Organometallic Chemistry in Industry

Organometallic Chemistry in Industry

A Practical Approach

With a Foreword by Robert H. Grubbs

Edited by
Thomas J. Colacot
Carin C.C. Johansson Seechurn

Editors

Thomas J. Colacot
Millipore Sigma (division of Merck
KGaA, Darmstadt, Germany)
6000 N Teutonia Avenue
Milwaukee, WI 53209
USA

Carin C.C. Johansson Seechurn
Johnson Matthey Plc
28 Cambridge Science Park
Milton Road, Cambridge CB4 0FP
United Kingdom

Library of Congress Card No.:
applied for

British Library Cataloguing-in-Publication Data
A catalogue record for this book is available from the British Library.

Bibliographic information published by the Deutsche Nationalbibliothek
The Deutsche Nationalbibliothek lists this publication in the Deutsche Nationalbibliografie; detailed bibliographic data are available on the Internet at <http://dnb.d-nb.de>.

Print ISBN: 978-3-527-34517-5
ePDF ISBN: 978-3-527-81917-1
ePub ISBN: 978-3-527-81919-5
oBook ISBN: 978-3-527-81920-1

Cover Design Formgeber, Mannheim, Germany
Typesetting SPi Global, Chennai, India
Printing and Binding Markono Print Media Pte Ltd, Singapore

Printed on acid-free paper

10 9 8 7 6 5 4 3 2 1

Contents

Foreword

In the late 1960s and through the 1970s, organometallic chemistry emerged from being a subfield of inorganic chemistry, where the interest was in boding and structure, to a field in its own right with chemists trained in inorganic or organic chemistry. The organic chemists brought reactivity to the field and helped to move organometallic chemistry into catalysis. The pioneering work of Collman, Vaska, and Halpern among others defined the basic mechanisms of the field and provided the basis for the application of this new field in organic transformations and organic synthesis. Now, most pharmaceuticals and natural product syntheses involve one, if not more, catalytic steps. The study of asymmetric hydrogenation and the ligands and mechanisms that controlled these processes paved the way for the discovery of a wide array of asymmetric processes. The structural flexibility of homogeneous catalysts and the wide array of ligands now available have resulted in most catalytic processes now being capable of producing products in high asymmetric purity. Heterogeneous catalysts, although they are generally favored for ease of processing, do not provide the flexibility required for more precise transformations. The rise of homogeneous catalysts has required the development of processes and methods that allow homogeneous catalysts to be exploited in practical large-scale processes.

Colacot (Millipore Sigma, a business of Merck KGaA) and Seechurn (Johnson Matthey), the editors of this book, have addressed these issues. After authoring the first chapter, which provides the historical background for the development of homogeneous catalysts and the basic mechanisms, they have chosen an outstanding group of authors to provide specific information about the practical aspects of the conversion of laboratory-scale reactions into real processes. Most of the processes are demonstrated by real examples. Themes of the chapters emphasize new developments in the pharmaceutical industry processes such as flow and continuous processes and the development of catalysts based on earth-abundant metals.

Chapters 2 and 3 discuss the advantages of continuous flow process. For example, the safe use of oxygen with organic solvents can be mitigated by the use of flow systems, and efficient processes can be developed for homogeneous reactions on scale. Particularly interesting is the use of the Buchwald–Hartwig reaction in a flow system with the efficient removal of the residual palladium catalysts. The second of the two chapters describes the methods for the use of low-temperature processes in the production of materials on a large scale, which involve reactive and environmentally sensitive reagents. These two chapters provide a detailed update on flow processes with the goal of increasing the use of flow processes in homogeneous processes. These processes regain some of the advantages that were traditional with heterogeneous catalysts while maintaining the selectivity of homogeneous processes. In a related process development, Chapter 8 describes the use of another "nanoreactor": micelles in water. In this chapter, the developments of traditional homogeneous cross-coupling reactions such as Heck and Suzuki–Miyaura in aqueous environments using a micelle environment are described. Carrying out the reactions in nanoreactors – micelles – results in interesting new selectivity and reactivity. From a process chemist's perspective, micelle-enabled processes can offer benefits such as the replacement of toxic organic solvents, reduced PMI value, improved reaction yields, high purity of API with reduced metal contents, and high cost efficiency.

As processes are scaled, the costs of the metal and ligands become more important. Chapters 4 and 5 describe the development of processes that are traditionally carried out using precious metals by rather employing either nickel or iron. These successful examples will encourage further development of efficient selective catalysts based on earth-abundant metals. In spite of potential costs, palladium catalysts have been shown to have a wide array of activities and selectivities. Chapter 6 demonstrates an outstanding example of the use of palladium in the commercial synthesis of beclabuvir utilizing the selectivity of palladium catalysts. Although, earth-abundant metals can take the place of palladium in a number of reactions, or rather complement Pd, the efficiency and selectivity of many palladium catalysts will ensure that it continues to be used in the pharmaceutical and fine chemical industry for many years to come.

Chapters 7, 9, and 10 cover specific reactions in process chemistry. The chapter on homogeneous hydrogenation provides a guide to the use of asymmetric hydrogenation in the synthesis of complex structures on a commercial scale. Asymmetric hydrogenation is one of the oldest and most used asymmetric processes in synthesis. This up-to-date guide provides the highlights of this field and helps to simplify the vast literature. In contrast, CH activation in complex synthesis is one of the newer areas of emphasis. For a number of years, there has been the recognition of the value of being able to functionalize C—H bonds directly, although C–H activation has not risen up like the cross-coupling reactions for industrial process. Therefore, the editors were conscientious enough to add a chapter on this topic. As is demonstrated in Chapter 7, this promise is now being realized as demonstrated by the use of a CH activation process in the synthesis of important compounds such as Merck's anacetrapib, sartans, etc. Olefin metathesis has been an important topic in academic synthesis for several decades; Phillips provides examples where this background of reactivity is now being translated into key structures for the pharmaceutical industry. He provides particularly good coverage of the important topics such as catalyst stability and removal that are required for the use of a homogeneous catalyst in a larger process.

The last chapter takes homogeneous catalysts outside of the applications in the pharmaceutical industry to the conversion of biomass-derived materials into chemical feedstocks. As many biomass sources are solids, a soluble catalyst is particularly suited for such applications. Although they focus on the conversion of carboxylic acids into olefins, the techniques and strategies would apply to many other such processes and can be developed for potential applications in industry.

It is particularly pleasing to see the evolution of organometallic chemistry into catalysts for extremely useful organic transformations. The basic principle and reaction mechanisms that were developed in the early decades of the area are now the basis for major processes that open the efficient synthesis of an amazing array of new chemical structures that have revolutionized how present-day bioactive materials are designed and prepared. Colacot and Seechurn have used their broad experience in new catalyst development, organic synthesis, and process chemistry involving homogeneous catalysts to assemble an outstanding team of authors from all over the world to highlight the important developments required to fulfill the promise of catalysis in organic synthesis for the twenty-first century. This is a very timely book for both academia and industry chemists and engineers to understand how academic concepts are translated into industries with a wide variety of important molecules as depicted in the cover of the book.

Robert Howard Grubbs
Division of Chemistry and Chemical Engineering
California Institute of Technology, Pasadena, CA 91125 USA
(626) 395 6003, rhg@caltech.edu

Prof. Grubbs Biography

B.A. and M.S. Chemistry, University of Florida, Gainesville, Florida, 1963 and 1965. Ph.D., Chemistry, Columbia University, New York, 1968. NIH Postdoctoral Fellow, Chemistry, Stanford University, 1968-69. He is the Victor and Elizabeth Atkins Professor of Chemistry at the California Institute of Technology, Pasadena, California, USA, and a faculty member since 1978. He was a faculty member at Michigan State University from 1969 to 1978.

The Grubbs group discovers new catalysts and studies their fundamental chemistry and applications. For example, a family of catalysts for the interconversion of olefins, the olefin metathesis reaction, has been discovered in the Grubbs laboratory. In addition to their broad usage in academic research, these catalysts are now used commercially. Other projects involve the design and synthesis of materials for use in medical applications. He has also been involved in the translation of technology through the founding of five companies.

His awards have included the Nobel Prize in Chemistry (2005) and 10 ACS National Awards. He was elected to the National Academy of Sciences (1989), Fellow of the American Academy of Arts and Sciences (1994), the Honorary Fellowship of the Royal Society of Chemistry (2006), Fellow of National Academy of Inventors, National Academy of Engineering (2015), and Foreign Member of the Chinese Academy of Sciences (2014) and of Great Britains's Royal Society (2017). He has 655+ publications and 160+ patents based on his research.

Preface

The perception that "there is probably no chemical reaction that cannot be influenced catalytically" was clearly stated by Wilhelm Ostwald even in the beginning of the past century. Many classic industrial processes, such as the production of ammonia by Haber process and sulfuric acid or nitric acid, require the use of heterogeneous catalysts such as finally divided transition metals.

Although industrial processes in general were dominated by heterogeneous catalysis, organometallic complexes emerged as a new class of compounds with a major function as homogeneous catalysts with more precision for the synthesis of organic chemicals. The earlier example of such a process is the manufacture of acetic acid by Cativa process, where methanol was subjected to carbonylation with the help of an Ir- or Rh-based organometallic complex.

The area of homogeneous catalysis literally became an emerging area for the synthesis of fine chemicals and pharmaceutical products after the discovery of a few important catalysts by Wilkinson in 1950s. Hence, Wilkinson should be recognized as the father of modern homogeneous catalysis. Wilkinson's work later inspired William Knowles to come up with a chiral Rh-based organometallic complex for a new area of homogeneous catalysis, called asymmetric hydrogenation. In 2001, William Standish Knowles and Ryōji Noyori shared the Nobel Prize for their work on enantioselective hydrogenation reactions, while the other half was awarded to K. Barry Sharpless for his work on enantioselective oxidation reactions. The chemistry community may remember the work of Kagan as well, although he did not get the Nobel Prize. Synthesis of numerous pharmaceutically relevant molecules for treatment of several diseases, agro-chemical compounds such as (S)-metolachlor, and fragrances in multi-ton quantities are some of the major applications of this area. The discovery and development of the olefin metathesis reactions in organic synthesis also led to the 2005 Nobel Prize to Yves Chauvin, Robert H. Grubbs, and Richard R. Schrock. Industries have started utilizing this method for special polymers, drug synthesis, as well as biological pesticides such as pheromones. There is no area in organic chemistry that has become as popular as the cross-coupling field, where the applications lie in the areas of drug synthesis, OLED, and related electronic as well as agrochemical applications. For this area, although the announcement was a bit delayed, the 2010 Nobel Prize in Chemistry was awarded jointly to Richard F. Heck, Professor Ei-ichi Negishi, and Professor Akira Suzuki for their work on palladium-catalyzed cross-coupling in organic synthesis. Some of the emerging areas in homogeneous catalysis are C–H activation and photoredox reactions where the organometallic complexes have a big impact on selectivity and activity. However, these areas have not quite reached the same stage as in the case of cross-coupling. Although a photocatalyst is required to generate an organic radical, a secondary catalyst such as a Ni-based organometallic compound is required to do some of the challenging coupling reactions such as sp^2–sp^3, sp^3–sp^3, and C–O/S coupling. The C–H activated process was also capable of functionalizing a sp^3 carbon. These chemistries in conjunction with flow processes are getting increasingly prominent in industrial processes and applications. Similarly doing chemistry with better *E*-factors is also becoming important in the industrial area. Inspired by the organometallic catalyzed reactions, particularly metal-catalyzed C–H activation, cyclopropanation, transfer hydrogenation, and emerging technologies such as "directed evolution" (2018 Nobel Prize in Chemistry where Frances Arnold shared the prize for generating highly active enzymes by mutation) and electrocatalysis are gaining momentum as potential technologies to be translated to industry.

We are fortunate enough to assemble a group of outstanding process chemistry researchers mostly from industry, with the exception of a few from academia, to write various chapters relevant to current-day developments in organic synthesis. We aimed to cover as many different topics as possible with the 11 chapters of the book. The introduction chapter sets the scene for the various reactivities of organometallic complexes that fundamentally enable all chemistries discussed in the subsequent chapters. Chapters 2 and 3 discuss the types of chemistry that have been found to be advantageously performed in a flow chemistry setting. Chapters 4 and 5 detail case studies where non-precious metal catalysis could be applied, with clear cost and sustainability benefits compared to using precious metal catalysts. In Chapters 6 and 7, C—H activation processes are described, which introduce a more atom-economical way of forming C—C, or C—X, bonds. In the first case, a palladium-catalyzed direct arylation reaction is discussed, and in the second case, a ruthenium-catalyzed directed CH-functionalization chemistry is detailed. Chapter 8 outlines the potential uses and advantages of carrying out conventional reactions with micellar catalysis in water, which is a very attractive, more environmentally friendly, option. Chapter 9 discusses homogeneous hydrogenation, which is possibly one of the most frequently seen applications for organometallic chemistry, or rather catalysis, that was recognized by the 2001 Nobel Prize. In Chapter 10, industrial applications of olefin metathesis, another Nobel Prize winning technology, are exemplified. Finally, Chapter 11 outlines a reasonably recent line of thought within the field of organometallic chemistry, converting biomass into chemically useful building blocks. This particular chapter focuses on the conversion of carboxylic acids into olefins. We thank all the authors for their scholarly contributions to make this book a unique one.

We thank Wiley-VCH especially, Dr. Elke Maase for giving this opportunity and for working with us patiently, as well as Prof. Grubbs for graciously taking time to write the foreword. We acknowledge all the reviewers although we would like to keep their names confidential. We also thank Millipore Sigma (a business of Merck KGaA, Darmstadt, Germany) and Johnson Matthey for their support on this collaborative project in helping science and technology to improve the quality of this planet and life in general through the utilization of organometallic chemistry. We are confident that this book will help chemists and chemical engineers in both academia and industry to improve their skills in organic synthesis.

August 9, 2019 *Thomas J. Colacot*
Milwaukee, USA and Cambridge, UK *Carin C.C. Johansson Seechurn*

1

Industrial Milestones in Organometallic Chemistry

Ben M. Gardner[1], *Carin C.C. Johansson Seechurn*[2], *and Thomas J. Colacot*[3]

[1] *Cambridge Display Technology Ltd, Unit 12 Cardinal Park, Cardinal Way, Godmanchester PE29 2XG, UK*
[2] *Johnson Matthey, 28 Cambridge Science Park, Milton Road, Cambridge CB4 0FP, UK*
[3] *Millipore Sigma (A Business of Merck KGAa Darmstadt, Germany), 6000N Teutonia Avenue, Milwaukee, WI 53209, USA*

1.1 Definition of Organometallic and Metal–Organic Compounds

Organometallic compounds can be defined as compounds that contain at least one chemical bond between a carbon atom of an organic moiety and a metal. The metal can be alkaline, alkaline earth, transition metal, lanthanide, or a metalloid such as boron, silicon, and phosphorus. Therefore, metal–phosphine complexes are also often included in this category, although they do not contain a typical metal–carbon bond – they are more commonly referred to as "metal–organic compounds." For the purposes of this book, applications of both organometallic and metal–organic compounds are discussed on the basis of "organometallic chemistry."

1.1.1 Applications and Key Reactivity

The three major types of applications of organometallic compounds in industry are in the areas of electronics, polymers, and organic synthesis. In organic synthesis, the organometallic compounds are used as either catalysts or stoichiometric reagents.

1.1.1.1 Electronic Applications

For electronic applications typically, the organometallic complex is subjected to chemical vapor deposition (CVD) to form an appropriate thin layer or subjected to organometallic chemical vapor deposition (OMCVD) where the deposition ultimately occurs via a chemical reaction at the substrate surface to produce a high-quality material. The production of thin films of semiconductor materials is used, for example, for LED applications via metal–organic vapor-phase epitaxy (MOVPE) where volatile organometallic Me_3E (E = Ga, In, Al, and Sb) compounds are used as precursors. They react with ultrapure gaseous hydrides in a specialized reactor to form the semiconducting product as a crystalline wafer [1–23].

Organometallic Chemistry in Industry: A Practical Approach, First Edition.
Edited by Thomas J. Colacot and Carin C.C. Johansson Seechurn.

1.1.1.2 Polymers

Another major application for organometallic complexes is in the polymer industry. Three common types of polymers produced via catalysis are particularly noteworthy. Polysiloxanes, also known as silicone, are polymers made up of repeating units of siloxane [4]. They have widespread application in a large number of different fields ranging from cookware to construction materials (e.g. GE silicone), medicine, and toys. Pt-based catalysts are commonly applied in the silicone industry for the production of a variety of products [5]. A milestone in the history of organometallic chemistry in the industry was the discovery of the Ziegler–Natta catalyst and its application in polymerization reactions [6]. Ziegler and Natta were awarded the Nobel Prize for their work in this field in 1963 [7]. Another area that has been recognized for its importance is olefin metathesis for which a Nobel Prize has been awarded to Grubbs, Schrock, and Chauvin. This has been applied to synthesize polymers via ROMP (ring-opening metathesis polymerization) [8].

1.1.1.3 Organic Synthesis

The focus of this book, however, is on the exploitation of organometallic compounds for organic synthesis, relevant to industry applications. One of the major applications in organic synthesis is catalysis.

In cases where the organometallic compound is used as a catalyst, for example in a process involving cross coupling, a precatalyst should be able to get activated to the active catalytic species to bind with the organic substrate(s), do the transformation, and release the product such that the active catalytic species returns to its original state in the catalytic cycle. During the organic transformation, the concentration of the catalyst can decrease with time because of poisoning. The efficacy and efficiency of the catalyst depend on how fast and how long it can retain its original activity. The turnover numbers (TONs) and turnover frequencies (TOF) are usually used to describe the activity of a catalyst. Organic chemists have started using organometallic compounds as catalysts to develop more efficient and practical processes [9–12].

The reactivity of organometallic complexes toward various reagents is the reason behind the widespread use of organometallic compounds as catalysts for a variety of organic transformations. The most important types of organometallic reactions are oxidative addition, reductive elimination, carbometalation, hydrometalation, β-hydride elimination, organometallic substitution reaction, carbon–hydrogen bond activation, cyclometalation, migratory insertion, nucleophilic abstraction, and electron transfer. In the following paragraphs, we will provide a brief overview of the basic theory with some selected applications.

Oxidative addition involves the breakage of a bond between two atoms X–Y. Splitting of H_2 with the formation of two new metal–H bonds is an example of an oxidative addition process (Scheme 1.1). Reductive elimination is the reverse of this process. In an oxidative addition process, the oxidation state of the metal is increased by 2, whereas in reductive elimination, oxidation state of the metal is decreased by 2. Both steps are crucial for metal-catalyzed cross-coupling reactions, as the first and the last steps of the catalytic cycle. Several factors can affect these two steps. The structure of the ligand (phosphine or other molecules coordinated with the metal), the coordination number of the metal in the complex, and the way in which the complex is activated to the catalytic species in the catalytic cycle, etc., can be modified and tailored to get the best outcome for a particular reaction [13]. The oxidative addition of H_2 onto Vaska's complex (Scheme 1.1) is a crucial step in metal-catalyzed hydrogenation reactions. The application of this methodology to industrially relevant molecules is further discussed in Section 1.3.3.

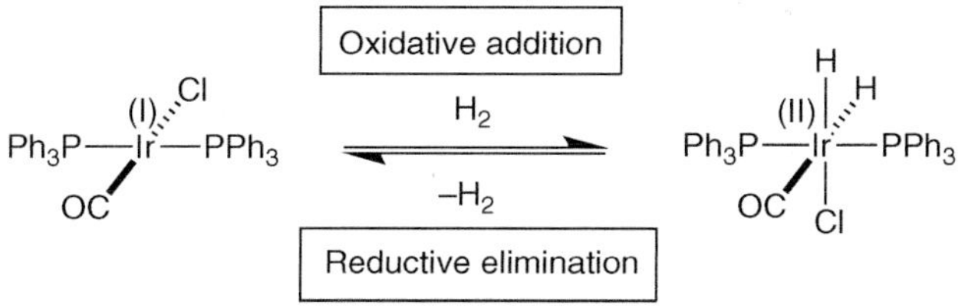

Oxidative addition
H_2
$-H_2$
Reductive elimination

Scheme 1.1 Oxidative addition and reductive elimination.

Carbometalation involves, as the name suggests, the simultaneous formation of a carbon–metal and a C—C bond. This is most commonly used to form a stoichiometric metal-containing reagent, such as the reaction between ethyllithium and bis-phenylacetylene in the synthesis of Tamoxifen™, a breast cancer drug (Scheme 1.2) [14].

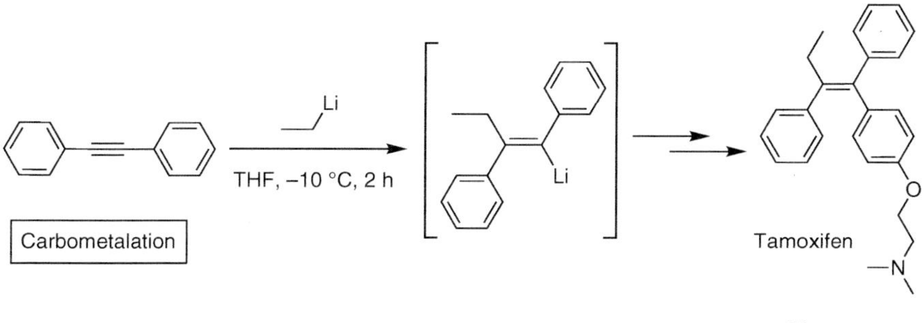

Li
THF, −10 °C, 2 h
Carbometalation
Tamoxifen

Scheme 1.2 Carbometalation as a key step toward the synthesis of Tamoxifen™.

Hydrometalation is similar to carbometalation, where, instead of a C—C bond, a C—H bond is formed alongside the carbon–metal bond. One such example is hydroalumination, where DIBAL (i-Bu_2AlH) is added across an alkyne (Scheme 1.3) [15]. This, similar to carbometalation, is most commonly a stoichiometric transformation with the aim of preparing an organometallic reagent that can be used as a reactant for subsequent desired transformations.

β-Hydrogen elimination, technically the reverse of hydrometalation, can in some cases result in the formation of undesired side products. In other cases, it is a "blessing" as the preferred reaction pathway. In Shell higher olefin process (SHOP), for the oligomerization to occur, a final β-hydrogen elimination reaction is performed to release the substrate from the catalyst (Scheme 1.4a) [16]. In the cross-coupling reaction between an aryl halide and an organometallic

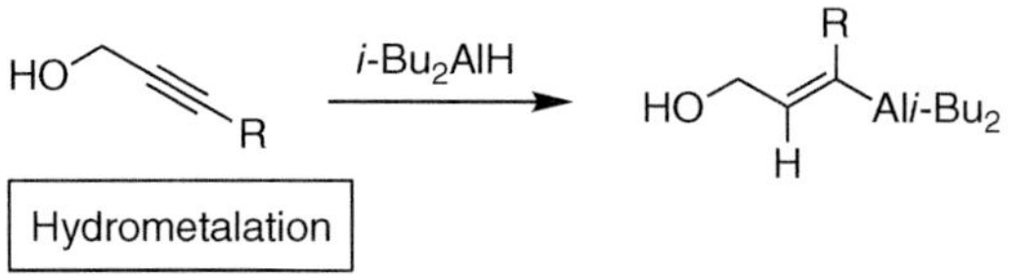

Hydrometalation

Scheme 1.3 Hydroalumination of alkynes.

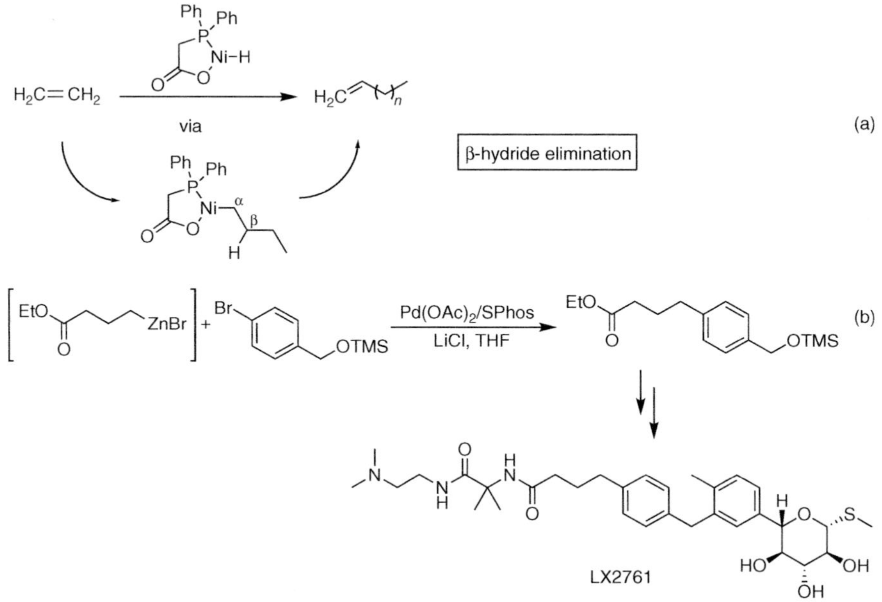

β-hydride elimination

LX2761

Scheme 1.4 a) β-hydride elimination is exploited in the Shell higher olefin process (SHOP). b) sp^2–sp^3 cross-coupling in the synthesis of a diabetes drug.

reagent containing β-hydrogens, this reaction can form the undesired alkene side products, hence detrimental. This is the reason why sp^2–sp^3 coupling and sp^3–sp^3 coupling become very challenging even today. However, a few success stories of these types of cross-coupling reactions have been reported, such as sp^2–sp^3 Negishi reaction for the synthesis of LX2761, a diabetes drug by Lexicon Pharmaceuticals (Scheme 1.4b) [17].

Organometallic substitution reactions can occur either via an associative or a dissociative substitution mechanism. This can be compared to S_N1 and S_N2 substitution mechanisms in organic chemistry. The overall outcome in either case is an exchange of a ligand on the organometallic complex. Scheme 1.5 illustrates an associative substitution mechanism to exchange Cl for X on Vaska's complex. This complex does not have any significant references to being employed in industry as a catalyst, but studies of its reactivity has been vital in providing the conceptual framework for homogeneous catalysis [18].

One of the reactions that has become increasingly exploited, particularly to complement the cross-coupling chemistry, is C–H activation. This is where the

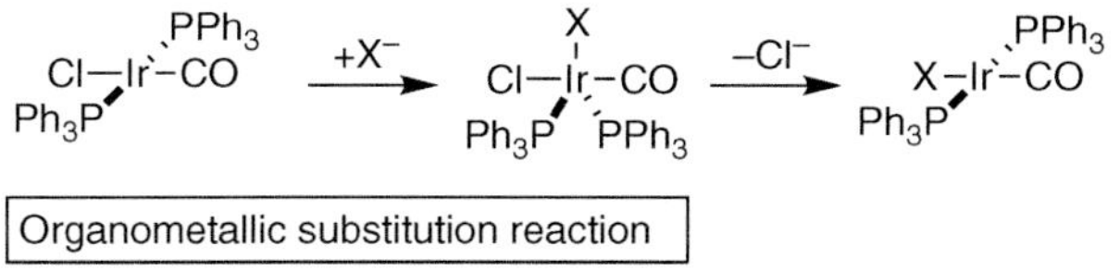

Organometallic substitution reaction

Scheme 1.5 Organometallic substitution reaction exemplified by Vaska's complex. Source: Wilkins 1991 [24]. Reproduced with permission of John Wiley & Sons.

metal gets inserted into a C—H bond of the substrate. There are many different pathways for this to happen; it can be promoted and directed to the site of choice by using a directing group, such as the amide exemplified in Scheme 1.6. Iridium-catalyzed direct borylation reactions can also be considered as a type of C–H functionalization reaction. This type of reactions is further discussed in Section 1.3.2.1.

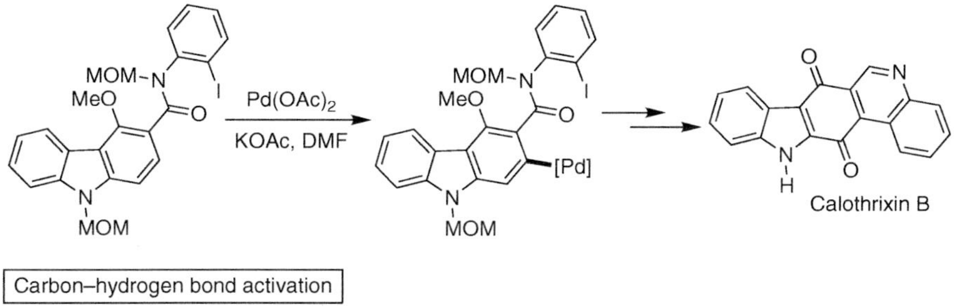

Carbon–hydrogen bond activation

Scheme 1.6 Carbon–hydrogen bond activation exemplified by the total synthesis of calothrixin B, which possesses various biological activities such as anti-malarial and anti-cance. Source: Ramkumar and Nagarajan 2013 [25]. Reproduced with permission of American Chemical Society.

In cyclometalation reaction, the strain of certain motifs is often exploited to insert the metal into C—C bonds. One example is the Rh-catalyzed insertion into cyclopropanes to form metallacyclobutanes (Scheme 1.7). This has been applied in the total synthesis of (±)-β-cuparenone [19]. Metallacyclobutanes also form a very crucial part of the metathesis olefination mechanism, as deduced by Chauvin [20].

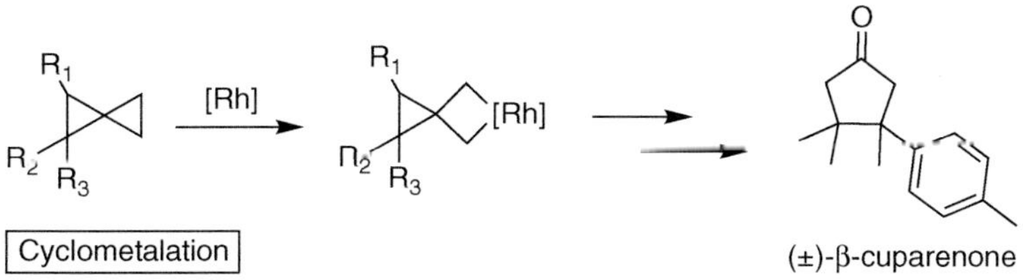

Cyclometalation

(±)-β-cuparenone

Scheme 1.7 Cyclometalation exemplified by the oxidative additions of Rh into a cyclopropane moiety.

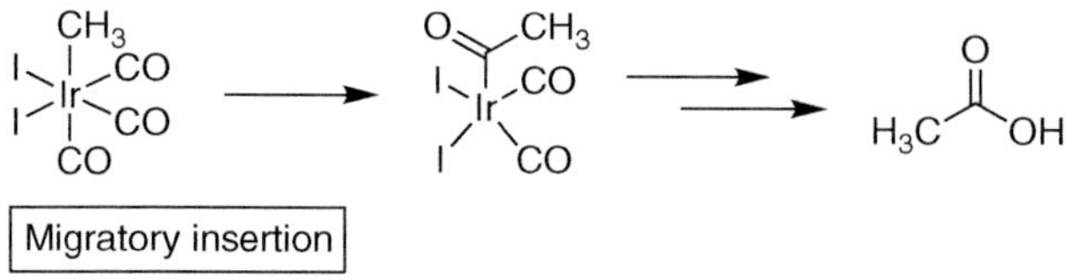

Migratory insertion

Scheme 1.8 Migratory insertion exemplified by a step in the Cativa process.

Migratory insertion is crucial for any carbonylation reaction and is illustrated in Scheme 1.8 by a step in the iridium-catalyzed Cativa process, where methanol is converted into acetic acid [21]. The migration involves the insertion of one ligand (CO) into the metal—C bond (Ir-Me). The reverse reaction, decarbonylation of aldehydes to form an alkane with the release of CO, is also a reaction known to be catalyzed by Rh complexes [22], such as Wilkinson's catalyst [23]. Migratory insertion is not restricted to CO alone but can also occur with SO_2, CO_2, and, most importantly, alkenes. The insertion of an alkene into an M—C bond is the key step in any oligo- or polymerization reaction, such as the Ziegler–Natta process [26].

Nucleophilic abstraction is a process when a ligand is fully or partly removed from the metal by the action of a nucleophile. In Scheme 1.9, the action of *n*-BuLi on a chromium-coordinated benzene ligand results in hydrogen abstraction [27]. Basically, the chemical reactivity of the ligand is altered when coordinated with a metal. This alters the reactivity of the ligated compound and may result in reactions that are not possible to carry out with the same non-ligated substrate.

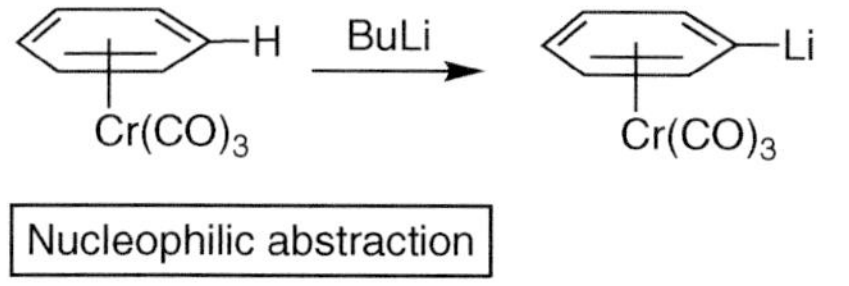

Nucleophilic abstraction

Scheme 1.9 Nucleophilic abstraction illustrated by hydrogen abstraction using *n*-BuLi.

Another important organometallic reaction to be discussed is electron transfer. The ability of certain organometallic complexes to initiate electron transfer reactions in combination with a visible light source has made some transformations possible that cannot be achieved using conventional chemistry. This is illustrated in Scheme 1.10 with one step in the photocatalytic Pschorr reaction using $Ru(bpy)3^{2+}$ as the photoredox catalyst [28, 29]. The phenanthrene formed can be further used for various purposes, such as in the manufacture of dyes, pharmaceuticals, etc. [30]. The potential of metal-catalyzed electron transfer reactions forms the basis for a new area in organic synthesis with lot of potentials [31].

Exploitation of the wide variety of "organometallic reactivity" has made the field of organometallics one of the most applied areas in process chemistry with particular importance to the pharmaceutical, agrochemical, polymer, and fine chemical industries.

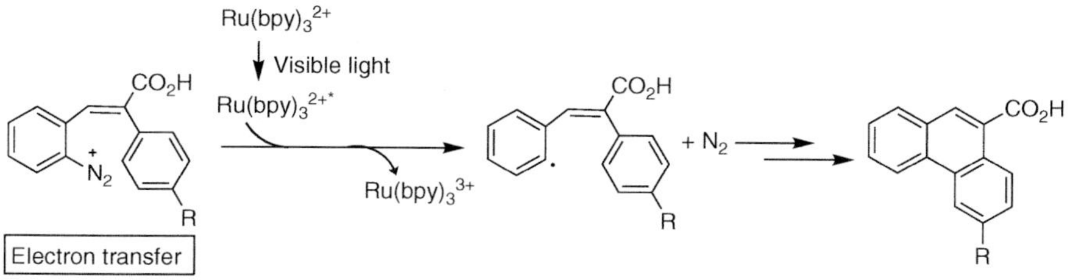

Scheme 1.10 Electron transfer illustrated by one step in the photocatalytic Pschorr reaction to form phenanthrene.

1.2 Industrial Process Considerations

Organometallic compounds are routinely prepared and used as stoichiometric reagents or catalysts for a range of synthetic processes on a multikilogram scale or even a ton scale.

In order to operate a commercially viable industrial chemical process, a reliable chemical synthesis route is needed as well as an understanding of how a process will behave during the scale up by taking into consideration factors such as heat and mass transfer, mixing, particle size, and filterability, etc. Air, moisture, and thermal sensitivity of some of the organometallic complexes or their intermediates needs to be addressed with proper handling techniques including inert conditions to achieve the maximum process efficiency and process safety. In addition, incorporation of environmental impact of the process is also very important, where exposure of chemicals and solvents and waste generation need to be minimized.

It is important to have a scalable chemical process, usually optimized on a bench scale to produce milligram to gram and then transferred to the pilot plant, typically to a kilogram scale. During this transfer, typically, one needs to readjust the rate of reagent addition to manage the exotherm, rate of agitation, rate of heating, degassing cycles, reaction time, etc. Identifying the optimal catalyst with the minimal loading especially when one uses *platinum group metals* (PGMs) in conjunction with expensive ligands is also important. Even for a well-known organic transformation such as a Pd-catalyzed cross-coupling, the process will not be economical if the reaction is poorly optimized, considering metal loss, purification of the products, and waste disposal. A proper understanding of the thermodynamics and kinetics is also important.

Experience in using DOE coupled with a "knowledge-based" process approach can accelerate the process development. It is important to involve both chemists and chemical engineers during the scale-up and process optimization, considering the equipment design, safety, raw material selection, etc. Even if the precatalyst is not sensitive to air, one has to conduct the reactions under inert conditions as the "active catalytic species" in the cycle might be sensitive to air. This can not only minimize the by-product formation but also increase the life cycle of the catalyst and hence the TONs and TOFs.

The kinetic control of an organometallic process can be another important factor. One example is low-temperature reactions involving organolithium reagents, where it is essential to avoid significant decomposition of thermally sensitive species, thermal promotion of undesired side reactions, and control the reactivity of exothermic processes.

Treatment of waste streams from organometallic processes must be considered carefully as they may contain precious metal or even other transition metal residues originating from the decomposition of the organometallic compounds. Apart from the well-documented environmental impact of PGM compounds, finely divided PGM particles, if allowed to dry out, pose a significant fire hazard. Because of the significant environmental hazards associated with heavy metal residues, predominantly arising from their persistence in the biosphere via bioaccumulation, generation of this type of waste stream on production scale should be avoided wherever possible, with environmental regulations strictly controlling the level of any emissions. Some common catalyst precursor complexes release harmful side products when activated or substituted. For instance, $[Pd(COD)(Cl)_2]$ releases 1,5-cyclooctadiene (COD) in the presence of phosphines, which, among its other chemical hazards, has a pungent odor even in low concentrations. Therefore, extreme care must be taken when dealing with process waste that contains it. Similarly, many of the metal carbonyl compounds can generate CO gas, which needs to be properly vented. Some of these carbonyl-based compounds undergo sublimation as well.

1.3 Brief Notes on the Historical Development of Organometallic Chemistry for Organic Synthesis Applications Pertaining to the Contents of this Book

Most organometallic processes have evolved and developed from seminal discoveries in the late 1800s or early 1900s. In some cases, it is easier to pinpoint the exact seminal reports, whereas in other cases, this task is not so easy. Sabatier's report of nickel-catalyzed hydrogenation can easily be identified as the discovery of metal-catalyzed hydrogenation reactions [32], for which he got the Nobel Prize in Chemistry in 1912. For the cross-coupling area, its origin is slightly more difficult to deduce precisely although our 2012 review articles and book provide a much better understanding of the area [9, 10, 13, 33, 34]. One could argue that it dates back to 1912 Nobel laureate Victor Grignard's discovery of RMgX reagents, where Grignard shared the Nobel Prize with Sabatier. Although both technologies (Grignard in 1912 and cross-coupling in 2010) got Nobel Prizes, the former is considered to be a "breakthrough innovation," whereas the latter is called "incremental innovation." The impact of cross-coupling in chemical processes shows its significance by being awarded the Nobel Prize, in comparison to many competing technologies.

In this section, we will briefly go through the origins of a few prominent areas within organometallic chemistry and how they relate to the current industrial applications with respect to the topics covered by the chapters in this book.

1.3.1 Synthesis of Stoichiometric Organometallic Reagents

1.3.1.1 Conventional Batch Synthesis

Arguably, the most important stoichiometric organometallic reagents are organolithium compounds, RLi. The studies of these reagents were pioneered by Karl Ziegler, Georg Wittig, and Henry Gilman [35]. Their relatively straightforward preparation, high basicity, and wide array of functionality provide convenient access to useful synthetic routes such as metalation, deprotonation, carbolithiation, and transfer or exchange of the nucleophilic organic fragment R^-.

In 1899, by substituting Mg for Zn in alkylation reactions, Philippe Barbier's student Victor Grignard (Figure 1.1) developed the RMgX alkylating agents that bear his name to this day. Being a less sensitive but more potent source of alkyl anions than their Zn-based counterparts, Grignard showed how they can efficiently alkylate carbonyl compounds, a discovery that proved to have huge impact in synthetic chemistry and earned him a Nobel Prize in 1912 [36].

Today, Victor Grignard is remembered as the father of organometallic chemistry. Organomagnesium compounds represent very useful alternatives to their lithium counterparts, exemplified by the widespread use of Grignard reagents, RMgX, for efficient alkylations and arylations. These reagents are now produced in multiton quantities. Organocalcium compounds are more reactive alkyl sources than Grignard reagents, but their applications are limited because of the increased difficulty of their preparation and the thermal instability they exhibit. Organocalcium compounds have also shown promise as hydroamination catalysts. In comparison to organolithium and organomagnesium, organoaluminum compounds, R_3Al reagents, are generally far less effective stoichiometric reagents but do add to alkenes and alkynes with high regio- and stereoselectivity via carboalumination. Importantly, however, they have found particular use as a vital component of the heterogeneous Ziegler–Natta polymerization process for the industrial-scale production of polyethylene and polypropylene. Aluminum

Figure 1.1 Victor Grignard. Source: https://commons.wikimedia.org/w/index.php?curid=545837. Licensed under CC BY 3.0.

alkyls are also widely utilized for group III–V chemistry for the production of electronic materials via CVD.

1.3.1.2 Organometallics in Flow

Industrial-scale organic synthesis for fine chemical applications, such as natural products or active pharmaceutical ingredients (APIs), and organometallic syntheses have traditionally been conducted in batch using large-volume (>100 l) reactors. In continuous flow processes, small amounts of reagent solutions are continuously pumped along a flowing stream to mix at a specific junction with resonance time to react them together to yield the product, which is being purified under the flow conditions and collected. In some cases, a cascade approach has been considered where multiple reagents have been mixed sequentially rather than performing reactions in different batch reactors. Industries have been using this technique for the manufacture of petrochemicals and bulk chemicals as this approach has proven to be not only most economical but also produce good-quality products consistently. The recent interest in flow chemistry in academia for the synthesis of more complex organic compounds has increased efforts to apply this rapidly burgeoning technology both in fine chemical and pharmaceutical industries. The advantages that flow processes can bring in a commercial context relative to batch production are shorter reaction times, greater temperature control, rapid optimization, shorter path length for photochemical reactions, and improved process safety. Chapter 2, authored by Joseph Martinelli of Eli Lilly, presents the design, development, and implementation of an API manufacturing route under continuous flow conditions to showcase the application of this technology in organic synthesis. Chapter 3 details the lithiation and borylation chemistry under flow, as developed by Joerg Sedelmeier and Andreas Hafner at Novartis. This chapter provides a snapshot of how this technology can also be applied for the synthesis of organometallic reagents.

1.3.2 Cross-coupling Reactions

Several years after Grignard's discovery of RMgX reagents, in 1941, Kharasch undertook the first systematic investigation of transition-metal-catalyzed sp^2–sp^2 carbon coupling, detailing the observation of homocoupling of Grignard reagents [37, 38]. Subsequent research from his group led to the earliest report of a cross-coupling reaction, where a cobalt-based catalyst was used to couple vinyl bromide with an aryl Grignard reagent [39]. This made him to be the father of cross-coupling reactions.

The metal catalysts in question are also organometallic complexes that mediate the coupling of two different hydrocarbon fragments for organic synthesis purposes in the fine chemical, agrochemical, and pharmaceutical industries. A simplified catalytic cycle is shown in Scheme 1.11. Many of the key reactivity steps that are characteristic for organometallic complexes are a prerequisite for these reactions to take place. Initial oxidative addition is followed by transmetalation (organometallic substitution) and finally reductive elimination to form the

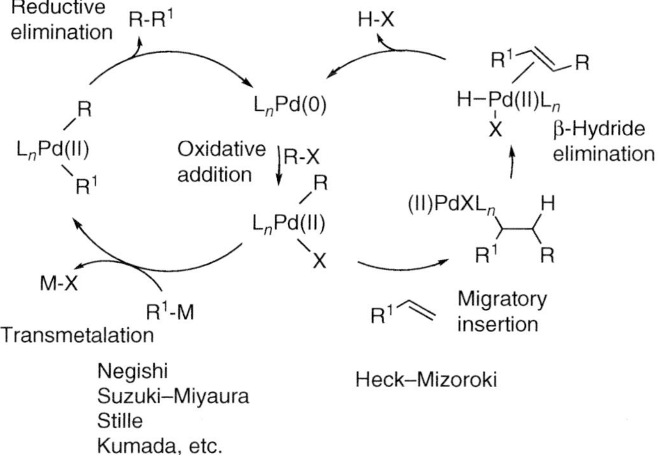

Scheme 1.11 Simplified catalytic cycles for cross-coupling reactions.

desired product and regenerate the catalyst. Each of these steps has been the subject of a number of studies to try and understand the exact nature of their mechanism. For Suzuki–Miyaura reactions, the transmetalation step has been the focus of attention of several research groups. The Denmark, Lloyd-Jones, and Hartwig groups have independently studied this step of the catalytic cycle for these types of cross-coupling reactions [40–42]. In Sonogashira reactions (sp–sp^2 bond formation), a Cu cocatalyst is commonly employed [43]. In many recent refinements of this reaction, however, the need for a cocatalyst has been circumvented by, for example, a careful choice of Pd catalyst and reaction conditions [44–46]. The mechanism of the Heck reaction differs from the other named cross-coupling reactions in that a β-hydrogen elimination is crucial for the formation of the final product.

Many pioneers have played a role in the development of this area and lent their names to the reactions they have discovered. The importance of cross-coupling to the field of chemistry was ultimately recognized in 2010 by awarding the Nobel Prize to Richard F. Heck, Ei-ichi Negishi, and Akira Suzuki for their research efforts in palladium-catalyzed cross-couplings in organic synthesis [9, 10, 13, 35, 36]. Cross-coupling is an example where incremental innovation is of equal importance to the breakthrough discovery, significant enough even for the award of the Nobel Prize.

Following from an earlier work by Fujiwara, in 1969, Richard Heck published the first examples of cross-coupling using stoichiometric palladium(II). Building on a separate work by Mizoroki, he proposed the first Pd(0)-mediated catalytic cycle for the cross-coupling of iodobenzene and styrene, opening the door for an explosion of discoveries in Pd-catalyzed cross-coupling chemistry. The traditional Mizoroki–Heck reaction forms a substituted alkene via cross-coupling of an unsaturated halide or pseudo-halide with an alkene under Pd catalysis and is frequently employed for C–C coupling in industrial settings. The Heck mechanism can also be accessed using nickel to mediate the catalysis. Under certain conditions, this brings advantages relative to the palladium version, such as higher activity. This is thought to be because of lower energy barriers to crucial steps in the catalytic cycle and greater selectivity to the desired product because of the greater efficiency of undesirable β-hydride elimination for Pd vs Ni. One example of successful implementation and scale-up of a nickel-catalyzed Mizoroki–Heck reaction at BI is discussed in detail in Chapter 4 by Jean-Nicolas Desrosiers and Chris H. Senanayake. This chapter is also relevant in terms of the emerging area of the use of base metal instead of precious metal catalysis.

The Suzuki–Miyaura reaction, where a boronic acid/ester is coupled to a halide or pseudo-halide precursor under palladium-mediated catalysis, is the most common C–C coupling reaction in industry. Suzuki couplings are advantageous at a large scale because of the mild reaction conditions, the commercially available and relative environmentally benign boronic acid/ester starting materials, and the comparative ease of disposal of boron-containing by-products compared to processes using other organometallic reagents.

The Kumada cross-coupling reaction utilizes a Grignard reagent and an organic halide as precursors and classically operates under palladium or nickel catalysis. Although demonstrated to be a largely efficient C—C bond formation strategy, Kumada couplings can be problematic in large-scale synthesis because of the high reactivity of Grignard reagents, which exhibit limited functional group tolerance. However, there is a new trend where palladium catalysts are substituted for those containing iron, advantages of which include lower cost of the metal because of higher earth abundance and lower toxicity. Although still relatively new, the iron-catalyzed Kumada coupling represents a rapidly growing area of organometallic synthesis with great industrial potential, which is discussed in Chapter 5, authored by Rakeshwar Bandichhor of Dr Reddy's Laboratories. Similar to the nickel-catalyzed Mizoroki–Heck reaction detailed in Chapter 4, Chapter 5 provides another example of the industrial trend in switching from precious metal to earth-abundant metal catalysis.

Overall, the breadth of catalytic cross-coupling reactions has found significant application in the field of organic synthesis for the production of pharmaceutically and agriculturally relevant molecules, often employed in several steps as part of a multistep synthesis. Under carefully optimized conditions, these catalysts typically offer advantages over stoichiometric alternatives such as high selectivity, mild reaction conditions, functional group tolerance, low loadings, and avoidance of protecting groups. In cases where the metal-catalyzed cross-coupling reaction takes place in the end game of a total synthesis, it is important to determine the levels of residual metal in the final API. There are well-defined limits for each metal and how much a drug can legally contain [47]. In general, homogeneous cross-coupling catalysts cannot be recycled, although the metal itself can be recovered from the waste stream.

1.3.2.1 C—H Bond Activation

In 1983, Robert Bergman and William Graham independently detail the first transition-metal-mediated intermolecular C–H activation of alkanes by oxidative addition by pentamethylcyclopentadienyl–iridium(I) complexes [48, 49]. This opened up the possibility of carrying out cross-coupling reactions where only one cross-coupling partner, or in rarer cases, neither of the two cross-coupling partners, is pre-functionalized as an aryl (pseudo)halide or an organometallic reagent. Noteworthy efforts within this field have been reported by the research groups of Keith Fagnou and coworkers [50], Melanie Sanford and coworkers [51], Christina White and coworkers [52], and Jin-Quan Yu and coworkers [53]. One class of reaction that falls under the C–H activation category is direct arylation of aromatics or heterocycles. In this type of chemistry, the heterocycle is nonfunctionalized and the successful reaction relies on the inherent nucleophilicity of the substrate or inherent acidity of certain C—H bonds in the molecule in order to achieve regioselectivity. Another breakthrough discovery in this area is the iridium/bipyridine-catalyzed direct borylation, first reported by Hartwig and coworkers [54]. This methodology was recently employed by Pfizer to form a nicotine analog (Scheme 1.12) [55].

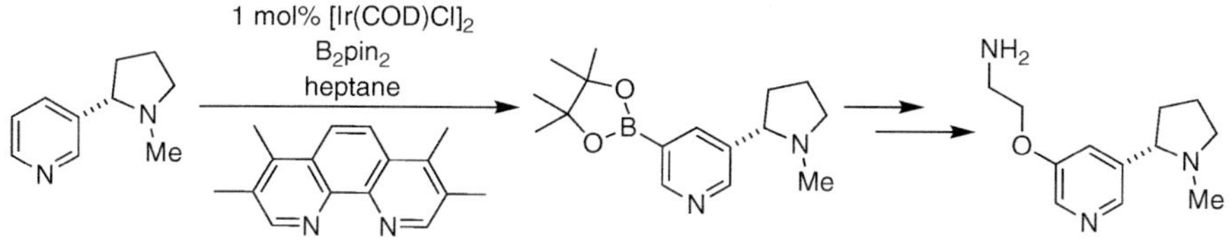

1 mol% $[Ir(COD)Cl]_2$
B_2pin_2
heptane

Scheme 1.12 Iridium-catalysed direct borylation in Pfizer's synthesis of a nicotine analog.

Several other metal complexes based on, e.g. Pd or Ru have been successfully employed in C–H activation or direct arylation reactions. Chapter 6, authored by Collin Chan, Albert J. DelMonte, Chao Hang, Yi Hsiao, and Eric M. Simmons from BMS describes an intramolecular Pd-catalyzed direct arylation reaction that was used in the manufacturing route of Beclabuvir. Ru complexes are also emerging as alternative, or complementary, catalysts to the Pd ones for the purpose of CH activation chemistry. Anita Mehta of Chicago Discovery Solutions describes her experience of Ru-catalyzed direct arylation reactions in water in Chapter 7.

1.3.2.2 Carbonylation

Transition-metal-catalyzed carbonylation (introduction of CO) of methanol has been employed very successfully in the large-scale manufacture of acetic acid in the Monsanto process, which uses an anionic Rh carbonyl catalyst [56]. This process has since been superseded by the more economical and environmentally friendly Cativa process, which relies on an analogous iridium-based catalyst system [21, 57]. Pd catalysts have also been employed in carbonylation chemistry, which is now a widely applied methodology in the pharmaceutical industry [58].

1.3.2.3 Catalysis in Water – Micellar Catalysis

Despite the widespread use of palladium-catalyzed cross-coupling both in research and in production, for some of the more problematic transformations such as the arylation of nitroalkanes, most or all of the process conditions can require organic solvents, a dry/inert atmosphere, elevated temperatures, and high catalyst loadings. A promising alternative to conventional cross-coupling conditions for which the former factors are requisite is micellar catalysis, whereby a surfactant is dispersed in aqueous media and the internal conditions of the micelles formed mimic organic solvents in which the substrates and the catalyst are compatible and the catalysis takes place. Care must be taken to select a surfactant that is environmentally benign and can interact with hydrophilic and hydrophobic functionalities (amphiphilic), whose polarity can accommodate the reagents to maximize the catalytic activity. The choice of catalyst itself is also important, as greater lipophilicity permits a lower catalyst loading. For organic synthesis on an industrial scale, micellar catalysis could confer major advantages in process safety, cost, and waste remediation, which is examined in Chapter 8 by Sachin Handa.

1.3.3 Hydrogenation Reactions

In 1900, French chemist Paul Sabatier first defined homogeneous and heterogeneous catalysis [59]. Together with Senderens, he discovered that the use of a trace amount of nickel as a catalyst promotes the hydrogenation of carbon compounds, and his work inspired others to develop the hydrogenation of fats, central to advances in the food industry [60]. He later shared a Nobel Prize with Grignard in 1912 for his work.

In 1922, Franz Fischer and Hans Tropsch reported the heterogeneously catalyzed production of linear alkanes and olefins from a mixture of CO and hydrogen (syngas) largely without oxygenated by-products. The process was so successful that it was later industrialized in 1925. In 1938, the German chemist Otto Roelen, during his investigation into the Fischer–Tropsch process, discovered that olefins can be catalytically converted to aldehydes in a process known as hydroformylation or the oxo process [61]. A $Co_2(CO)_8$ catalyst was used to affect the conversion with the H_2/CO syngas feedstock. Now, the industrial-scale production of most useful synthetic aldehydes from alkenes is achieved via this process, traditionally under Rh- or Co-catalyzed conditions. The products can be hydrogenated to produce alcohols or used directly as starting materials in the organic synthesis of complex pharmaceutical agents [62]. Another process employing olefin as a raw material is the Wacker process. This involves the catalytic oxidation of ethylene to acetaldehyde and is one of the first Pd-catalyzed homogeneous processes on an industrial scale [63, 64].

Continuing the hydrogenation thread, in 1965, Wilkinson and Coffey reported the catalytic activity of $[RhCl(PPh_3)_3]$ for the homogeneous hydrogenation of alkenes [65, 66]. This discovery formed the basis for the development of asymmetric homogeneous hydrogenation as exemplified by the chiral Noyori-type ruthenium catalysts [67]. In 2001, William Knowles, Ryoji Noyori, and Karl Barry Sharpless were awarded the Nobel Prize in Chemistry for their work on enantioselective metal-catalyzed hydrogenation and oxidation reactions [68].

Asymmetric hydrogenation – the addition of two H atoms preferentially to one of the two faces of a prochiral unsaturated molecule – is a very important tool in the manufacture of bioactive pharmaceuticals and agrochemicals. The enantioselectivity derives from the chirality of the ligand bound to the metal (commonly ruthenium or iridium) center in the active catalyst. The substrate is oriented relative to the chiral ligand in the catalyst–substrate complex to minimize steric interactions, and hydrogen is delivered to the sterically least hindered face. Asymmetric hydrogenation was first successfully applied to the industrial-scale manufacture of L-Dopa (Scheme 1.13) [69], and a treatment for Parkinson's disease aided by the explosion in the availability of chiral catalysts in the 1990s has since been used in the production of a range of pharmaceuticals, such as the calcium channel blocker *mibefradil* (Scheme 1.13) [70]. In Chapter 9, authored by Stephen Roseblade of Johnson Matthey, the aspects of asymmetric homogeneous hydrogenation and details of its application in several industrial cases are discussed.

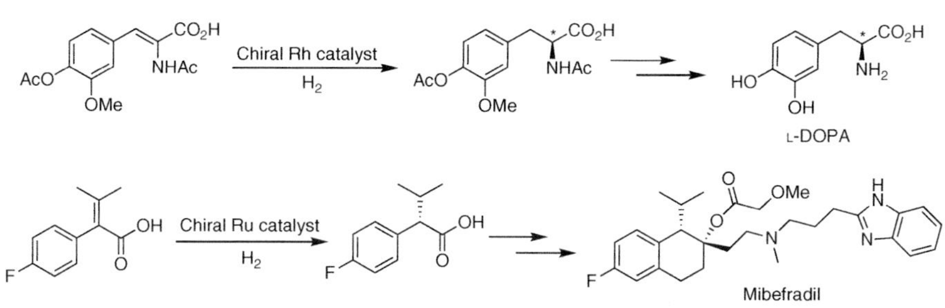

Scheme 1.13 The application of asymmetric hydrogenation in the syntheses of L-DOPA and mibefradil.

1.3.4 Olefin Formation Reactions

1.3.4.1 Wittig Reaction

In 1954, Georg Wittig discovered the olefination reaction that bears his name [71]. The transformation used a new nucleophilic carbon source, a dipolar phosphonium ylide, to convert carbonyl compounds into alkenes (Scheme 1.14).

$$R_2C=O \quad + \quad P_3''R=CR'_2 \longrightarrow R_2C=CR'_2 \quad + \quad P_3''R=O$$

Scheme 1.14 Wittig olefination.

Wittig's work later earned him the Nobel Prize in Chemistry in 1979. To this date, no real viable solution has been presented to turn this into a catalytic version that could eliminate the problem of the large amount of phosphine oxide side product that is generated.

1.3.4.2 Metathesis Reactions

In 2005, Yves Chauvin, Robert Grubbs, and Richard Schrock won the Nobel Prize in Chemistry for the development of the metathesis method in organic synthesis. Most commonly, the catalysts used for metathesis reactions are based on ruthenium or molybdenum. The key intermediates in the catalytic cycle are a series of metallacyclobutanes initiated from the reaction between the carbene-containing catalyst and one of the starting olefins (Scheme 1.15).

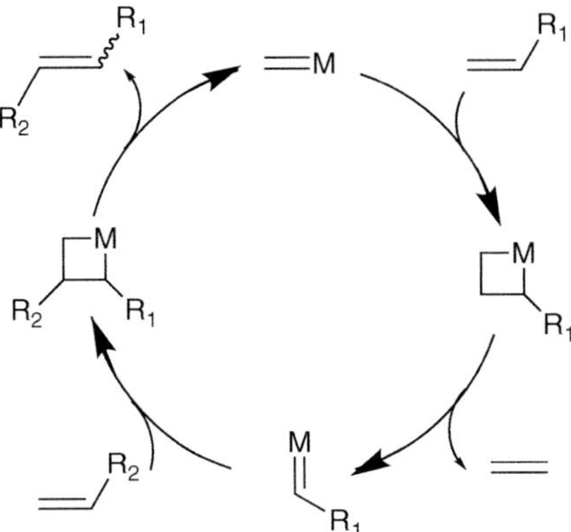

Scheme 1.15 General catalytic cycle for Metathesis reactions.

Olefin metathesis is finding increasing use in drug discovery and process chemistry on a commercial manufacturing scale, and with developments in catalyst systems, highly functionalized substrates can now be targeted. Exemplified by the first (publicly disclosed) case in the pharmaceutical industry to apply metathesis at scale, namely the hepatitis C drug candidate *ciluprevir* [72], chemists are able to access increasingly diverse chemistry on scale. Industrially relevant metathesis chemistry is discussed by John Phillips of Materia in Chapter 10.

1.3.4.3 Dehydrative Decarbonylation

Because of the ubiquity of carboxylic acids in industrial feedstocks obtained from biomass processing, such as the fatty acids, citric acid, and itaconic acid, an efficient method of transforming these groups into useful alkene functionalities, particularly linear olefins for the production of important polymers, is an attractive synthetic objective. Such a process – dehydrative decarbonylation – is traditionally achieved with palladium, rhodium, iridium, or nickel catalysts and can provide value-added odd-numbered alpha olefins desirable in the large-scale production of polymers, lubricants, and surfactants, from renewable even-numbered natural fatty acids. Dehydrative decarbonylation is discussed in by Alex John at California State Polytechnic University in Chapter 11.

1.3.4.4 Olefins as Starting Materials

Nucleophilic addition chemistry to an alkene is regarded as a fundamental methodology for organic synthesis; however, unactivated alkenes do not demonstrate this chemistry because of their lack of polarity. Substitution of the alkene with a carbonyl functionality such as in α,β-unsaturated carbonyl compounds renders the β position electrophilic by resonance and amenable to reaction with nucleophiles. With an appropriately selected nucleophile, conjugate addition results in C—C bond formation, saturation of the former olefin, and retention of the carbonyl group for further reactivity, thus is a very useful transformation in the synthesis of complex organic molecules relevant to the fine chemical and pharmaceutical industries. As stereocontrol is so important, practical catalytic methods that confer this are highly desirable for large-scale processes, and rhodium systems incorporating chiral ligands have shown particular utility in affecting enantioselective conjugate additions [73].

1.3.5 Poly- or Oligomerization Processes

In 1955, German Karl Ziegler and Italian Giulio Natta developed a process to produce stereoregular polymers from 1-alkenes on a commercial scale [74]. The catalyst class that bears their name generally comprises silica-supported titanium halides and alkyl aluminum activators, capable of polymerizing propylene and other 1-alkenes to highly crystalline isotactic polymers.

The Ziegler–Natta (ZN) polymerization process has been widely industrialized to provide almost the entire global supply of polyethylene and polypropylene. By contrast to the more dominant heterogeneous ZN system, which employs silica-supported titanium tri- or tetrachloride with added organoaluminum activators, homogeneous ZN catalysis generally employs a metallocene catalyst of the type $[Cp^R{}_2MCl_2]$ (Cp^R = cyclopentadienide derivative, M = Ti, Zr, and Hf) alongside a methylaluminoxane (MAO) activator. The homogeneous ZN system enables a structural control at the molecular level that is not available to heterogeneous ZN catalysts via the constrained geometry of the organometallic complex. Non-metallocene-based ZN catalysts incorporate a wide range of alternative ligand types bearing heteroatoms that impart different steric constraints at the metal center, leading to divergent polymerization regimes.

1.3.6 Photoredox Catalysis for Organic Synthesis

In recent years, there has been an increasing interest in the development of synthetic applications of photoinduced electron transfer chemistry [75, 76]. This technology has also been adopted by industrial teams and demonstrated in a number of research publications [77, 78]. The catalysts of choice for this type of reactions are in general Ru- or Ir-based organometallic complexes with bipyridine- or phenylpyridine-based ligands.

1.4 Conclusion and Outlook

The use of organometallic complexes as catalysts or stoichiometric reagents has revolutionized the way organic molecules are being synthesized in the pharmaceutical, agrochemical, and fine chemical industries. This has not been an overnight change but a very gradual development starting from an initial discovery followed by incremental improvements and modifications. Cross-coupling reactions such as Suzuki–Miyaura are good examples of this.

Without any doubt, the growing requirements for greener, more efficient, and streamlined processes in industry will lead to further development and refining of current technologies. In some cases, well-established organometallic reactivity can be exploited for the purpose of carrying out organic transformations in a new manner. The emerging area of photoredox catalysis in cross-coupling reactions is a good example. Another area, where similar developments are ongoing, is that of electrochemistry in organic synthesis. In other cases, it may be chemical engineering breakthroughs that make a major contribution to more efficient processes. Continuous processes carried out in flow reactors showcase examples of this emerging technology.

There is also a growing trend of working toward displacing precious metals by earth-abundant, metal-based catalysts. Although this is an achievable target for certain transformations, this is still to be established as a general viable alternative. One thing is certain; organometallic chemistry in the industry, whether as catalysts or stoichiometric reagents or whether transition metals or base metals, will remain the strongest pillar in organic synthesis especially from the perspective of process chemistry.

Biography

Carin C.C. Johansson Seechurn carried out her undergraduate studies at U.M.I.S.T. (University of Manchester Institute of Science and Technology), including a year abroad at CPE (l'Ecole Superieure Chimie, Physique, Electronique de Lyon) in France to receive her MChem with French in 2003. She then joined the University of Cambridge, UK, where she received her PhD. degree in the area of organocatalysis working under Dr Matthew Gaunt. After the completion of the Ph. D studies in 2006, she continued research with Dr. Gaunt, now in the area of palladium catalysis. In 2008, she joined the Catalysis and Chiral Technologies division of Johnson Matthey in the United Kingdom, based at Royston. In 2019, she moved to Cambridge. Carin is involved in the development, scale-up, and commercialization of new homogeneous metal catalysts. She is a coauthor of a number of publications, book chapters, and patents on the topic of homogeneous catalysis.

Ben M. Gardner, having obtained a first-class undergraduate masters' degree in Chemistry in 2008, Ben was awarded his PhD degree from the University of Nottingham in 2012 under the supervision of Stephen T. Liddle, his thesis entitled "The Chemistry of Uranium Triamidoamine Complexes." He continued to work in the Liddle group as a postdoctoral researcher investigating various aspects of nonaqueous synthetic actinide chemistry until 2016 when he took up the position of research chemist in the Homogeneous Catalysts division of Johnson Matthey plc. based in Royston, UK, where he worked in the development and scale-up of precious metal catalysts. In 2018, Ben took up his current position of scientist at Cambridge Display Technology Ltd., where he is developing new conductive materials for organic electronics applications.

Thomas J. Colacot is working at Millipore Sigma (a business of Merck KGaA, Darmstadt, Germany) since May 2018 in the United Sates as an R&D Fellow, Director – Global Technology Innovation (Lab and Specialty Chemicals). His expertise is in the areas of new product and technology development; process R&D; scale-up; and tech transfers of organic, organometallic, and fine chemicals relevant for pharmaceutical, electronic, and biological applications with very strong interactions/partnership with customers. Before this, he worked as a global R&D manager/technical fellow in Homogeneous Catalysis of Johnson Matthey. He is considered to be one of the leading industrial experts in cross-coupling. He is a coauthor of about 100 articles, which include reviews, peer-reviewed publications, and books such as "New Trends in Cross-Coupling: Theory and Applications" (RSC) and "Organometallics in Process Chemistry – Topics in Organometallic Chemistry" (Springer, in process). He has also given over 400 presentations in national and international conferences, universities, and chemical and pharmaceutical companies. He is a member of many national committees and a visiting professor at IIT Bombay, Mumbai. His contributions to the field have resulted in many awards and accolades, some of which are the 2017 Catalysis Club of Philadelphia award for the outstanding contributions in catalysis, 2015 American Chemical Society National Award in Industrial Chemistry, 2015 IPMI Henry Alfred Medal, 2016 Chemical Research Society of India CRSI medal, 2016 IIT Madras Distinguished Alumnus Award, and Royal Society of Chemistry 2012 Applied Catalysis Award and Medal. In 2018, Merck KGaA recognized him as globally one of the most outstanding researchers. Thomas holds a PhD degree in Chemistry with an MBA degree and is a fellow of the Royal Society of Chemistry.

References

1 Ostroverkhova, O. (2016). *Chem. Rev.* 116: 13279–13412.

2 For seminal publication, see: Tang, C.W. and Vanslyke, S.A. (1987). *Appl. Phys. Lett.* 51: 913–915.

3 Singh, D., Nishal, V., Baghwan, S. et al. (2018). *Mater. Des.* 156: 215–228.

4 Mark, J.E., Schaefer, D.W., and Lin, G. (2015). Types of polysiloxanes. In: *The Polysiloxanes*, 32–67. New York: Oxford University Press.

5 Lewis, L.N., Stein, J., Gao, Y. et al. (1997). *Platinum Met. Rev.* 41: 66.

6 Hoff, R. and Mathers, R.T. (eds.) (2010) *Handbook of Transition Metal Polymerization Catalysts*, Onlinee. Wiley.

7 Claverie, J.P. and Schaper, F. (2013). *MRS Bull.* 38: 213–218.

8 Choinopoulos, I. (2019). *Polymers* 11: 298–329.

9 Johansson Seechurn, C.C.C., DeAngelis, A., and Colacot, T.J. (2015). Introduction to new trends in cross-coupling. In: *New Trends in Cross-Coupling – Theory and Applications* (ed. T.J. Colacot), 1–19. Cambridge: RSC.

10 Gildner, P.G. and Colacot, T.J. (2015). *Organometallics* 34: 5497–5508.
11 Hagen, J. (2015). Homogeneous catalysis with transition metal catalysts. In: *Industrial Catalysis: A Practical Approach*, 17–46. Weinheim: Wiley-VCH.
12 Klein, A., Goldfuss, B., and van der Vlugt, J.-I. (2018). *Inorganics* 6: 19–22.
13 DeAngelis, A. and Colacot, T.J. (2015). *New Trends in Cross-Coupling – Theory and Applications*, 20–90. Cambridge: RSC.
14 McKinley, N.F. and O'Shea, D.F. (2006). *J. Org. Chem.* 71: 9552–9555.
15 Wilke, G. and Müller, H. (1956). *Chem. Ber.* 89: 444.
16 Keim, W. (2013). *Angew. Chem. Int. Ed.* 52: 12492–12496.
17 Sirois, L.E., Zhao, M.M., Lim, N.-K. et al. (2019). *Org. Process. Res. Dev.* 23: 45–61.
18 Vaska, L. and DiLuzio, J.W. (1962). *J. Am. Chem. Soc.* 84: 679–680.
19 Matsuda, T., Tsuboi, T., and Murakami, M. (2007). *J. Am. Chem. Soc.* 129: 12596–12597.
20 Chauvin, Y. (2006). *Angew. Chem. Int. Ed.* 45: 3740–3747.
21 Sunley, G.J. and Watson, D.J. (2000). *Catal. Today.* 58: 293–307.
22 Kreis, M., Palmelund, A., Bunch, L., and Madsen, R. (2006). *Adv. Synth. Catal.* 348: 2148–2154.
23 Osborn, J.A. and Wilkinson, G. (1967). *Inorg. Synth. Inorg. Synth.* 10: 67.
24 Wilkins, R.G. (1991). *Kinetics and Mechanism of Reactions of Transition Metal Complexes*, 2e. Weinheim: VCH.
25 Ramkumar, N. and Nagarajan, R. (2013). *J. Org. Chem.* 78: 2802–2807.
26 Sinn, H. and Kaminsky, W. (1980). *Adv. Organomet. Chem.* 18: 99–149.
27 Spessard, G. and Miessler, G. (2010). *Organometallic Chemistry*, 2e, 285–289. Oxford University Press.
28 Cano-Yelo, H. and Deronzier, A. (1984). *J. Chem. Soc., Perkin Trans.* 2: 1093–1098.
29 Cano-Yelo, H. and Deronzier, A. (1987). *J. Photochem.* 37: 315–321.
30 Francisco de Azeredo, S.O. and Figueroa-Villar, J.D. (2014). *World J. Pharm. Pharm. Sci.* 3: 1362–1379.
31 Narayanam, J.M.R. and Stephenson, C.R.J. (2011). *Chem. Soc. Rev.* 40: 102–113.
32 Sabatier, P. and Senderens, J.B. (1902). *Compt. Rend.* 134: 514–516.
33 Johansson Seechurn, C.C.C., Kitching, M.O., Colacot, T.J., and Snieckus, V. (2012). *Angew. Chem. Int. Ed.* 51: 5062–5085.
34 Li, H., Johansson Seechurn, C.C.C., and Colacot, T.J. (2012). *ACS Catal.* 2: 1147–1164.
35 Eisch, J.J. (2002). *Organometallics* 21: 5439–5463.
36 G. S. Silverman, P. E. Rakita, *Kirk-Othmer Encyclopedia of Chemical Technology*, doi:https://doi.org/10.1002/0471238961.0718090719091222.a01, 2000.
37 Kharasch, M.S. and Fields, E.K. (1941). *J. Am. Chem. Soc.* 63: 2316–2320.
38 Kharasch, M.S. and Reinmuth, O. (1954). *Grignard Reactions of Nonmetallic Substances*, 1146–1132. New York: Prentice-Hall.
39 Kharasch, M.S. and Fuchs, C.F. (1943). *J. Am. Chem. Soc.* 65: 504–507.
40 Thomas, A.A., Wang, H., Zahrt, A.F., and Denmark, S.E. (2017). *J. Am. Chem. Soc.* 139: 3805–3821.

41 Lennox, A.J.J. and Lloyd-Jones, G.C. (2013). *Angew. Chem. Int. Ed.* 52: 7362–7370.
42 Carrow, B.P. and Hartwig, J.F. (2011). *J. Am. Chem. Soc.* 133: 2116–2119.
43 Chinchilla, R. and Nájera, C. (2011). *Chem. Soc. Rev.* 40: 5084–5121.
44 Selected example of Cu-free Sonogashira reaction:Pu, X., Li, H., and Colacot, T.J. (2013). *J. Org. Chem.* 78: 568–581.
45 Selected example of Cu-free Sonogashira reaction:Handa, S., Smith, J.D., Zhang, Y. et al. (2018). *Org. Lett.* 20: 542–545.
46 For proposed mechanism of Cu-free Sonogashira:Gazvoda, M., Varant, M., Pinter, B., and Košmrlj, J. (2018). *Nat. Commun.* 9: 4814–4823.
47 ICH guideline Q3D on elemental impurities, EMA/CHMP/ICH/353369/2013, https://www.ema.europa.eu (accessed 4 March 2019).
48 Janowicz, A.H. and Bergman, R.G. (1982). *J. Am. Chem. Soc.* 104: 352–354.
49 Hoyano, J.K. and Graham, W.A.G. (1982). *J. Am. Chem. Soc.* 104: 3723–3725.
50 For selected publication, see: Caron, L., Campeau, L.-C., and Fagnou, K. (2008). *Org. Lett.* 10: 4533–4536.
51 Neufeldt, S.R. and Sanford, M.S. (2012). *Acc. Chem. Res.* 45: 936–946.
52 For selected publication, see: Chen, M.S., Prabagaran, N., Labenz, N.A., and White, M.C. (2005). *J. Am. Chem. Soc.* 127: 6970–6971.
53 For selected publication, see: Xia, G., Weng, J., Liu, L. et al. (2019). *Nat. Chem.* 11: 571–577.
54 Ishiyama, T., Takagi, J., Ishida, K. et al. (2002). *J. Am. Chem. Soc.* 124: 390.
55 Sieser, J.E., Maloney, M.T., Chisowa, E. et al. (2018). *Org. Process. Res. Dev.* 22: 527–534.
56 Cheung, H., Tanke, R.S., and Torrence, G.P. (2002). Acetic acid. In: *Ullmann's Encyclopedia of Industrial Chemistry*. Weinheim: Wiley-VCH.
57 Jones, J.H. (2000). *Platinum Met. Rev.* 44: 94–105.
58 Barnard, C.F.J. (2008). *Organometallics* 27: 5402–5422.
59 E. K. Rideal (2019), Paul Sabatier – Obituary notices. https://royalsocietypublishing.org/doi/pdf/10.1098/rsbm.1942.0006 (accessed 12 February 2019).
60 Normann, W. (1903). Process for converting unsaturated fatty acids or their glycerides into saturated compounds, DE141029.
61 Ojima, I., Tsai, C., Tzamarioudaki, M., and Bonafoux, D. (2004). *Org. React.* https://doi.org/10.1002/0471264180.or056.01.
62 Whiteker, G.T. and Cobley, C.J. (2012). *Top Organomet Chem.* 42: 35–46.
63 Jira, R. (2009). *Angew. Chem. Int. Ed.* 48: 9034–9037.
64 Elschenbroich, C. (2006). *Organometallics*. Weinheim: Wiley-VCH.
65 Osborn, J.A., Wilkinson, G., and Young, J.F. (1965). *Chem.Commun.*
66 Coffey, R.S. (1968). Hydrogenation process, GB1121643.
67 Noyori, R., Ohkuma, T., Kitamura, M. et al. (1987). *J. Am. Chem. Soc.* 109: 5856.
68 Ault, A. (2002). *J. Chem. Educ.* 79: 572.
69 Knowles, W.S. (2002). *Angew. Chem. Int. Ed.* 41: 1998–2007.
70 Jacobsen, E.N., Pfaltz, A., and Yamamato, H. (eds.) (1999). *Comprehensive Asymmetric Catalysis*, 1443–1445. Berlin; New York: Springer.
71 Wittig, G. and Schöllkopf, U. (1954). *Chemische Berichte* 87: 1318.
72 N. Yee, X. Wei, C. Senanayake Challenge and opportunity in scaling-up metathesis reaction: synthesis of Ciluprevir (BILN 2061), In *Sustainable*

Catalysis, Eds. P. J. Dunn, K. K. (Mimi) Hii, M. J. Krische, M. T. Williams, pp 215-232, 2013, *Wiley*

73 Burns, A.R., Lam, H.W., and Roy, I.D. (2017). *Org. React.* 93: 1.

74 Cecchin, G., Morini, G., and Piemontesi, F. (2003). Ziegler–Natta catalysts. In: *Kirk-Othmer Encyclopedia of Chemical Technology*. Wiley-VCH.

75 Stephenson, C. and Yoon, T. (2016). *Acc. Chem. Res.* 49: 2059–2060, special edition "Photoredox Catalysis in Organic Chemistry".

76 Shaw, M.H., Twilton, J., and MacMillan, D.W.C. (2016). *J. Org. Chem.* 81: 6898–6926.

77 For selected example, see: Huff, C.A., Cohen, R.D., Dykstra, K.D. et al. (2016). *J. Org. Chem.* 81: 6980–6987.

78 For selected example, see: Hsieh, H.-W., Coley, C.W., Baumgartner, L.M. et al. (2018). *Org. Process Res. Dev.* 22: 542–550.

2

Design, Development, and Execution of a Continuous-flow-Enabled API Manufacturing Route

Alison C. Brewer[1], *Philip C. Hoffman*[1], *Timothy D. White*[1], *Yu Lu*[1], *Laura McKee*[1,*], *Moussa Boukerche*[1,*], *Michael E. Kobierski*[1], *Nessa Mullane*[2], *Mark Pietz*[1], *Charles A. Alt*[1,†], *Jim R. Stout*[3], *Paul K. Milenbaugh*[3], *and Joseph R. Martinelli*[1]

[1] *Small Molecule Design and Development, Lilly Research Laboratories, Eli Lilly and Company, 1200 W Morris Street, Indianapolis, IN 46285, USA*
[2] *Eli Lilly Kinsale Limited, Dunderrow, P17 NY71 Kinsale, Co. Cork, Ireland*
[3] *D&M Continuous Solutions, LLC, 8496 Georgetown Rd, Indianapolis, IN 46268, USA*

It is a long and arduous journey from an initial hit in a biological activity screen to a medicine being used to improve the lives of people. Each molecule that makes this journey faces myriad challenges along the way. Some of these challenges are inherent to the properties of the molecule itself, whereas others arise from the process being used. Complications can also arise from numerous other sources, some of which are more practical concerns such as the source and supply of raw materials, the logistics of moving materials from one manufacturing site to another, or equipment cleaning protocols. Other complications are less obvious, such as the time pressures that can mount as development efforts are stopped and restarted to match clinical milestones or the challenges associated with transferring technical packages to contract manufacturing organizations (CMOs) located all over the world. This collection of challenges must be navigated during development and throughout manufacturing campaigns such that the resulting commercial process is robust and reliably produces high-quality active pharmaceutical ingredients (APIs) that provide the desired therapeutic benefit.

Chemistry manufacturing and controls (CMCs) refers to the collection of development activities related to the drug substance and drug product, including the manufacturing process and control strategy to ensure high quality, which is regulated. The control strategy is composed of unit operations and analytical methods by which the API or drug product is shown to have sufficient quality, and generally, this requires limiting all impurities to very low levels. The thresholds for impurities in a drug substance are outlined by the International Conference on Harmonisation (ICH) [1] and are shown in Table 2.1 [2]. Herein, the terms drug substance and API will be used interchangeably, as defined by

*Current address: AbbVie 1401 Sheridan Road, North Chicago, IL 60064, USA.
†Currently retired from Eli Lilly and Company.

Organometallic Chemistry in Industry: A Practical Approach, First Edition.
Edited by Thomas J. Colacot and Carin C.C. Johansson Seechurn.

Table 2.1 Thresholds for impurities in a drug substance.

Maximum daily dose	Reporting threshold	Identification threshold	Qualification threshold
≤2 g/d	0.05%	0.10% or 1 mg/d(lower of the two applies)	0.15% or 1.0 mg/d (lower of the two applies)
>2 g/d	0.03%	0.05%	0.05%

the (ICH) guidance on Good Manufacturing Practices (GMP) for API manufacturing [3]. Developing manufacturing processes capable of routinely achieving this level of control is the main tenant of CMC organizations, which requires a sophisticated level of process understanding.

Typically, the level of process understanding increases with the phase of development, and this chapter will describe the developmental activities associated with a molecule in early-phase studies at Lilly as a potential anticancer agent. The work described was conducted in preparation for a scale-up campaign ("campaign 2") executed at four different sites in three different countries around the world. This chapter will provide an overview of the general approach to route design a high-level description of the impurity control strategy and highlight the importance of organometallic chemistry in API synthesis by detailing the development of two transition-metal-catalyzed cross-coupling steps, both of which were carried out under GMP conditions, as shown in Scheme 2.1.

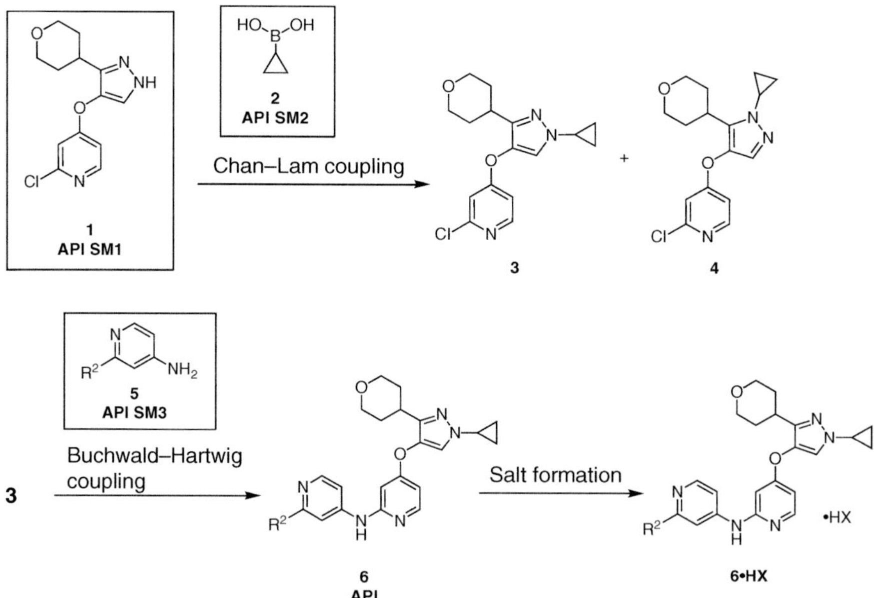

Scheme 2.1 Final steps conducted under GMP for the scale-up of API **6•HX**.

2.1 Continuous-flow-Enabled Synthetic Strategy

Rapidly restarting development of a manufacturing process for **6** required to achieve aggressive project timelines and necessitated initiation of an API campaign within six months of the CMC team re-engaging on the project. The first step for the team was to assess the synthetic route and determine if it was suitable to eventually implement. Ultimately, the team needed to design a synthetic route that would afford an enduring and robust process with an associated control strategy to ensure the reliable production of high-quality API.

The synthetic route used for the previous scale-up campaign ("campaign 1") is shown in Scheme 2.2. This route featured many desirable attributes and positioned the team to move rapidly. The route had already been successfully scaled-up to make approximately 19 kg of **6•HX**. This route also offered numerous options for developing a robust control strategy because it featured several

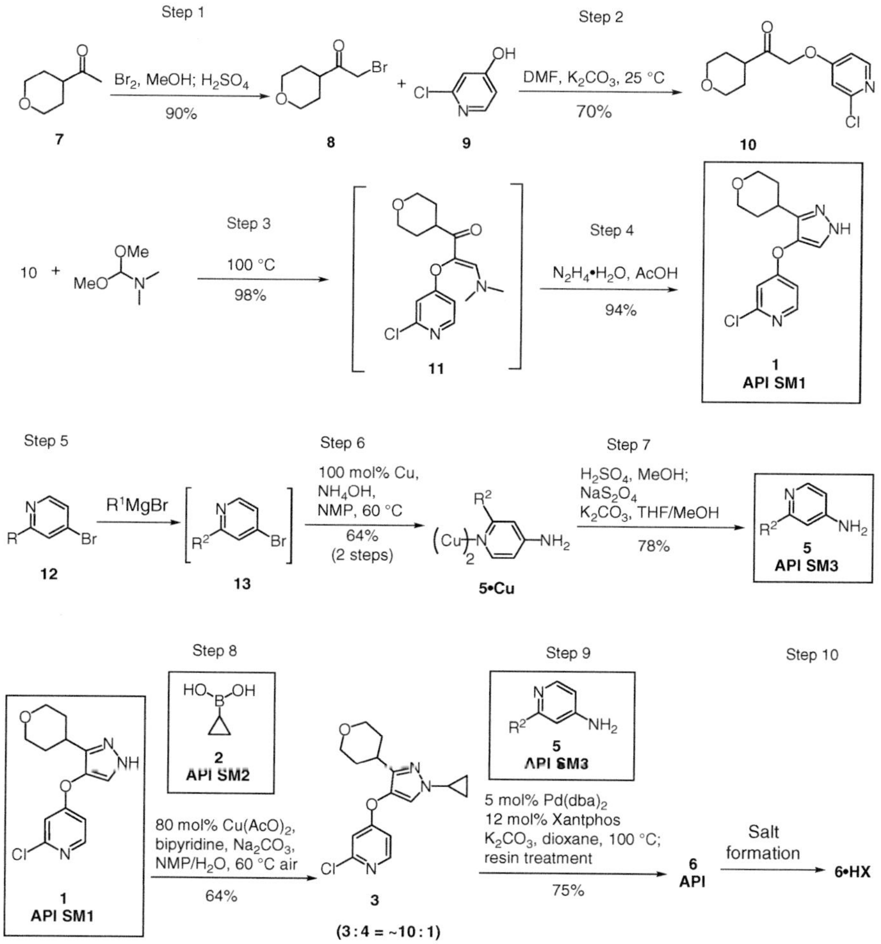

Scheme 2.2 Synthetic route used for campaign 1.

isolated intermediates that could be used as control points. In addition, the intermediates in the route flagged in our *in silico* genotoxicity assessment tested negative in AMES assays and were therefore not a source for potential genotoxic impurities (GTIs) in the final API. Despite a number of attractive features, there was one outstanding process, safety concern, with regard to the use of oxygen as a reagent in step 8. Aerobic processes on scale are seldom used in the pharmaceutical industry because of safety concerns; however, Lilly has been spearheading the use of continuous flow systems with dilute oxygen in nitrogen to enable aerobic oxidations in pharmaceutical manufacturing. Although our technical team felt confident that a continuous flow process could be developed for step 8, other synthetic approaches were examined.

Several approaches to **6** were examined and the main three strategies are summarized in Scheme 2.3. Although numerous other approaches and variations of the approaches shown in Scheme 2.3 were considered, the team focused on these prioritized options for the sake of rapid implementation.

The first option, SR1, is the same synthetic approach used in campaign 1, as shown in Scheme 2.2. Despite its length, this approach has several attractive features including a proven ability to deliver kilogram-scale quantities of API, opportunities for achieving improvement for the next campaign, no GTI intermediates, is amenable to continuous flow process development, and includes ample isolations in the endgame, allowing for development of a robust control strategy.

The other synthetic approaches are generalized as the two retrosynthetic approaches shown in Scheme 2.3, SR2 and SR3. Although these routes were generally shorter, they faced various issues. Route SR2 attempted to solve the regiochemical issues of the step 8 Chan–Lam coupling by using

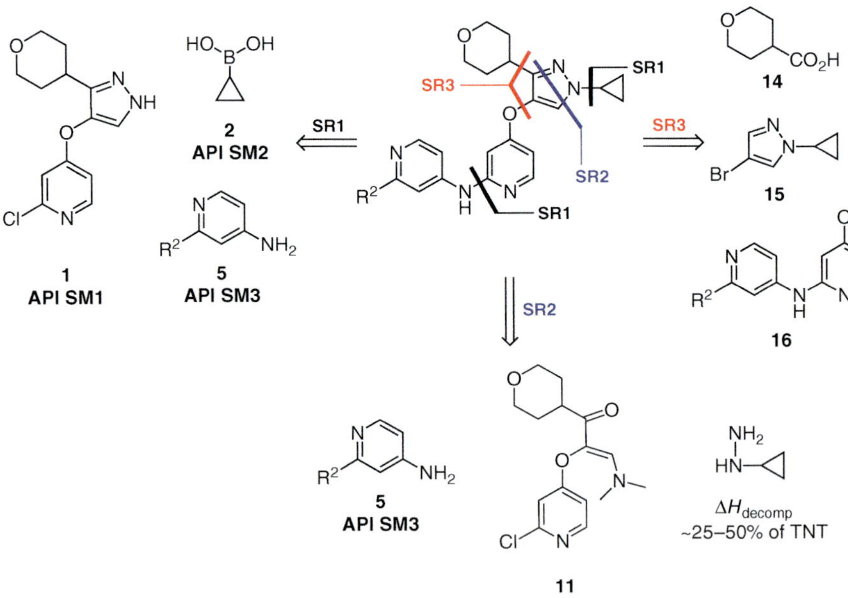

Scheme 2.3 Some retrosynthetic approaches to 6.

cyclopropylhydrazine in the step 4 condensation reaction. However, there were two major problems with this strategy. The first problem was related to process safety: cyclopropylhydrazine has an enthalpy of decomposition in the range of 25–50% of that of TNT. Although this is a major issue, there are approaches for on-demand generation and consumption that can be employed with continuous flow processing to allow for the use of such hazardous reagents [4]. More significantly, this approach did not improve the regioselectivity issue as originally hypothesized. When the experiment was run under conditions similar to those typically used in step 4, the ratio was in favor of the undesired isomer rather than the desired isomer (Scheme 2.4). Given these limitations, this synthetic approach was abandoned.

The synthetic route SR3 was a significantly more convergent and efficient approach. This route featured two key transformations: an Ullmann coupling with a hydroxypyridine derivative, such as **16**, and a Minisci-type decarboxylative C–H functionalization. Various iterations of this route were explored, but it quickly became apparent that the Ullmann coupling with the desired hydroxypyridine would not be trivial. Further investigations with a model system showed that the Ullmann coupling of compounds **17** or **18** with **9** or 4-hydroxypyridone was unproductive, Scheme 2.5.

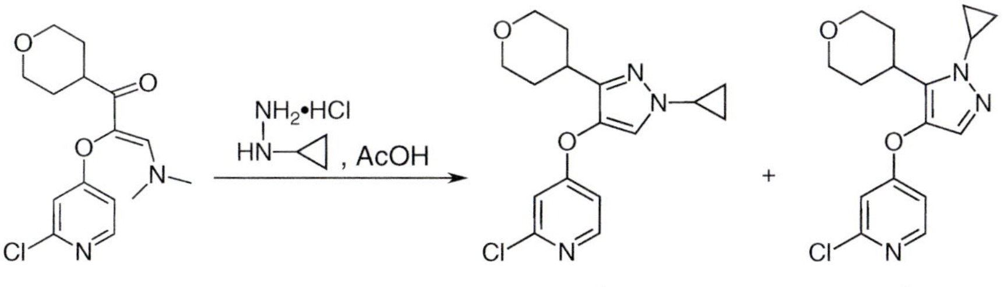

Ratio of 3 : 4 = 1 : 3
products not isolated

Scheme 2.4 Cyclopropylhydrazine condensation to test the viability of SR2.

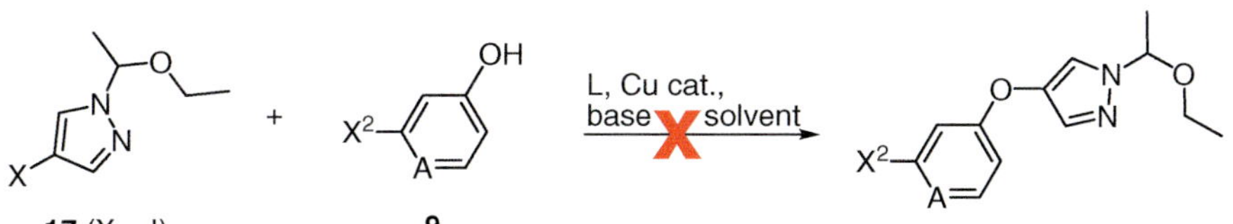

17 (X = I)
18 (X = Br)

9
(X^2 = Cl, A = N)
4-hydroxypyridine
(X^2 = H, A = N)
phenol
(X^2 = H, A = C–H)

Scheme 2.5 Model ullmann couplings evaluated for SR3.

Control reactions using phenol as a coupling partner in the Ullmann reaction were also unproductive, Scheme 2.5. Despite the promise of a more convergent route approach, the short timeline to deliver the material required that the team deprioritize this route.

After careful consideration of the options described above, the team decided to design a manufacturing strategy around SR1 in order to maximize flexibility and began to put efforts to optimize the route used for early-phase material deliveries. The main drawback to this approach was the process safety issue associated with oxygen in step 8. However, as stated above, Lilly has demonstrated using high pressures of dilute oxygen in nitrogen as a viable safe continuous flow aerobic process. This methodology was envisioned to enable this synthetic approach for the upcoming scale-up campaign.

It is worth noting that all of the proposed routes heavily feature applied organometallic chemistry. SR1 features two late-stage transformations of this sort: step 8 Chan–Lam and step 9 Buchwald–Hartwig cross-coupling. SR2 and SR3 also feature applied organometallic chemistry. SR2 featured the same Ullmann chemistry to make **API SM3** and a similar Buchwald–Hartwig reaction for the endgame. SR3 was proposed using a variety of transition-metal-catalyzed reactions including the Ullmann coupling and some version of a contemporary C–H functionalization.

In order to develop this process, several issues needed to be addressed. The development work done to improve the processes for making **1** and **5** will not be discussed in this chapter. The overall synthetic approach used for both remained unchanged and throughput was improved by implementing process improvements related to reagent addition order, reaction workup, and product isolation. Process development for the Chan–Lam coupling (Section 2.2) and the Buchwald–Hartwig coupling (Section 2.3) will be discussed in detail. Additionally, as stated above, because API quality is a primary consideration for CMC development, an impurity control strategy for this route will be presented (Section 2.4).

2.2 Design and Scale-up of Chan–Lam Coupling

A Chan–Lam coupling between **1** and cyclopropylboronic acid (**2**) proved to be an efficient means of preparing **3**, the penultimate of the API (Scheme 2.6) [5]. Typical literature conditions [6] were used initially, employing stoichiometric copper acetate in combination with sodium carbonate in 1,2-dichloroethane under an atmosphere of air (Scheme 2.6, conditions A). Two regioisomers can form in the cross-coupling reaction depending on which nitrogen in the pyrazole ring is functionalized, and when bipyridine was used as the ligand, formation of the desired regioisomer **3** was favored in a 10 : 1 ratio over **4**.

Following this positive proof of concept for the use of a Chan–Lam coupling to construct the key C—N bond, second-generation conditions eliminated the undesirable dichloroethane solvent [7] and used sub-stoichiometric quantities of copper (Scheme 2.6, conditions B). Under these conditions, concerns over the

1 + 2 → Conditions → 3 (Desired isomer) + 4

Conditions A	Conditions B
1.1 equiv $Cu(OAc)_2$	0.8 equiv $Cu(OAc)_2$
1.1 equiv 2,2′-bipyridine	0.8 equiv 2,2′-bipyridine
2 equiv K_2CO_3	3.2 equiv Na_2CO_3
1,2-dichloroethane	NMP/H2O (2 : 1)
Bubbling air	Bubbling air
0 °C	60 °C

Scheme 2.6 Cu-catalyzed Chan–Lam coupling to the desired isomer, **3**.

use of air were overcome by using a mixture of NMP and water as the reaction solvent. Under 1 atm of air, a 2 : 1 mixture of NMP/H_2O exhibited no flash up to 150 °C. Bubbling air could safely be used as the oxidant without the risk of an ignition event because the reaction operated at >25 °C below the flash point of the mixture. These conditions were employed in a 25 kg manufacturing campaign to prepare 19 kg of **3** (Scheme 2.6, conditions B).

Despite the mitigation of flammability concerns over the use of O_2 by the NMP/H_2O conditions, there were a number of additional challenges with the process. The transformation was sensitive at scale; hence, the process required much higher catalyst loadings to achieve good conversion when scaled up. Almost a stoichiometric amount of Cu catalyst (0.8 equiv) was required for the 25-kg campaign even though the chemistry was carried out in a series of continuously stirred tank reactors (CSTRs) to minimize scale-up effects. Because of the poor solubility of the Cu catalyst in the solvent matrix and the use of inorganic base, the reaction mixture was heterogeneous. Additionally, **3** tended to form an oil layer as the reaction progressed. This complex multiphase system suffered from clogging in transfer lines between the CSTRs on both 1.5 and 25 kg scale.

An alternative approach to the design and implementation of the Chan–Lam coupling was needed based on the issues outlined above. Although other routes to **3** were entertained (*vide supra*), development of the Chan–Lam coupling was prioritized to enable rapid delivery of the next campaign. We were particularly focused on designing a process that allowed for safe use of O_2 while improving robustness and eliminating scale sensitivity. Although the previously implemented NMP/H_2O solvent system did provide a strategy for the safe use of air as a source of O_2, the tendency of **3** to phase separate as an oil in the presence of water was a liability. Although the combination of air and organic solvents is a significant safety hazard because of flammability, it has been demonstrated that operating below the LOC of the reaction solvent by using diluted air (e.g. 5% O_2 in N_2) is a practical approach to safe operation [8]. Elevated pressures and constant gas flow are usually required to ensure that sufficient levels of O_2 are present when employing diluted air. Continuous processing can readily accommodate these pressure and gas flow requirements, and our previous experience with vertical vapor–liquid pipes-in-series continuous reactors led us to believe that this class of reactors would be particularly well suited to this aerobic oxidation [9].

Table 2.2 Summary of screening conditions for the oxidative coupling of **1** and **2**.

1 + **2** (3 equiv) → **3** + **4**; 25 mol% $Cu(OAc)_2$, 38 mol% ligand, 3 equiv $i Pr_2NEt$, Solvent, 1000 psig 6.25% O_2/N_2, 60 °C

Entry	Ligand	Solvent	Additive[a]	%Yield 3	3 : 4	Homogeneous?
1	Bipy	DMF	—	91	13.1	Yes
2	Bipy	DMSO	—	62	11.2	No
3	Bipy	Acetone	—	85	13.3	No
4	Bipy	NMP	—	74	13.0	No
5	Bipy	THF	—	53	12.0	No
6	Bipy	2-MeTHF	—	43	12.0	No
7	Pyridine	2-MeTHF	—	83	6.6	No
8	4,4′-OMe_2 bipy	2-MeTHF	—	30	13.3	No
9	4,4′-$^tBu_2$ bipy	2-MeTHF	—	84	12.3	No
10	Bipy	THF	Myristic acid	48	12.8	No
11	Bipy	2-MeTHF	Myristic acid	48	13.2	No
12	Bipy	THF/DMSO (1 : 1)	Myristic acid	41	14.6	Yes
13	Bipy	2-MeTHF/ DMSO (1 : 1)	Myristic acid	47	15.0	No
14[b]	Bipy	THF/DMSO (1 : 1)	Myristic acid	89	14.9	Yes

a) The reaction was run at 75 °C.
b) 2 equiv relative to $Cu(OAc)_2$.

2.2.1 Development of Homogeneous Conditions

Based on previous work demonstrating that the vertical pipes-in-series reactor can accommodate both development- and manufacturing-scale vapor–liquid reactions,[1] our goal was to develop homogeneous conditions for the Chan–Lam coupling that would allow utilization of the vertical pipes-in-series reactor. To accomplish this, it was necessary to identify a reaction solvent that fully solubilized the starting materials and product as well as the copper catalyst. The insoluble carbonate base was also an issue and needed to be replaced, and soluble organic bases were the most logical replacements. Starting materials and products (**1**, **2**, and **3**) all had high solubility in NMP, DMF, DMSO, THF, 2-MeTHF, and acetone (>250 mg/ml). These solvents were used for the initial reaction screening along with several organic-soluble amine bases. In addition to high conversion and a favorable ratio of **3** to **4**, having a fully homogeneous reaction mixture was also prioritized (Table 2.2).

Amine bases were found to be a generally suitable, soluble replacement for the carbonate base used in earlier conditions.[2] The use of DMF as the solvent and iPr_2NEt as the base provided a proof of concept for a fully homogeneous Chan–Lam coupling with good conversion (Table 2.2, entry 1). However, DMF was not viewed as a long-term option because it is included on the REACh substance of very high concern list [10]. Other solvents, although showing moderate reactivity and comparable selectivity for the desired product, did not result in fully homogeneous reaction mixtures and were therefore not suitable for the pipes-in-series reactor (Table 2.2, entries 2–6). In addition to bipyridine, several other ligands were screened. The use of the monodentate ligand pyridine led to deterioration in the regioselectivity (Table 2.2, entry 7) while both 4,4′-OMe_2bipy and 4,4′-$^tBu_2$bipy showed a similar regioselectivity to bipy (Table 2.2, entries 8 and 9).

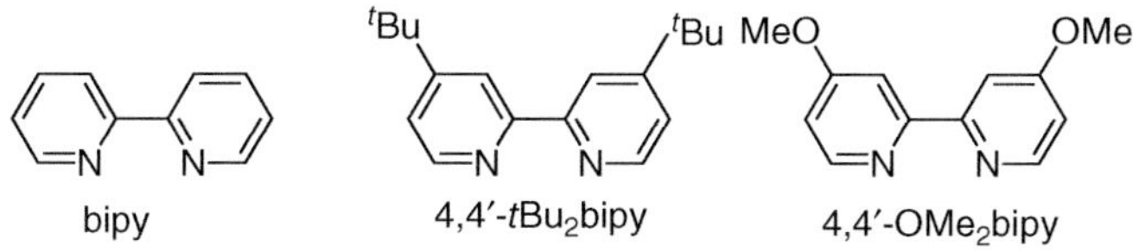

bipy   4,4′-tBu$_2$bipy   4,4′-OMe$_2$bipy

In a 2001 report, Buchwald and coworkers demonstrated that addition of myristic acid to copper-catalyzed Chan–Lam couplings increased the reaction rate [11]. The authors propose that this improvement is due to coordination of the myristic acid to the copper center, which increases the solubility of the copper catalyst in organic solvent. Hypothesizing that this strategy might improve the copper solubility in our system, a series of reactions were designed and executed using 2 equiv of myristic acid relative to copper. Although the solubility was improved, the system was not fully homogeneous with a precipitate forming at the end of the reaction. The addition of DMSO as a cosolvent prevented precipitation as the reaction reached completion and a 1 : 1 mixture of THF/DMSO

1 For details on the custom online HPLC described, see the supporting information of Ref. [12c].
2 iPr_2NEt, NEt_3 and 2,6-lutidine all showed acceptable performance in initial screening.

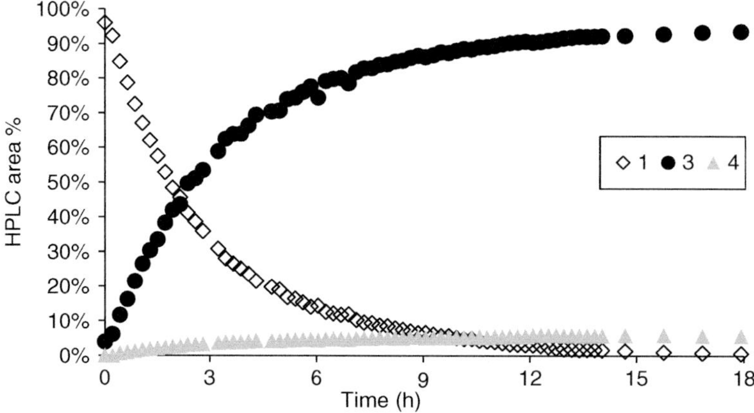

Figure 2.1 Reaction kinetics for the conversion of **1–3** using $Cu(OAc)_2$/bipy/myristic acid catalyst system.

allowed for a fully homogeneous solution at both the beginning and end of the reaction (Table 2.2, entry 12). The addition of DMSO as a cosolvent slowed conversion, so the reaction temperature was increased to 75 °C. This improved the rate without negatively affecting the reaction outcome (Table 2.2, entry 14). The reaction was then scaled to 1 g in a 100 ml autoclave using 1000 psi of 6.25% O_2/N_2 as the oxygen source and monitored by online high-performance liquid chromatography (HPLC); complete conversion of **1** was observed after 12 hours (Figure 2.1).

2.2.2 Application of a Platform Technology to Aerobic Oxidation

The homogeneous conditions were then demonstrated in a vertical pipes-in-series continuous reactor. A schematic of the pipes-in-series reactor is shown in Figure 2.2. Vapor and liquid travels in the same direction through the reactor, up through large diameter pipes and down through smaller diameter jumper tubes. There are two flow regimes: segmented flow in the small diameter tubing and bubble flow in the vertical pipes. This innovative design affords several practical and safety advantages:

1. The reactor allows for the use of high gas pressures and flow rates; the reaction can readily be run below the LOC of the solvent while maintaining an excess stoichiometry of O_2.
2. The pipes-in-series reactor design can be easily and readily scaled from small-volume lab scale (22 ml) up to large-volume manufacturing scale (>300 l).[3] Allowing rapid process development and facilitating process transfer into manufacturing.
3. The configuration of the pipes-in-series reactor allows for excellent vapor/liquid mass transfer rates.

---

3 Eight percentage O_2 was selected for initial experiments before the limiting oxygen concentration was known for the solvent system. However, there is a sufficient basis of safety in the lab scale reactor since flames cannot propagate in the jumper tubes due to the small internal diameter (0.0225″).

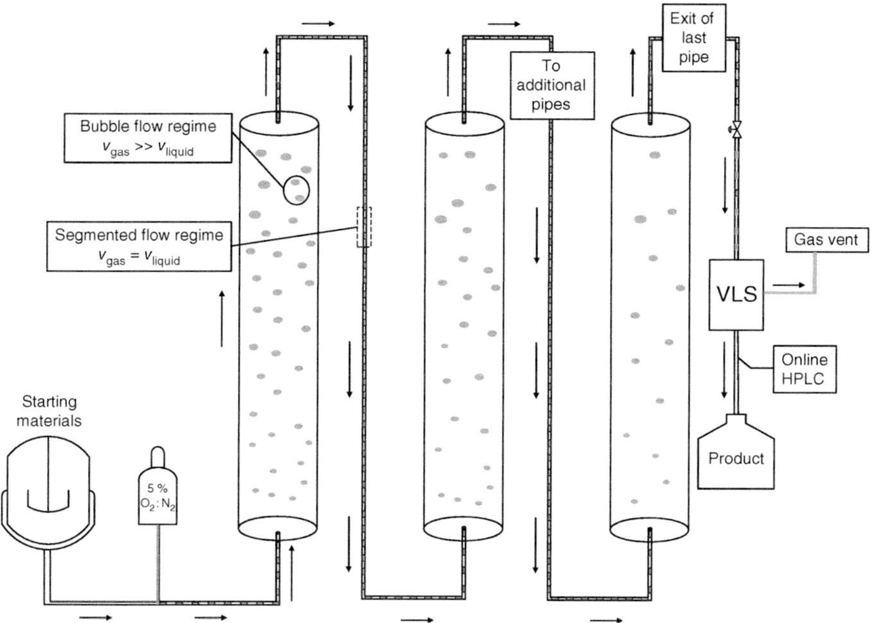

Figure 2.2 Pipes-in-series reactor design schematic.

With homogeneous reaction conditions in hand, we began exploring the transformation in a 75 ml research-scale vertical pipes-in-series reactor. There were two liquid feed streams and one gas feed stream into the reactor: (i) **1** and **2** in THF/DMSO; (ii) $Cu(OAc)_2$, bipyridine, myristic acid, and iPr_2NEt in THF/DMSO; and (iii) 8% O_2 in N_2.[3] Figure 2.3 shows the block flow diagram for the reaction. The product stream from the reactor was monitored by online HPLC.[1] Based on the reaction rate observed in batch reactions, liquid and gas flow rates were chosen to give a 12 hours tau.

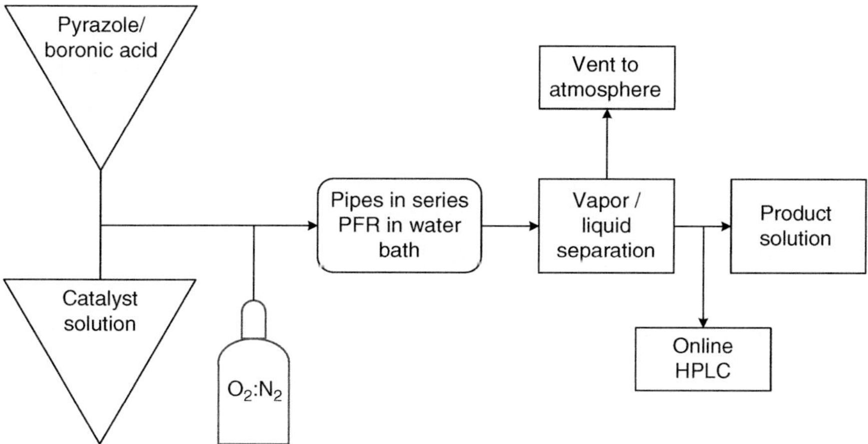

Figure 2.3 Setup of vapor-liquid pipes-in-series reactor.

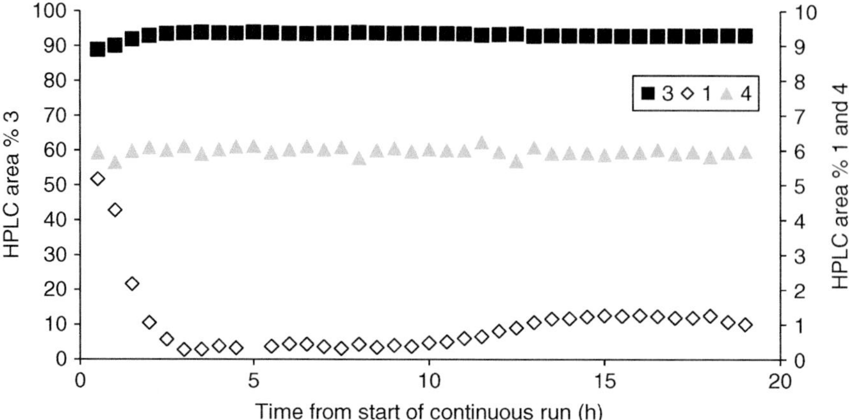

Figure 2.4 Online HPLC data trend for continuous run with 12 hours τ at 75 °C using 3 equiv of O_2 under 400 psig 8% O_2/N_2.

The initial results in the pipes-in-series reactor were comparable with batch experiments. Using 3 equiv of O_2, controlled by a flow gas rate of 8% O_2/N_2 at 400 psi, the reaction achieved steady state at 99% conversion and produced **3** with a 15 : 1 ratio relative to regioisomer **4** at 75 °C (Figure 2.4). Sampling at several points along the continuous reactor allows for reaction kinetics to be obtained. Intermediate samples were taken from pipes 11, 21, and 31 in addition to the final reaction sample taken after pipe 40 and the vapor–liquid separator. When using a 12 hours reaction time (τ), the samples from different pipes represented different reactions times: pipe 11 corresponded to a three hour reaction time, pipe 21 corresponded to a six hour reaction time, and pipe 31 corresponded to a nine hours reaction time. The samples at these intermediate points are analogous to sampling at different times in a batch reaction because they were all taken at

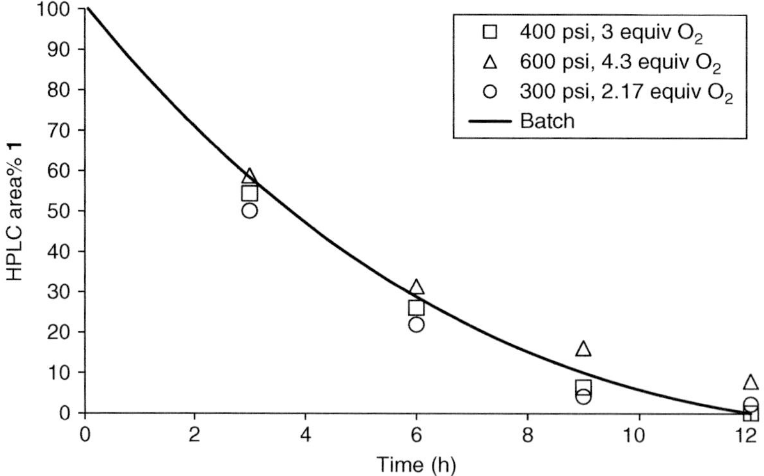

Figure 2.5 Reaction kinetics in the 75 ml pipes-in-series reactor compared to a 100 ml Parr.

the same time after the reactor reached steady state. The kinetic profile observed in flow was comparable with the profile observed in batch runs (Figure 2.5).

A series of pressures and O_2 stoichiometries were explored. Surprisingly, the reaction rate was only minimally affected by different combinations of O_2/N_2 pressures and flow rates. Although aerobic oxidation reactions are often sensitive to O_2 partial pressure and mixing, these conditions did not appear to be limited to mass transfer and were nearly independent of O_2 pressure.[4] Additional screening work at a 75 ml scale was done using 400 psi of 8% O_2/N_2 and 3.0 equiv of O_2 because these conditions were robust and afforded high conversion at steady state.

Next, a series of reaction conditions were examined in flow. A range of reaction temperatures was explored and the reaction was found to be considerably faster at 90 °C as compared to 75 °C. Minimal changes to the impurity profile of the crude reaction mixture between the two temperatures were observed (Figure 2.6).[5] **3** was formed with a 92% *in situ* yield in less than six hours using 12.5 mol% $Cu(OAc)_2$ with 18.75 mol% bipyridine, 25 mol% myristic acid, and 3 equiv of iPr_2NEt in 20 vol THF/DMSO (50/50).

2.2.3 Optimization of Reaction and Workup Parameters

Given the initial success carrying out a selective oxidative coupling between **1** and **2** in the continuous reactor, reaction workup conditions were explored. It became quickly evident that the myristic acid would be difficult to purge. Myristic acid is poorly soluble in water even as a carboxylate salt and was not removed from **3** using aqueous washes. Myristic acid was also not purged by crystallization. The use of an $EtOH/H_2O$ crystallization had significant advantages to the impurity control strategy, but the low solubility of myristic acid in $EtOH/H_2O$ mixtures led to poor rejection in the crystallization.

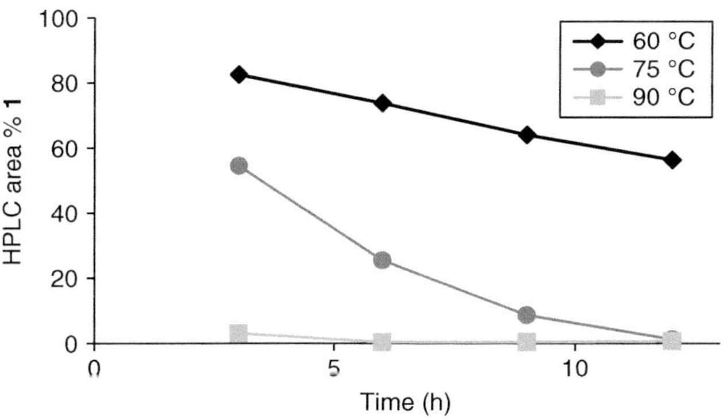

Figure 2.6 Reaction kinetics plots of temperature effect on conversion.

4 In a series of batch runs, doubling the O_2 partial pressure did not change the reaction rate, consistent with the results observed in flow.
5 The ratio of **3** : **4** decreased from 15.2 : 1 to 13.4 : 1 when transitioning from 75 to 90 °C which represents <1% decrease in the in situ yield of **3**.

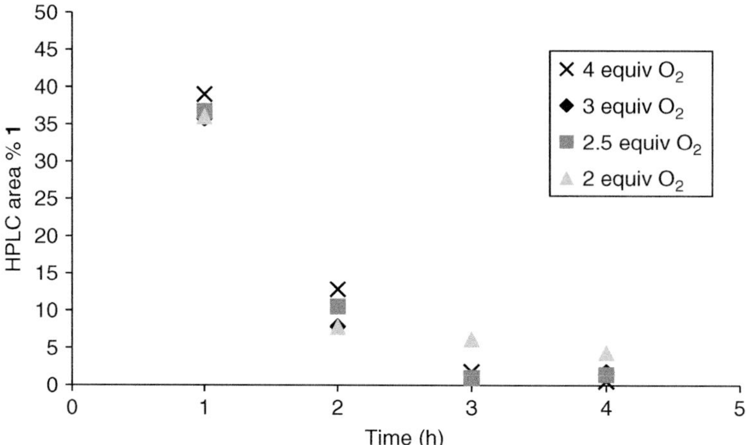

Figure 2.7 The effect of O_2 equivalents on conversion.

Other long-chain carboxylic acids were explored as alternatives to myristic acid in an effort to simplify the workup while maintaining homogeneity of the catalyst system. The addition of 2 equiv of valeric acid relative to $Cu(OAc)_2$ was similarly able to solubilize the copper catalyst and led to a comparable reaction profile.[6] Because valeric acid has significantly higher water solubility than myristic acid, it could be easily separated from **3** by aqueous washes.[7]

The reaction rate was not dependent on the concentration of iPr_2Net, which could be reduced to 1.5 equiv with no change in the reaction profile. The optimized conditions employed 12.5 mol% $Cu(OAc)_2$, 18.75 mol% bipyridine, 25 mol% valeric acid, and 1.5 equiv of iPr_2NEt in THF/DMSO (50/50) at 90 °C. Using 400 psi of 6.25% O_2/N_2 in the continuous reactor, the effect of O_2 stoichiometry was found to be minimal in the range of 2.5–4.0 equiv of O_2 (Figure 2.7)[8] and 3 equiv of O_2 was used moving forward.

With robust reaction conditions in hand, the workup and isolation were examined. Initially, the crude reaction mixture was diluted with 10 volumes of 2-MeTHF and subsequently washed with 1N HCl, 5 wt% aqueous $NaHCO_3$, and brine. This series of washes purged iPr_2NEt, valeric acid, and DMSO from the organic layer, but residual copper remained in the organic layer with **3** and this was problematic in the crystallization. Aqueous solutions commonly employed to remove copper salts such as NH_4Cl, NH_4OH, and EDTA were unsuccessful in this case. Next, treatment of the washed organic solution with a series of copper scavenger resins was explored. Many of the resins tested were effective for lowering the amount of copper in the solution and QuadraSil® TA looked especially promising. This particular resin contains a diethylenetriamine functional group and, given the miscibility of diethylenetriamine with water, we thought that it

6 Later in development, trifluoroacetic acid was also identified as a suitable additive for achieving a homogeneous reaction under the conditions described.
7 At 20 °C, the solubility of myristic acid in water is 20 mg/l and the solubility of valeric acid in water is 50 g/l.
8 The reaction stalled when 2.0 equiv of O_2 was used.

Table 2.3 Comparison of aqueous wash sequences for workup.

Wash	[Cu] in aqueous (ppm)	[Cu] in organic (ppm)
(A)		
1N HCl	795	4
5 wt% $NaHCO_3$ (aq.)	2.5	7
Water	2.5	8
(B)		
15 wt% Diethylenetriamine (aq.)	1260	<2.5
1N HCl	3.87	<2.5
5 wt% $NaHCO_3$ (aq)	<2.5	<2.5

might be possible to use an aqueous solution of diethylenetriamine to extract copper from the organic product solution into the aqueous layer. This approach would be less expensive and simpler than a resin treatment. This hypothesis was tested by diluting the crude reaction mixture with 10 volumes of 2-MeTHF and then washing with 10 volumes each of 15 wt% aqueous diethylenetriamine, 1N HCl, and 5 wt% aqueous $NaHCO_3$. The diethylenetriamine wash was effective in reducing the concentration of copper in the organic phase to below the limit of detection. The color was also notably lighter after the amine wash. After executing this series of aqueous washes, residual diethylenetriamine, iPr_2NEt, valeric acid, and DMSO were not detected in the resulting 2-MeTHF solution of **3**. The workup proved to be robust across a range of scales and afforded solutions of **3** with Cu levels below the limit of detection (Table 2.3).

The ethanol/water crystallization used in the previous campaign exhibited excellent rejection of regioisomer **4** and, with minor modifications, was successfully used again. The solution of **3** in 2-MeTHF was solvent exchanged with ethanol and diluted with water. The resulting solution of **3** in 3 volumes of ethanol and 1 volume of water at 45 °C was further diluted with antisolvent via slow addition of 3.5 volumes of water. The crystallization was seeded with 1% (w/w) **3** at 45 °C and the mixture was cooled to 5 °C. The product was isolated by filtering the resulting slurry, afforded **3** in good yield, and high purity with <0.05% **4**.

2.2.4 Safety Considerations for Aerobic Oxidation on Scale

Data were collected to ensure that appropriate process safety considerations had been addressed before scaling up the aerobic oxidation. The LOC of 50/50 THF/DMSO (%O_2 needed for an ignition event) was measured over a range of temperatures and pressures (Figure 2.8).[9] Pressure had only a small effect on

9 The LOC testing was conducted by Fauske & Associates, LLC.

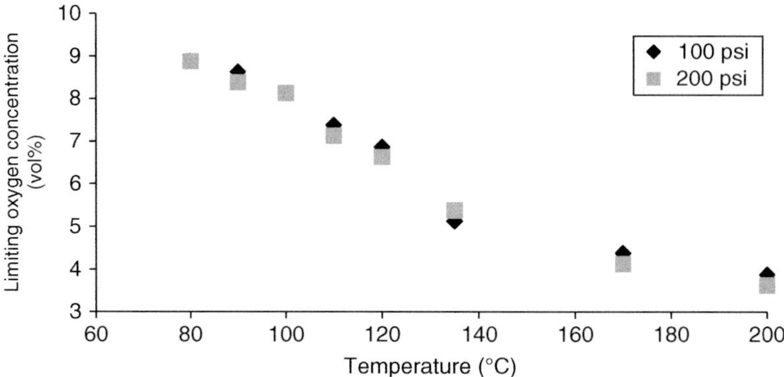

Figure 2.8 LOC data for THF/DMSO (50/50) under 100–200 psi at 80–200 °C.

the LOC while the temperature had a more notable effect [5]. At the planned operating temperature of 90 °C, the LOC at 100 and 200 psi was found to be 8.63% and 8.38%, respectively. We chose to implement the use of 5% O_2/N_2 going forward to operate within a sufficient margin of safety [5].[10]

Another safety consideration was the propensity of THF to form peroxides under oxidative conditions. The crude reaction solution of **3** and several intermediate organic-phase solutions from the workup were tested for the presence of THF peroxide by GCMS; no peroxide was detected in these samples.[11] Finally, ARC data were obtained for the starting and final reaction solutions and DSC was obtained for **1**, **2**, and **3**. For all samples, the exothermic decompositions were found to be well above the reaction temperature and the process was deemed to be safe based on these evaluations.

2.2.5 Continuous Scale-up and Manufacturing

After demonstrating the optimal reaction conditions in the 75 ml pipes-in-series reactor (Scheme 2.7), process scalability and robustness were tested in a 2 l reactor before transferring the process to manufacturing. Understanding the key differences in the reactors across a range of scales allowed for the development of a process that could be reliably and reproducibly taken from lab scale to manufacturing scale. Some of the notable differences in the reactor dimensions and characteristics across different reactor scales are given in Table 2.4. Some of the characteristics are necessarily different to accommodate higher throughput; others are the result of design to minimize the impact of scale changes on the overall

---

10 For processes where the oxygen concentration is continuously monitored, NFPA 69 requires a minimum safety margin of 2 vol% below the LOC when the LOC is ≥5 vol%. Our contract manufacturer used a more stringent margin of safety (60% of LOC), which also provides a wider safety margin in case of a temperature excursion. See: *NFPA 69: Standard on Explosion Prevention Systems*; National Fire Protection Association: Quincy, MA, 2014.

11 Typically, a colorimetric test in the form of a titration or test strip would be used to detect peroxides. In this case however, because the reaction solutions were dark and opaque, it was impractical to use a technique which relies on a color change for detection; As little as 5 ppm 2-hydroperoxytetrahydrofuran could be detected in fresh, unstabilized THF.

1 + 2

12.5% $Cu(OAc)_2$
18.75% 2,2′-bipyridine
25% valeric acid
1.5 equiv iPr_2NEt

THF/DMSO (1 : 1)
300 psi 5% O_2/N_2
90 °C

3

Scheme 2.7 Optimized conditions for Cu-catalyzed Chan–Lam coupling to prepare **3**.

Table 2.4 Characterization of vertical pipes-in-series reactor sizes.

	Lab scale (75 ml reactor)	**Research scale (2 l reactor)**	**Manufacturing scale (200 l reactor)**
Number of pipes	40	20	24
Length of pipes	2.67 cm	1.07 m	3.78 m
ID of pipes	0.69 cm	1.09 cm	5.35 cm
Length of jumper tubes	3.05 m	1.22 m	4.46 m
ID of jumper tubes	0.057 cm	0.18 cm	0.75 cm
Pressure drop across reactor	14–52 psi	22–25 psi	~100 psi

performance of the reactor. Specific goals of the scale-up were demonstrating the scalability of the process in terms of reaction rate, impurity profile, and robustness across variables that could not be readily explored in batch, including ranges of oxygen pressures, O_2/N_2 compositions, and residence times.

Despite the differences in the different-sized reactors, scale-up to the 2 l reactor was readily accomplished using conditions previously demonstrated in the 75 ml reactor with a comparable kinetic profile (Figure 2.9). The scalability of the

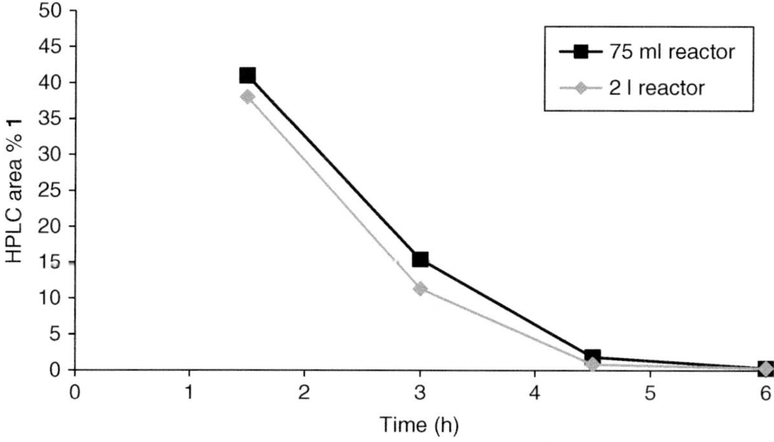

Figure 2.9 Reaction rate comparison for the 75 ml and 2 l reactors.

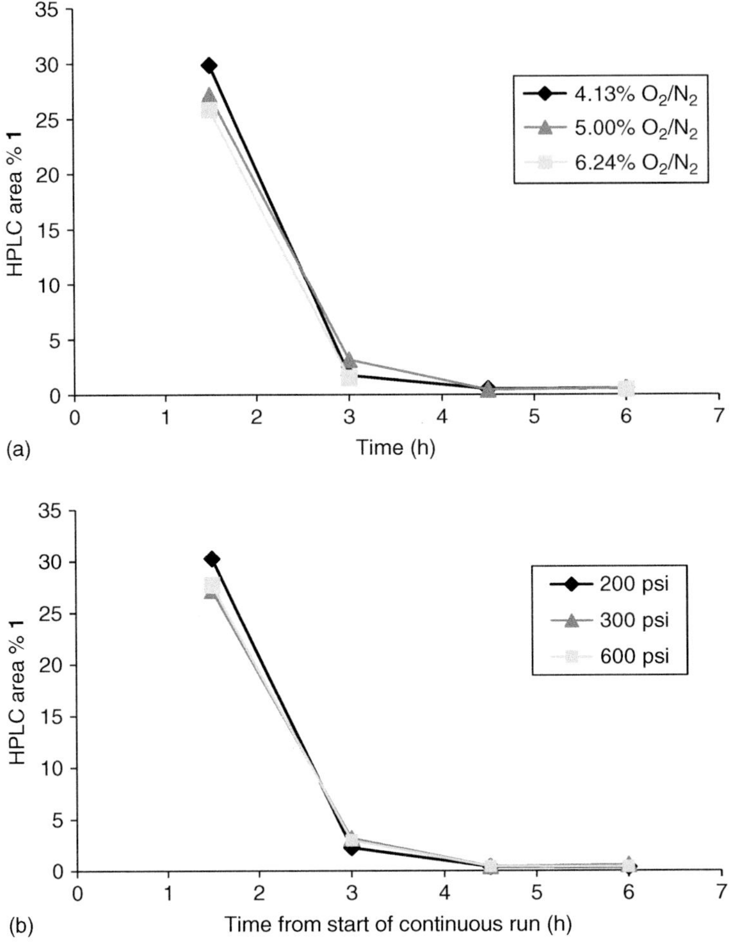

Figure 2.10 Reaction kinetics in 21 pipes in series reaction run with 3 equiv of O_2 using: (a) 300 psi of 4.13–6.24% O_2/N_2 and (b) 200–600 psi of 5.0% O_2/N_2.

reactor design is reflected in the ease of scale-up; a 26-fold increase in scale did not require optimization of either the reaction conditions or the reactor.

Minimal variations were observed when comparing reaction rates using 300 psi of gas mixtures ranging from 4.13% to 6.24% O_2/N_2 (Figure 2.10a). The negligible pressure drop across the 75 ml reactor as compared to a much larger pressure drop at manufacturing scale (Table 2.4) is of particular note when comparing different reactor sizes[12] and ensuring that the reaction was robust with respect to O_2 partial pressure, which was also a key factor when preparing for scale-up and manufacturing. Complete conversion was observed within six hours when 200–600 psi of 5% O_2/N_2 was used with no notable variation in the kinetic profile

12 A 100 psi pressure drop was measured in the 200 l reactor during the campaign. Because we did not know the magnitude of the pressure drop prior to the campaign, robust performance across a wide range of gas pressures was desirable.

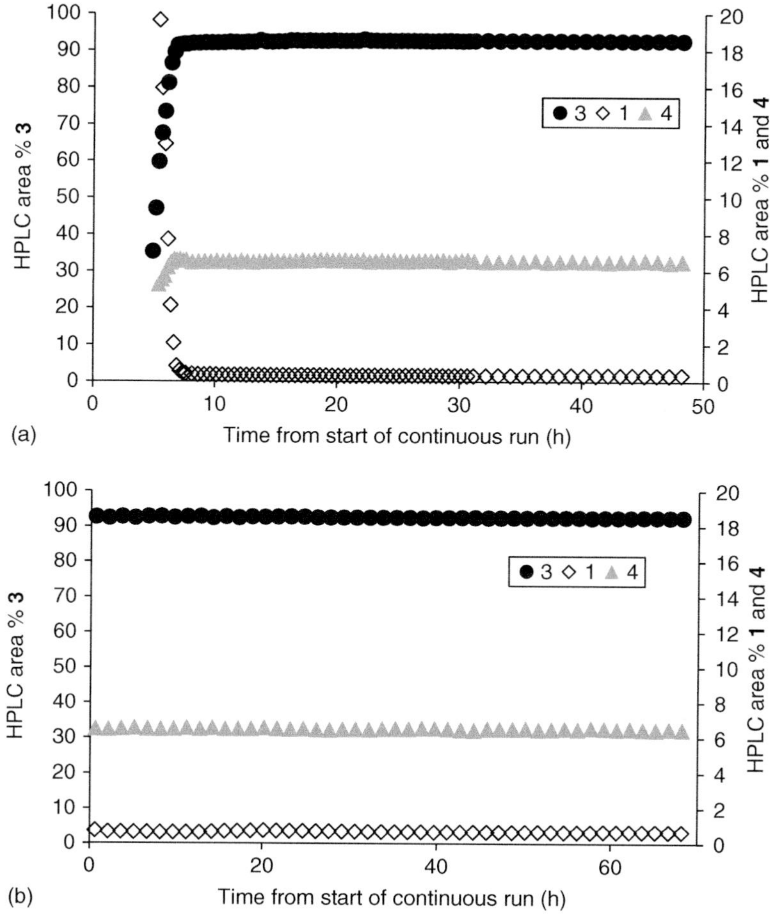

Figure 2.11 Steady state profiles for reactions run in 2 l reactor using (a) 6 hours tau (starting from a solvent filled reactor) and (b) 18 hours tau (starting from a product filled reactor).

(Figure 2.10b). Because the manufacturing scale reactor (200 l) was oversized for the scale of the planned campaign, a longer tau would be used in manufacturing to allow for more time to correct any unanticipated excursions while minimizing the risk to material. Before the campaign, we wanted to ensure that the reaction was stable under reaction conditions for longer times, so reactions were run using both 12 and 18 hours tau. No change in the steady-state profile was observed, further indicating process robustness (Figure 2.11).

On manufacturing scale, a 75 kg campaign was carried out in a 200 l vertical pipes-in-series reactor. About 290 psi of 5.0% O_2 in N_2 was used with a gas stoichiometry of 3.0 equiv of O_2. The reaction temperature was 90 °C with a tau of 8.45 hours. The reactor was run for 114 hours to utilize all of the starting material feed. At the steady-state conversion of **1**, the measured HPLC was 99.8% (Figure 2.12). The entire contents of the continuous run were treated as one batch. Thus, the entire reaction mixture was carried through the extractive workup,

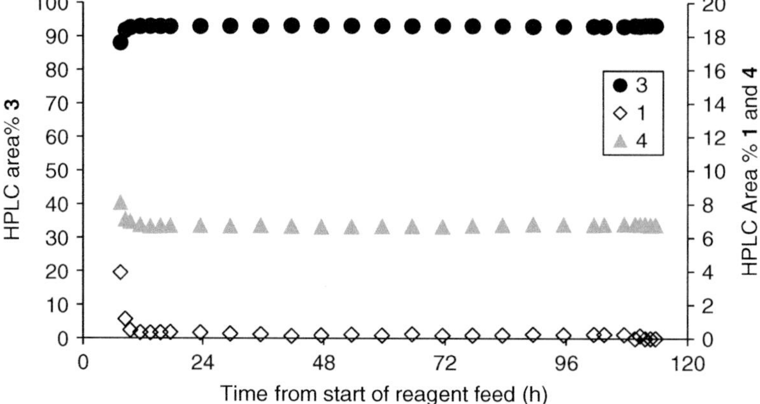

Figure 2.12 Steady state profile for manufacturing scale (2001) reaction.

solvent exchange, and crystallization as described above to afford a 69% isolated yield of **3** with 100% potency and 99.6% purity by HPLC. It is also noteworthy that isomer **4** was not detected in the final product (<0.05%).

2.3 Design and Scale-up of a Buchwald–Hartwig Cross-coupling

For the synthesis of API freebase **6**, a Buchwald–Hartwig amination reaction of 2-chloropyridine **3** and 4-aminopyridine **4** was developed for the desired C–N coupling. Originally, in the medicinal chemistry synthetic route, sodium phenoxide (NaOPh) was used as the base for this transformation. A major impurity was identified as the product of C–O coupling between **3** and NaOPh, resulting in suboptimal conversion of **3** to the desired product **6**. Additionally, chromatography was performed to purify product **6** in a moderate isolated yield on gram scale.

During the initial round of process development, several organic and inorganic bases were screened in order to avoid the C–O coupling impurity. Potassium carbonate (K_2CO_3) provided the best *in situ* conversion and reaction profile among the screened bases. After further optimization, including catalyst loading and solvent selection, the optimized conditions were implemented in campaign 1 (Scheme 2.8).

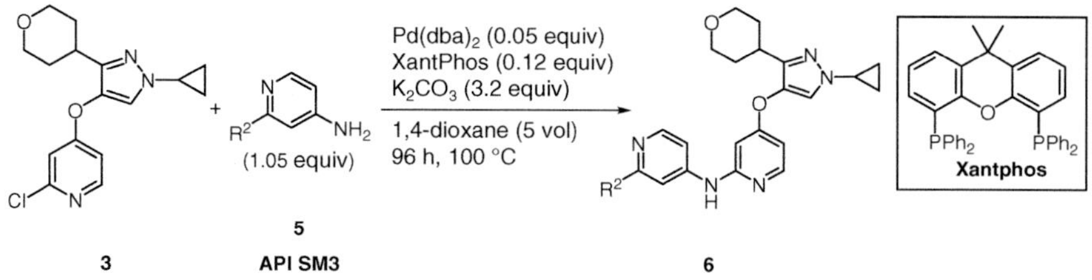

Pd(dba)$_2$ (0.05 equiv)
XantPhos (0.12 equiv)
K_2CO_3 (3.2 equiv)
1,4-dioxane (5 vol)
96 h, 100 °C

(1.05 equiv)

3 + **5** **API SM3** → **6**

PPh$_2$ PPh$_2$ **Xantphos**

Scheme 2.8 Buchwald–Hartwig amination conditions from campaign 1.

Campaign 1 was successful in delivering 13.2 kg of product **6** with 74.5% yield and 98.9% HPLC purity. However, several aspects of this process required further optimization. The chemical transformation employed a high loading of the catalyst and ligand (5 mol% and 12 mol%, respectively), required long reaction times (96 hours) at high temperature (100 °C), and used 1,4-dioxane, an undesirable solvent [12]. Development was focused on lowering catalyst loading, screening reagents, and solvents so that full conversion of **3** could be achieved under less demanding and more process-friendly conditions. Additionally, the workup procedure in previous campaign was elaborate, including extensive acid/base extractions, resin treatment to remove Pd, and a crystallization from EtOH/H_2O, which resulted in a 15% product loss to the mother liquor. Moving forward, development included streamlining the extraction, screening Pd scavengers, and developing a crystallization process that would produce **6** with the required purity while minimizing mother liquor loss.

2.3.1 Initial Screening

Literature on mechanistic studies of the Buchwald–Hartwig amination reaction revealed that a 1 : 1 ratio of Pd(0):bidentate phosphorus (P^2) ligand is optimal for the C—N bond formation, and especially important when XantPhos is used [13]. Therefore, in the reaction screening where $Pd_2(dba)_3$ was used as the source of Pd(0), a 1 : 1 ratio of Pd:P^2 was adopted. Preliminary reaction screening showed that DPEPhos is a better ligand than XantPhos. Under otherwise identical reaction conditions, DPEPhos gave much higher conversion than XantPhos (Table 2.5, entry 1 vs entry 2). Given that the cost of DPEPhos is substantially lower than XantPhos, it was established as the standard ligand for further ligand screening [14]. With regard to the replacement of 1,4-dioxane as the solvent, although both toluene and THF showed good reaction conversion, *t*-AmOH was identified as the optimal solvent (Table 2.5, entry 3) because of its benign character, superior reaction profile, and the opportunity to explore soluble amylate salts as bases.

During the initial round of base screening, it was noted that K_2CO_3 could be replaced by NaO*t*-Am, which gave higher conversion under otherwise identical conditions (Table 2.5, entry 4 vs entry 5). One additional benefit of using NaO*t*-Am as the base is that it is soluble in *t*-AmOH, a feature especially beneficial on production scale where efficient mixing of slurries becomes more challenging and the solubility of inorganic bases can be affected by numerous factors (e.g. particle size, agitation, and water content). Further investigation into the base revealed that both NaOPh and aq. NaOH gave similar but slightly lower reaction conversion than NaO*t*-Am (Table 2.5, entries 5–7). From the aspect of future development, one key consideration was the homogeneity of the reaction mixture. Although both NaO*t*-Am and NaOPh are soluble in the reaction mixture, the NaCl by-product precipitates as the reaction progresses. Attempts to add water into reaction as a cosolvent (*t*-AmOH:H_2O = 5 : 1) did successfully solubilize NaCl at the end of reaction (EOR), but the resulting reaction mixture was biphasic. Additionally, the reactions required longer reaction times and the reaction profile did not compare favorably to the reaction without added water.

Table 2.5 Summary of screening conditions for the formation of **6**.

3 + R^2-pyridine-NH_2 (1.1 equiv) **5** **API SM3** → ($Pd_2(dba)_3$, Ligand, base, solvent, 16 h, 100 °C) → **6**

Entry	Ligand	Base (equiv)	Solvent	HPLC area% of 6	HPLC area% of 3
1	XantPhos[a]	K_2CO_3 (3.2)	Toluene[b]	30.1	40.2
2	DPEPhos[a]	K_2CO_3 (3.2)	Toluene[b]	89.6	1.9
3	DPEPhos[a]	K_2CO_3 (3.2)	*t*-AmOH[b]	96.7	0.1
4	DPEPhos[c]	K_2CO_3 (0.9)	*t*-AmOH[d]	89.3	6.1
5	DPEPhos[c]	NaO*t*-Am (1.1)	*t*-AmOH[d]	94.3	ND
6	DPEPhos[c]	NaOPh (1.1)	*t*-AmOH[d]	93.2	5.4
7	DPEPhos[c]	NaOH (1.1)	*t*-AmOH[d]	83.4	9.2
8	DPEPhos[c]	DBU (1.1)	*t*-AmOH[d]	31.5	42.5
9	DPEPhos[c]	DBN (1.1)	*t*-AmOH[d]	15	51.5
10	DPEPhos[c]	TMG (1.1)	*t*-AmOH[d]	12.1	42
11[e]	DPEPhos[c]	Phosphazene (1.1)	*t*-AmOH[d]	12.2	54.5
12	P(*c*-hex)$_3$[f]	K_2CO_3 (0.9)	*t*-AmOH[d]	1.1	58.8
13	P(*o*-tol)$_3$[f]	K_2CO_3 (0.9)	*t*-AmOH[d]	0.5	64.1
14	JohnPhos[g]	NaO*t*-Am (1.1)	*t*-AmOH[d]	16.8	30.7
15	*t*-BUXPhos[g]	NaO*t*-Am (1.1)	*t*-AmOH[d]	12.6	32.5
16	TrippyPhos[g]	NaO*t*-Am (1.1)	*t*-AmOH[d]	8.3	30.1
17	BippyPhos[g]	NaO*t*-Am (1.1)	*t*-AmOH[d]	18.4	30
18	AmPhos[c]	NaO*t*-Am (1.1)	*t*-AmOH[d]	2.5	50.8
19	Naph-NHC[c]	NaO*t*-Am (1.1)	*t*-AmOH[d]	3.3	50
20	Allyl-NHC[c]	NaO*t*-Am (1.1)	*t*-AmOH[d]	3.3	48.4
21	(*S*)-BINAP[c]	NaO*t*-Am (1.1)	*t*-AmOH[d]	90.6	3.3
22	(*R*)-BINAP[c]	NaO*t*-Am (1.1)	*t*-AmOH[d]	90.9	4.1

ND, Not detected.

a) 5 mol% $Pd_2(dba)_3$ and 10 mol% ligand.
b) 15 vol solvent.
c) 0.3 mol% $Pd_2(dba)_3$ and 0.6 mol% ligand.
d) 10 vol *t*-AmOH.
e) 60 °C, 16 hours.
f) 5 mol% $Pd_2(dba)_3$ and 20 mol% ligand.
g) 0.3 mol% $Pd_2(dba)_3$ and 1.2 mol% ligand.

Some organic bases, including DBU, DBN, TMG, and phosphazene, were also studied (Table 2.5, entries 8–11). Although a homogeneous reaction solution was observed in each case, all showed much lower reaction conversion.

After preliminary reaction screening data showed the favorable comparison of DPEPhos and XantPhos, it was desired to conduct a broader ligand screen. In this set of experiments, it was noted that monodentate ligands ($P(c\text{-hex})_3$, $P(o\text{-tol})_3$, JohnPhos, *t*-BuXPhos, TrippyPhos, BippyPhos, and AmPhos; Table 2.5, entries 12–18) and *N*-heterocyclic carbene (NHC) ligands (naphthylquinone-NHC and allyl-NHC; Table 2.5, entries 19 and 20) all showed inferior results, with less than 20% product **6** (HPLC area%) formed in all cases. Notably, a higher amount of dimer by-product **23** derived from homocoupling of **3** was observed. As much as 2.7% (HPLC area%) **23** was observed when monodentate ligands were employed. Interestingly, the only ligands that showed comparable conversion to DPEPhos were (*S*)-BINAP and (*R*)-BINAP (Table 2.5, entries 21 and 22). Because the BINAP ligands did not show advantage over DPEPhos in either conversion rate or cost, BINAP was not further pursued.

2.3.2 Synthesis and Isolation of Pd(dba)DPEPhos Precatalyst

For the generation of the precatalyst, a common practice is to form the Pd–ligand complex *in situ* before the addition of substrates, which provides higher conversion than addition of the Pd source and ligand separately to the substrates [15]. A survey of aromatic solvents (toluene, anisole, *o*-xylene, *m*-xylene, and *p*-xylene) and *t*-AmOH indicated that toluene is preferable for the formation and long-term storage of the precatalyst solution.[13] Therefore, toluene was selected as the solvent for the precatalyst formation going forward into campaign 2.

In addition to the *in situ* generation of the preformed complex in solution, isolation of this type of species (such as Pd(dba)[(*R*)-BINAP]) and its application in amination reactions have been documented in the literature [16]. In our study, the isolated precatalyst Pd(dba)DPEPhos was prepared and characterized by NMR and single-crystal X-ray diffraction.[14] Its long-term stability has also been monitored by ^{31}P NMR. Although the Pd(dba)DPEPhos solid stored in the glove box was unchanged by ^{31}P NMR over several months, the solid kept in a vial on the benchtop was slowly oxidized.[15] Because there was no direct evidence that the isolated catalyst offered better activity in this amination reaction as compared to the preformed catalyst solution in toluene, isolation of the precatalyst was not implemented for the production of **6** in campaign 2.

With the reaction screening completed at this stage, the reaction conditions were further examined on different scales, from 1 to 500 g (Scheme 2.9). In all cases, HPLC monitoring of reaction profile showed high conversion to

13 (i) While 10 vol toluene is sufficient to dissolve the pre-catalyst, up to 400 vol *t*-AmOH is required to make a homogeneous solution of the pre-catalyst; (ii) in a related study, a pre-catalyst solution in toluene was used four months after its preparation and showed no sign of diminished activity.

14 The CCDC reference code for Pd(dba)DPEPhos is 1885714.

15 Studies also showed that the pre-catalyst was partially oxidized after filtration in air during preparation.

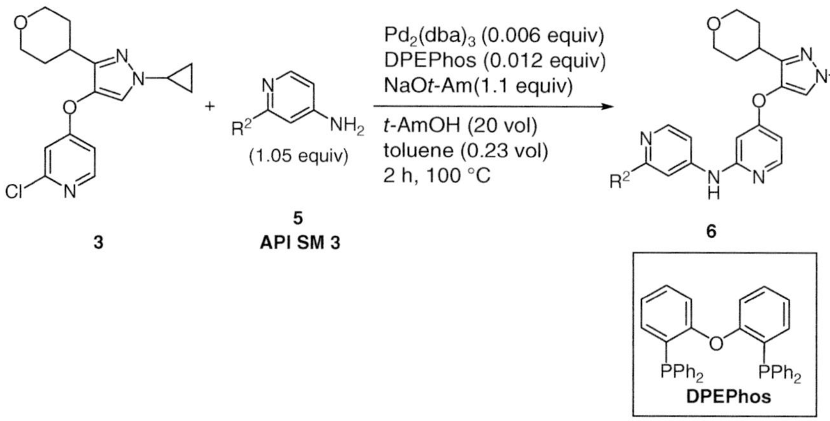

$Pd_2(dba)_3$ (0.006 equiv)
DPEPhos (0.012 equiv)
NaO*t*-Am(1.1 equiv)
t-AmOH (20 vol)
toluene (0.23 vol)
2 h, 100 °C

3 + **5** **API SM 3** (1.05 equiv) → **6**

DPEPhos

Scheme 2.9 Optimized reaction conditions for the formation of **6**.

the desired product (>90% of **6** in HPLC area%; typical isolated yield ~89%) with a reaction time of two hours at 100 °C. Therefore, these optimal reaction conditions were established and ultimately applied in the manufacturing campaign.

2.3.3 Workup Procedure, Metal Removal, and Crystallization

The workup and isolation procedure used in campaign 1 involved several unit operations: concentration of the reaction mixture, addition of aq. HCl, aqueous-phase washes with dichloromethane (DCM) and ethyl acetate (EtOAc) to remove impurities, pH adjustment of the aqueous solution with K_2CO_3 to precipitate the product as a freebase, product dissolution in ethanol, depletion of Pd with Si-thiol scavenger, and crystallization by addition of H_2O. In campaign 1, it was noted that the loss of product in DCM extraction was high, nearly 15%. More importantly, when this purification procedure was applied on the new reaction conditions for campaign 2 (Scheme 2.9), intense gassing was noted by addition of aq. K_2CO_3, concomitant with the formation of a sticky solid resulting from oiling of the product. Because of these complications in the workup and isolation, this isolation process was abandoned.

Instead of acid–base extraction followed by resin treatment to remove Pd, the initial development strategy focused on the aqueous wash to remove the NaCl formed in the reaction, combined with an extractive method involving chelating agents capable of Pd sequestration. Metal scavengers were screened, including hexamethylenetetramine, trisodium hydrogen ethylenediaminetetraacetate hydrate (NaEDTA), D-glucosamine, sodium diethyldithiocarbamate (NaDETC), ammonium pyrrolidinedithiocarbamate (APDTC), and 1,3,5-Triazine-2,4,6-trithiol trisodium salt (NaTMT). Among them, NaTMT (as 15% aq. solution; commercially available) demonstrated the highest efficacy in removing Pd and was chosen for further development (Table 2.6).

Based on these initial results with NaTMT, an extractive workup involving four rounds of treatment with aq. NaTMT under refluxing conditions followed by a

Table 2.6 Initial development of NaTMT for Pd removal.

Layer	Metal scavenger	Pd in solution[a)] (ppm)
Organic	None	87.8
Organic	NaTMT	4.4
Aqueous		82.4

a) Pd level in solution was determined through ICP (inductively coupled plasma) analysis of Pd.

final wash with 20% aq. NaCl was designed. Although approximately 97% of the total Pd amount was rejected using this process, the remaining Pd in the final isolated product was still ~300 ppm, well above the specification of 100 ppm. An additional filtration through a filter aid helped to further remove Pd level down to 138 ppm, but this was still above the 100 ppm specification.

In a separate set of experiments, Siliabond Thiol ("Si-Thiol") was identified as an excellent option for Pd depletion. Other resins were evaluated, including QuadraSil TA, DMT, TMT, MTU, cysteine, TAACOH, THU, and TAACONa, but none were effective. Then, a combination of the extractive aq. NaTMT method followed by Si-Thiol treatment was tested. The Pd level was reduced to 11 ppm after three extractions with aq. NaTMT (15% w/w) at reflux followed by 20% brine wash and stirring with 10% (w/w) Si-Thiol at reflux. However, an elevated level of S was noted in the isolated product (418 ppm). This was attributed to the partial miscibility of *t*-AmOH and water, and leaching sulfur-containing aqueous into the organic phase resulted in S contamination upon drying the isolated product.

To avoid contamination of **6** with sulfur residue, it was decided to utilize Si-Thiol as the exclusive Pd scavenger for campaign 2, eliminating the NaTMT washes. This allows for a simple workup procedure, which starts with two sequential washes with 10 volume of water to remove inorganic by-products. The water washes were done at 80 °C to prevent the desired product from precipitating and to allow a distinct phase separation. Then, the organic-phase solution was stirred with Si-Thiol (30% w/w relative to a theoretical yield of **6**) at 80 °C for four hours, and the resin was removed by hot filtration. This stream was then concentrated to 10.5 and 4.5 volume of heptane was slowly added as an antisolvent over two hours. Finally, after cooling to 10 °C for three hours, the desired product was collected as a solid. The streamlined workup conditions consistently provided high efficacy for Pd rejection in lab demo batches, resulting in <100 ppm Pd residue in the isolated **6** each time. Eventually, it was successfully implemented in campaign 2 to provide the API freebase **6** with 12 ppm Pd.

During the course of process development, three different crystal forms of **6** were obtained and characterized: *t*-AmOH solvate, anhydrate, and monohydrate. A thorough comparison of the purity profiles among the three forms showed a slightly better profile for the *t*-AmOH solvate, with an overall higher rejection of impurities. However, a slurry bridge experiment showed that the monohydrate form is the most thermodynamically stable form in either *t*-AmOH as a single solvent or a mixture solvent of *t*-AmOH/heptane (7 : 3, v/v), when the water content is over 1% w/w. Because of the miscibility of water and *t*-AmOH, the risk of some residual water remaining after the water washes and distillation operations was high. Thus, if either *t*-AmOH solvate or anhydrate forms were desired, there was a high burden on the distillation to achieve a very low water content. As a result, it was decided to target the monohydrate crystal form for campaign 2. In order to ensure a robust and reproducible crystallization process, three criteria were set: (i) maintaining a water content of 3–5% (w/w) after distillation, (ii) adding 0.5% (w/w) seed crystals of monohydrate, and (iii) careful drying of the wet product at 40–50 °C under vacuum of ~50 mbar. Once the IPC criteria were fulfilled (KF: 3.6–4.6%, LOD <1% w/w), the drying was ended.

2.3.4 Scale-up and Manufacturing

In manufacturing campaign 2, the step 9 Buchwald–Hartwig coupling was carried out in three batches. The optimized process was executed as described above to afford an average isolated yield of 80.6% **6**•monohydrate with 99.5% potency and 99.6% purity by HPLC. The results of this campaign are summarized in Table 2.7. It is worth noting that the yields for batch 1 and batch 2 are slightly lower because of the fact that the distillation end points were missed, resulting in excess *t*-AmOH in the crystallization and slightly higher than normal mother liquor losses. Finally, batch 3 benefitted from the recovery of all the material from the dryer, which resulted in an artificially high isolated yield.

2.4 Impurity Control

As stated above, maintaining and ensuring API quality is a core deliverable for any CMC team. In order to do so, the team must develop a deep understanding of the origin and fate of all relevant process impurities. Federal regulations laid by the Food and Drug Administration (FDA) [17], in addition to guidance documents from the FDA and ICH, detail the requirements and expectations of regulatory agencies in the USA and around the world. It is up to the CMC team to outline the relevant potential impurities for a given synthetic route for an API and develop an

Table 2.7 Results for the scale-up batches of step 9.

	Average	Batch 1	Batch 2	Batch 3
Yield	80.6%	69.8%	71.3%	101.4%
Mass	62.6 kg	18.3 kg[a)]	18.7 kg	25.6 kg

a) One kilogram of material was retained as seed, accounting for approximately 3.8% of the yield.

appropriate and robust impurity control strategy. This section outlines the initial work done to develop an impurity control strategy for **6**.

Impurities in the final API originate from species entering with the raw materials used for the process (i.e. starting-material-related impurities) or species that form during the reaction as a result of undesired side-reactions (i.e. process-related impurities). Ultimately, CMC teams cultivate a full understanding of the source and fate of all impurities throughout the synthesis, thus allowing the establishment of suitable limits and controls. Further process understanding is gained by evaluating each parameter of each reaction and gauging the risk it poses to API quality, often referred to as a risk assessment (RA). This process is typically carried out in alignment with ICH Q9, Quality Risk Management [18]. The output of the RA is then used to design experiments to determine the acceptable ranges (proven acceptable ranges or PARs) for each parameter and the associated design space for each step [19].

Building up this level of process understanding for a multistep synthesis requires a tremendous amount of data. The industry standard for this type of data is HPLC analysis, typically of reaction samples or isolated products (before and after drying). The industry is also undergoing a shift to utilize process analytical technology (PAT), whereby information about a process can be gained in real time through on-line analysis (direct analysis of a process stream). Examples of PAT would include an *in situ* FTIR (Fourier transform infrared) probe or an FBRM (focused beam reflectance measurement) probe for detecting a particular chemical species and for estimating particle size and distribution, respectively. The use of PAT is well aligned with the FDA initiative for quality by design (QbD). The principles of QbD will not be discussed in detail in this chapter but are vitally important to modern CMC development and the expectation is that every effort will be taken to ensure API quality by using technology to fundamentally build quality into the process.

As projects mature and progress to later phases of development, greater levels of process understanding are gained. For this project, in early-phase development, the team was only starting to build the foundation for the impurity control strategy. The key to this process is to understand the formation and fate of each impurity in the sequence. The rejection of the impurities through each synthetic step (i.e. purity upgrades via crystallization or other purification technique) is the basic approach to establish impurity control. Impurity fate and purge information along with qualification of known impurities through toxicology studies, and maximum daily dosing levels of the drug in the drug product, enable setting specifications on the API. Typically, impurities are limited by ICH guidelines as listed in Table 2.1. Thus, it is critical to understand the amount of a given impurity that can be carried forward from one step to the next, such that when processed under representative conditions, API of sufficient quality is obtained.

A series of experiments were conducted to begin building up this data set. Because of time constraints and the early phase of development, the team focused on the final two bond-forming steps of the synthetic sequence. The initial control strategy was built through a combination of spiking studies and solubility studies. An impurity spiking study is an experiment in which a known amount of an impurity is added to the reaction, and the fate of that impurity is determined by evaluating the isolated product and other relevant process stream, e.g. mother liquors. Spiking studies can be done by adding the actual impurity of interest to the reaction mixture at the end of the reaction, to specifically study the fate of the impurity through the product isolation and thereby avoid any changes to the impurity under the reaction conditions. Alternatively, the experiment can be done by adding the source of the impurity to initial reaction mixture and generating the impurity under the reaction conditions. The team must determine the best approach for each impurity and the approach for steps 8 and 9 will be discussed below.

Metal control is another critical aspect to consider for API synthesis, especially in the context of this text and the ever-increasing popularity of transition-metal-catalyzed reactions in complex molecule synthesis. The later sections of this chapter provide more details on the removal of metal from each step. Table 2.8 outlines the thresholds for Cu and Pd in a drug substance according to ICH Q3D and the associated specification for the intermediate tested.

2.4.1 Solubility and Impurity Spiking Studies

In step 8, there are numerous impurities to consider, as shown in Scheme 2.10. As stated previously, because of time pressures and the early phase of development, only two impurities (**1** and **4**) were studied to gain greater process understanding and begin to build an impurity control strategy. All other impurities were routinely controlled to ≤0.15% through the crystallization process (see Section 2.2.3).

In step 8, control of 1 and 4 was established using a combination of solubility data and spiking data. Evaluation of the solubility of 1, 3, and 4 in the crystallization solvent system revealed much higher solubility for both 1 and 4 at process-relevant temperatures (Table 2.9). These data were an excellent starting point for understanding how to remove the isomer 4 and the unreacted starting material 1 from the desired product 3, assuming a fully equilibrated and thermodynamic isolation via crystallization. The team did not have sufficient

Table 2.8 Thresholds for residual Cu and Pd in a drug substance adapted from ICH Q3D and specification.

Element	Class[a]	Oral PDE[b] (μg/d)	Specification[c] (intermediate tested)
Cu	2B	3000	100 ppm (3)
Pd	3	100	100 ppm (6)

a) See Section 2.4 of ICH Q3D for full definition of classes; abbreviated definitions: class 2B elements can be excluded from risk assessments unless intentionally added to the process. Class 3 elements have relatively low toxicities when ingested orally.
b) Permitted daily exposure.
c) Specification for Cu in intermediate **3** and Pd in **6**.

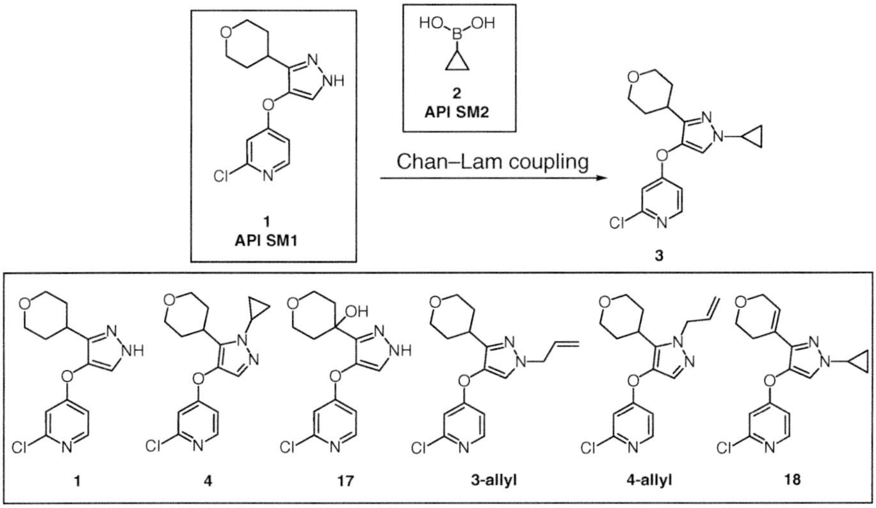

HO B OH
2
API SM2
1
API SM1
Chan–Lam coupling
3
1 4 17 3-allyl 4-allyl 18

Scheme 2.10 Some impurities observed in step 8.

Table 2.9 Solubility of compounds 3, 4, and 1 in 40 : 60 ethanol:water.

Temperature	Solubility 3 (mg/g)	Solubility 4 (mg/g)	Solubility 1 (mg/g)
25 °C	11.6	73.7	72.6
5°C	3.1	45.6	16.6

time to fully derive an understanding of crystallization kinetics; rather, time had to be spent to design away from a very complex oiling issue that will not be covered in this chapter.

A single spiking study was run to assess the rejection of **1** from the desired product **3**. This experiment was done by spiking a known amount of **1** into the EOR mixture and subsequent processing through the workup and isolation. A 4.2% spike of **1** was rejected down to 0.51% in the isolated solid (**3**), corresponding to an 88% rejection. Given that reaction conversions in this step were routinely >98%, the team felt confident that any unreacted starting material would be removed during the isolation. At the time, there was no supply of authentic **4** to do a corresponding spiking study, but the solubility data indicated that removal of **4** by crystallization should be favored. Because ~6.5% of **4** is routinely generated in step 8, control of this impurity was the key. Indeed, the >2.5× solubility of **4** vs **1**, and 15× solubility of **4** vs **3**, in the final crystallization mixture at 5 °C provided the desired control of these key impurities for the current scale-up campaign.

Step 9 is the final bond-forming step in this synthesis and the isolation of **6** is a key control point for impurities. There are five process-derived impurities related to Buchwald–Hartwig amination reaction that have been

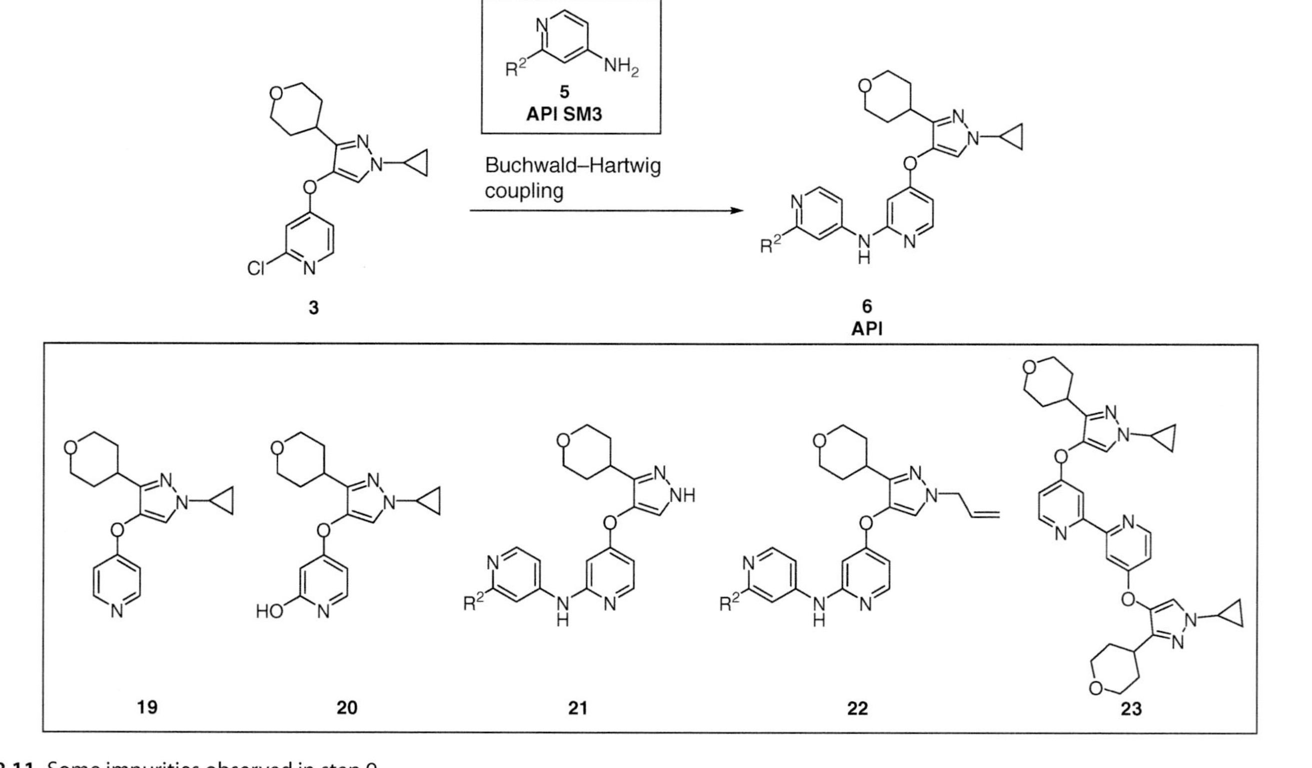

Scheme 2.11 Some impurities observed in step 9.

identified (Scheme 2.11). All impurities were unambiguously confirmed by the syntheses of the reference standard samples and analyses of their NMR and HPLC/LC-MS data.

As discussed above, it was desired to study the rejection of these impurities during the isolation process, so spiking studies were performed for each impurity. These studies were done by spiking the authentic impurity at the EOR. After impurity spiking, the reaction mixture was subjected to the standard workup and crystallization procedure to determine the relative rejection rate[16] of the specific impurity (Figure 2.13). As indicated in Figure 2.13, both aqueous washes and treatment with Pd scavenger Si-thiol ("org1" through "org3") did not contribute to the rejection of any impurities. However, the impurities were fairly well rejected by the crystallization process. This worked particularly well for **19**, **20**, and the dimer impurity **23**, with over 80% relative rejection achieved for each (Table 2.10). Much lower rejection was observed for the des-cyclopropyl impurity **21** (32% relative rejection rate) and the *N*-allyl impurity **22** (16% relative rejection rate) (Table 2.10).

Although greater process understanding would be required in later phase development, the initial data gathered for steps 8 and 9 provided a foundation for establishing an impurity control strategy to ensure API quality. The process and impurity control designed for campaign 2 successfully delivered the material of very high quality at each step, and the phase-appropriate nature of the strategy allowed timely delivery of the material to support the overall program.

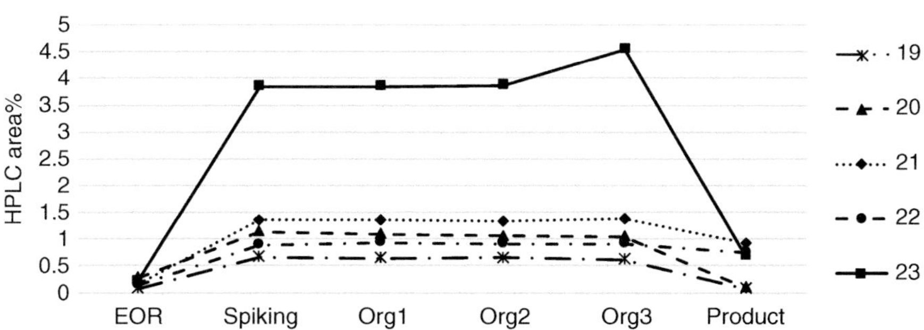

Figure 2.13 Spiking studies for some impurities observed in step 9.

Table 2.10 Relative rejection rate for some impurities observed in step 9.

Impurity	19	20	21	22	23
Relative rejection rate	88%	91%	32%	16%	82%

16 Relative rejection rate = (HPLC area% at spiking – HPLC area% in product)/(HPLC area% at spiking) * 100%.

2.5 Conclusions

The case study mentioned above detailed the development and optimization of an API manufacturing route enabled by continuous processing. The development of homogeneous (single liquid phase) conditions for the step 8 Chan–Lam coupling was critical for three reasons. First, this development allowed for the use of dilute air in a continuous process, which is critical from a safety perspective as it allows the reaction to be run under nonflammable conditions. Second, these conditions mitigated robustness issues associated with the heterogeneous mixtures typically associated with this reaction. And third, turning the step 8 Chan–Lam coupling into a safe and robust process obviated the need to redesign the synthesis, thus saving the team significant time and money. The final bond-forming step of this synthesis was a Buchwald–Hartwig coupling, which also featured an organometallic catalyst. As described above, these final bond-forming steps are critical to the development of the overall impurity control strategy, which is a vitally important element to ensure API quality. The execution of these steps at kilogram scale and under GMP conditions highlights the significant role that organometallic chemistry plays in API synthesis and underpins the ubiquity of organometallic chemistry in modern organic synthesis.

Biography

Alison C. Brewer is a senior research scientist at Eli Lilly and Company, where she leads the small molecule design and development catalysis group. Alison received an undergraduate degree from Lafayette College and subsequently attended the University of North Carolina, where she earned her Ph.D. degree under the supervision of Professor Michel R. Gagné in 2009. From 2009 to 2011, she was an NIH postdoctoral fellow with Professor Shannon S. Stahl at the University of Wisconsin.

Philip C. Hoffman obtained his B.S. degree in Chemical Engineering from Purdue University in 1998. He has worked for Eli Lilly and Company for 21 years, first as a process engineer in Solvent Recovery followed by 18 years in Small Molecule Design and Development (SMDD), where he has gained extensive experience in API process development and scale-up.

Timothy D. White received his B.Sc. in Chemistry from the University of Illinois at Urbana-Champaign in 1995. He then attended the University of Utah, and under the guidance of Frederick West, he earned his M.S. degree in Organic Chemistry in 1998. Tim then joined the process research and development group at Pfizer in Groton, Connecticut, where he worked on the development of scalable processes including tofacitinib citrate. In 2007, he moved to the Chemical Process R&D Department at Eli Lilly and has worked on flow chemistry steps that have been implemented in manufacturing. In addition to green and flow process improvements in the small-molecule space, he was worked with large molecules to generate antibody drug conjugates and solution-phase and solid-phase peptide synthesis.

Yu Lu was born and raised in Jing Dezhen, a city of Jiang Xi Province in southern China. He attended Fudan University in Shanghai to obtain his B.S. degree in Chemistry in 2004. From there, he moved to the United States and obtained M.S. degree in Chemistry from the University of Missouri, Columbia, in 2007, where he did research with Prof. Susan Lever. Yu Lu then joined Prof. Michael Krische's lab at the University of Texas at Austin, where he did research in transition-metal-catalyzed allylation reactions, and accomplished the total synthesis of Bryostatin 7. After obtaining his Ph.D. degree in Chemistry from UT-Austin in 2012, he joined the discovery chemistry group at Dow AgroSciences (now Corteva Agriscience) and did research in fungicide and herbicide projects. In 2016, Yu Lu joined Eli Lilly as a process chemist and has been supporting and leading multiple projects through early to late phase developments.

Laura McKee, P.E., is graduated with a double major in Chemical Engineering and Biomedical Engineering from Carnegie Mellon University in 2011. She then worked at Eli Lilly and Co. in Small Molecule Research and Development until 2017. Laura now works as a technical operations specialist in Process Research and Development at AbbVie, where she is responsible for technical transfer of small molecules and antibody drug conjugates to the API Pilot Plant and commercial facilities. She has worked on development and scale-up of multiple unit operations both for batch and continuous processes. Her specialties include multiphase reactions, extraction, distillation, and filtration.

Moussa Boukerche received his Ph.D. in Chemical Engineering in 2005 from Université Claude Bernard Lyon I (France) under the supervision of Pr. Jean Paul Klein. He developed experimental methodologies for the screening of polymorphs to directly nucleate the most stable form from solution. Moussa is currently a member of the Center of Excellence for Isolation and Separation Technologies (CoExIST) at AbbVie Inc. (USA), where he is responsible for the design and development of crystallization processes in API manufacturing. Before AbbVie Inc., Moussa worked in the field of industrial crystallization for several companies, including Eli Lilly (USA), Pfizer (UK), Aughinish Alumina (Ireland), and SANOFI (France).

Michael E. Kobierski received his Bachelor of Science degree in Chemistry from Gannon University in 1988 and his Master of Science degree in Organic Chemistry from Case Western Reserve University in 1991. He joined Eli Lilly and Company in 1991 and has been working in Process Research and Development for 28 years at Lilly. Mike has contributed to numerous projects during his tenure starting from early phase development at the discovery interface all the way up through commercial phase product launch, including a stint in the manufacturing group at Kinsale, Ireland. Mike was instrumental in the commercial phase development of both the Alimta and Olumiant manufacturing processes. He has spent most of his career in small-molecule research but is currently involved in solid-phase peptide synthesis. Mike was also the recipient of the 2015 ACS Technical Achievements in Organic Chemistry award.

Nessa Mullane has nine years of experience in the medical device and pharmaceutical industry. She is currently a process chemist at Eli Lilly and Company, where she has been responsible for commercial manufacturing, new product introduction, and technology transfer. Nessa is a graduate of the University of Limerick, Ireland, where she received a B.Sc. degree in Industrial Chemistry and a Ph.D. degree in Organic Chemistry.

Mark Pietz is a graduate of Michigan Technological University, where he received his B.S. degree in Chemical Engineering in 1995. He has 24 years of experience in API process development and technology transfer in the pharmaceutical industry and is currently a consultant engineer with Eli Lilly and Company, where he has spent the past 16 years.

Charles A. Alt earned his B.S. degree in Chemistry from Indiana University – Bloomington in 1984, where he conducted undergraduate research under the guidance of Prof. Paul A. Grieco. From there, Charles earned his M.S. degree in Chemistry from Butler University in Indianapolis in 1987. In 1984, Charles started his career as a process chemist at Eli Lilly and Company working in agricultural, medicinal, and pharmaceutical research, retiring in 2017.

Jim R. Stout received his B.Sc. degree in Chemistry from Indiana University, Bloomington, in 1988. Following graduation, Jim joined the research chemistry group at Reilly Industries as a research chemist, where he worked on the development and scale-up of processes for a variety of specialty chemicals. While working, Jim continued his education receiving his M.S. degree in Organic Chemistry in 1995 from Butler University. In 2001, Jim joined the chemical process R&D group at Eli Lilly working on developing scalable GMP processes where he completed several GMP manufacturing campaigns. Jim joined D&M Continuous Solutions in 2010, where he worked on the development of chemical processes in flow. His role at D&M has also included the engineering and development of flow chemistry equipment including high-pressure gas reagent equipment and MSMPRs in series. Currently, Jim is working on solid-phase peptide synthesis (SPPS) and on the development of equipment for SPPS.

Paul K. Milenbaugh began his career at Reilly Industries in April 1986 as a Chemical Operator where he became skilled at processing distillations, reactions, hydrogenations, and crystallization extractions. He left Reilly as a Team Coordinator after 15 years to join Eli Lilly in 2001. He became an expert with pilot plant equipment and operations, including complex setups and rarely used equipment operations. Paul became the co-owner of D&M Continuous Solutions in February 2008. D&M builds and provides support in the operation of continuous processing equipment for pharmaceutical companies globally. D&M has gained recognition in the development and building of continuous processing equipment for research through manufacturing scale (including GMP Operations).

Joseph R. Martinelli earned his B.S. degree in Chemical Engineering from the University of Wisconsin – Madison in 2002. From there, Joe moved to MIT and conducted graduate studies under the guidance of Prof. Stephen L. Buchwald. After earning his Ph.D. in 2007, Joe moved to Eli Lilly and Company as a process chemist. Currently, Joe sits on a commercialization team, overseeing the development of half of Lilly's small- and medium-molecule pipeline.

References

1 International Council for Harmonisation of Technical Requirements for Pharmaceuticals for Human Use. http://www.ich.org/home.html (accessed 21 February 2019).

2 ICH Q3A(R2) Impurities in New Drug Substances. http://www.ich.org/fileadmin/Public_Web_Site/ICH_Products/Guidelines/Quality/Q3A_R2/Step4/Q3A_R2__Guideline.pdf (accessed 21 February 2019).

3 ICH Q7 Good Manufacturing Practice Guide for Active Pharmaceutical Ingredients. http://www.ich.org/fileadmin/Public_Web_Site/ICH_Products/Guidelines/Quality/Q7/Step4/Q7_Guideline.pdf (accessed 21 February 2019).

4 For an example of on-demand generation and consumption of hazardous regents enabled by flow chemistry, see: (a) Mastronardi, F., Gutmann, B., and Kappe, C.O. (2013). Continuous flow generation and reactions of anhydrous diazomethane using a Teflon AF-2400 tune-in-tube reactor. *Org. Lett.* 15: 5590–5593. (b) Movsisyan, M., Delbeke, E.I.P., Berton, J.K.E.T. et al. (2016). Taming Hazardous Chemistry by Continuous Flow Technology. *Chem. Soc. Rev.* 45: 4892–4928.

5 For Chan–Lam couplings with cyclopropylboronic acid, see: (a) Tsuritani, T., Strotman, N.A., Yamamoto, Y. et al. (2008). *N*-Cyclopropylation of indoles and cycli amides with copper(II) reagent. *Org. Lett.* 10: 1653–1655. (b) Benard, S., Neuville, L., and Zhu, J. (2008). Copper-mediated *N*-cyclopropanation of azoles, amides, and sulfonamides by cyclopropylboronic acid. *J. Org. Chem.* 73: 6441–6444. (c) Benard, S., Neuville, L., and Zhu, J. (2010). *Chem. Commun.* 46: 3393–3395. (d) Racine, E., Monnier, F., Vors, J.-P., and Taillefer, M. (2013). Direct *N*-cyclopropanation of secondary acyclic amides promoted by copper. *Chem. Commun.* 49: 7412. (e) Mudryk, B., Zheng, B., Chen, K., and Eastgate, M.D. (2014). Development of a robust process for the preparation of high-quality dicyclopropylamine hydrochloride. *Org. Process Res. Dev.* 18: 520–527.

6 For recent reviews, see: (a) Qiao, J.X. and Lam, P.Y.S. (2011). Copper-promoted carbon-heteroatom bond ross-coupling with boronic acids and derivatives. *Synthesis*: 829–856. (b) Rao, K.S. and Wu, T.-S. (2012). Chan–Lam coupling reactions: synthesis of heterocycles. *Tetrahedron* 68: 7735–7754.

7 1,2-dichloroethane is a suspect carcinogen and subject to authorization under the REACh directive in the EU. See: https://echa.europa.eu/candidate-list-table (accessed 21 February 2019).
8 Osterberg, P.M., Niemeier, J.K., Welch, C.J. et al. (2015). Experimental limiting oxygen concentrations for nine organic solvents at temperatures and pressures relevant to aerobic oxidations in the pharmaceutical industry. *Org. Process Res. Dev.* 19: 1537–1543.
9 (a) Johnson, M.D., May, S.A., Haeberle, B. et al. (2016). Design and comparison of tubular and pipes-in-series continuous reactors for direct asymmetric reductive amination. *Org. Process Res. Dev.* 20: 1305–1320. (b) May, S.A., Johnson, M.D., Buser, J.Y. et al. (2016). Development and manufacturing GMP scale-up of a continuous Ir-catalyzed homogeneous reductive amination reaction. *Org. Process Res. Dev.* 20: 1870–1898. (c) Johnson, M.D., May, S.A., Calvin, J.R. et al. (2016). Continuous liquid vapor reactions part 1: design and characterization of a reactor for asymmetric hydroformylation. *Org. Process Res. Dev.* 20: 888–900. (d) Abrams, M.L., Buser, J.Y., Calvin, J.R. et al. (2016). Continuous liquid vapor reactions part 2: asymmetric hydroformylation with rhodium-bisdiazaphos catalysts in a vertical pipes-in-series reactor. *Org. Process Res. Dev.* 20: 901–910.
10 The candidate list for Substances of Very High Concern (SVHC) is updated twice a year and can be found at http://echa.europa.eu/candidate-list-table (accessed 21 February 2019).
11 Antilla, J.C. and Buchwald, S.L. (2001). Copper-catalyzed coupling of arylboronic acids and amines. *Org. Lett.* 3: 2077–2079.
12 1,4-dioxane is classified as a Group 2B carcinogen: "possibly carcinogenic to humans" by IARC (International Agency for Research on Cancer). See: https://monographs.iarc.fr/wp-content/uploads/2018/06/mono71-25.pdf (accessed 21 February 2019).
13 (a) Shekhar, S., Ryberg, P., Hartwig, J.F. et al. (2006). Reevaluation of the mechanism of the amination of aryl halides catalyzed by BINAP-ligated palladium complexes. *J. Am. Chem. Soc.* 128: 3584–3591. (b) Klingensmith, L.M., Strieter, E.R., Barder, T.E., and Buchwald, S.L. (2006). New insights into Xantphos/Pd-Catalyzed C—N bond forming reactions: a structural and kinetic study. *Organometallics* 25: 82–91.
14 Cost information obtained by searching SciQuest Enterprise Reagent Manager: DPEPhos (CAS# 166330-10-5) ~\$500/kg on bulk quote; XantPhos (CAS# 161265-03-8) ~\$1000/kg on bulk quote (accessed 1 February 2018).
15 (a) Muci, A.R. and Buchwald, S.L. (2002). Practical palladium catalysts for C—N and C—O bond formation. *Topics Curr. Chem.* 219: 131–209. (b) Hartwig, J.F. (2003). Palladium-catalyzed α-arylation of carbonyl compounds and nitriles. *Acc. Chem. Res.* 36: 234–245.
16 Wolfe, J.P. and Buchwald, S.L. (2000). Scope and limitations of the Pd/BINAP-catalyzed amination of aryl bromides. *J. Org. Chem.* 65: 1144–1157.
17 https://www.fda.gov/Drugs (accessed 21 February 2019).
18 https://www.ich.org/fileadmin/Public_Web_Site/ICH_Products/Guidelines/Quality/Q9/Step4/Q9_Guideline.pdf (accessed 21 February 2019).

19 For examples of case studies on the development of control strategies:
(a) Thomsom, N.M., Seibert, K.D., Tummala, S. et al. (2015). Case studies in the applicability of drug substance design spaces developed on the laboratory scale to commercial manufacturing. *Org. Process Res. Dev.* 19: 925–934.
(b) Thomsom, N.M., Singer, R., Seibert, K.D. et al. (2015). Case studies in the development of drug substance control strategies. *Org. Process Res. Dev.* 19: 935–948.

3

Continuous Manufacturing as an Enabling Technology for Low-Temperature Organometallic Chemistry

Andreas Hafner[1] *and Joerg Sedelmeier*[2]

[1] *Bayer CropScience Schweiz AG, Rothausstrasse 61, 4132 Muttenz, Switzerland*
[2] *F. Hoffmann-La Roche Ltd, Pharmaceutical Division, Small Molecules Technical Development, Process Chemistry & Catalysis, Grenzacherstrasse 124, 4070 Basel, Switzerland*

3.1 Introduction

Over recent years, flow chemistry and continuous manufacturing have drawn a lot of attention as an enabling technology for performing organic synthesis as demonstrated by a steadily increasing number of publications in this field [1]. Today, numerous chemistries have been performed under continuous flow conditions, and multiple benefits have been demonstrated such as process intensification, increased process safety, reduction/removal of headspace, enhanced control of process parameters, cost reductions, and higher yield and selectivity with good quality in addition to minimize the waste streams [2]. Its implementation affects the way we design the synthetic routes and allows for a wider operational window to conduct organic synthesis. Continuous technologies mitigate concerns related to strongly exothermic transformations and handling of unstable or toxic intermediates and hazardous reagents, thereby addressing safety concerns, which are inherent in traditional batch mode [3]. Chemistries, which are traditionally highly undesired or impossible to be performed in conventional batch vessels, can now be considered in a retrosynthesis planning stage [4]. These circumstances are a great opportunity for the pharmaceutical industry to reduce time-to-market timelines and to develop more sustainable and cost-effective processes [5]. Although flow chemistry as an enabling tool for synthetic chemistry is nowadays well established in academia, industrial applications on larger scale are still scarce and remain infrequently described in the literature. The reasoning for the hesitant uptake in industry is that organic synthesis has been performed in classical batch mode for decades by process chemists or engineers who are well experienced in the development and scale-up of batch processes that do not change as they lack experience in the field of continuous manufacturing. Furthermore, a lack of clear selection criteria of reaction classes that benefit from operating in continuous vs batch and the justification for return of investment might be hindering the successful implementation of flow technology. Accelerated project timelines, limited

Organometallic Chemistry in Industry: A Practical Approach, First Edition.
Edited by Thomas J. Colacot and Carin C.C. Johansson Seechurn.

resources, and high project attrition rates further limit the opportunities to introduce new technologies with some risks.

This chapter reviews the present state of continuous processing and provides first-hand experiences on the process development and scale-up of organolithium- and organomagnesium-based organic synthesis in continuous flow mode at lower temperature with examples from a practical point of view.

3.2 Organo-Li and Mg Processes in Flow Mode

Organometallic species (Met = Li and Mg) originating from either deprotonation or halogen–metal exchange reactions serve as useful carbanion equivalents in organic synthesis and play a pivotal role in the creation of new carbon–carbon bonds [6]. Reactions involving organometallic chemistry are predominantly fast and highly exothermic with reaction enthalpies in the range of 300 kJ/mol and pose an inherent process safety risk because of their high reactivity and strong basic nature of the reagents and intermediates [7]. In general, organometallic species are generated by a simple two-step procedure: firstly, the desired aromatic compound is metalated by deprotonation, halogen–metal exchange, or via direct metal insertion. Subsequently, the unstable organometallic reaction intermediate is quenched by an electrophile yielding the desired product (Scheme 3.1). Both steps are fast and highly exothermic; hence, the yield and selectivity are strongly influenced by the reaction temperature, residence time, kinetics, effective heat transfer management, and rate and type of mixing.

In classical batch operations, these challenging reactions are typically conducted in specialized batch equipment under cryogenic temperatures to mitigate the formation of undesired by-products, to facilitate the safe handling of energetic and reactive intermediates, and most importantly to prevent potential runaway scenarios. A slow dosing regimen of one reaction partner ensures that accurate control of the inner temperature, however, leads to long hold-up times of the metalated species because of the limited heat removal capacity of the batch reactor (Figure 3.1a). These challenges become more pronounce with the increase of scale. A solution to address these concerns is to apply continuous manufacturing technologies, while operating with minimized hold-up volumes and short hold-up times (Figure 3.1b) [8].

3.2.1 Technological Advantages of Flow Technology Compared to Traditional Batch Operation

While in a batch process only one step can be performed at a time, multiple steps can be performed simultaneously under continuous flow conditions.

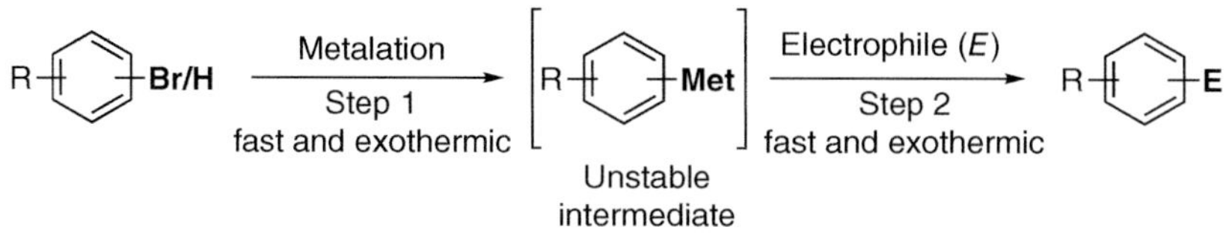

R–Ar–Br/H → (Metalation; Step 1; fast and exothermic) → [R–Ar–Met] (Unstable intermediate) → (Electrophile (*E*); Step 2; fast and exothermic) → R–Ar–E

Scheme 3.1 Metalation/quench sequence.

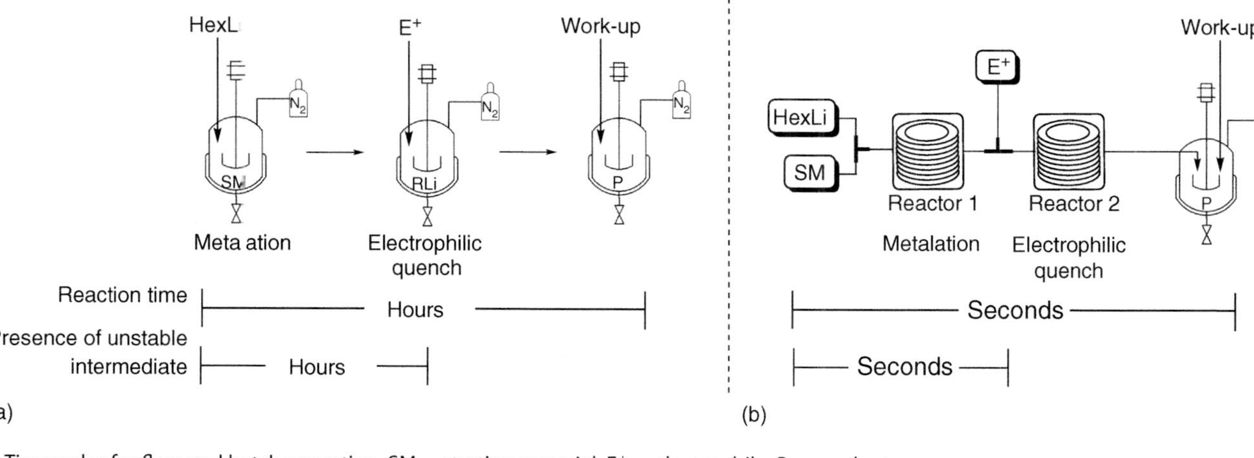

Figure 3.1 Timescales for flow and batch operation. SM = starting material; E^+ = electrophile; P = product.

Organometallic reactions are generally controlled by mixing, and reaction times can be dramatically reduced when efficient mixing performance is ensured. Considering the enhanced mass and heat transfer capabilities of flow reactors in comparison to batch reactors and the high reactivity of organolithium compounds, reaction times in the range of seconds or even milliseconds can be realized. Organomagnesium species (Grignard reagents) are slightly less reactive in comparison to organolithiums resulting in reaction times in the low minute range, whereas organozinc species show reactivities in the range of a couple of minutes, although in some cases higher temperatures are required. Applying a continuous make and consume flow approach (Figure 3.1) allows for significant reduction of residence times and thus the process can be performed in shorter time frames compared to conventional batch operations. Residence times are narrowly defined, which is of paramount advantage when handling unstable or short-lived organometallic moieties. The half-life of organometallic reagents is temperature-dependent and hence decomposition occurs over time especially when increased temperatures are applied. The make and consume concept allows for cleaner and more robust processes with cost savings and improved space/time yields (cycle times), which is highly desirable from a safety and product-quality perspective.

When considering the abovementioned factors and criteria, it appears that the reaction class of low-temperature reactions involving Li- and Mg-based organometallics is ideal to be run in a continuous flow mode [9]. Although flow processes are superior when reactions are controlled by mixing or when temperature-sensitive and unstable intermediates are present, conventional batch operation is more suited for reactions involving prolonged heating and mixing and solids are precipitated with time. Ideally, the two technologies are considered complementary to each other, and the most applicable technology needs to be selected for a certain transformation.

3.2.2 Temperature Profile of Continuous Flow Reactions

Flow chemistry enhances the control over reaction parameters such as mixing efficiency, mixing time, residence time, back-pressure, stoichiometry, and temperature on a microscale level. In the discipline of low-temperature organometallic chemistry, two general heat management strategies can be applied during development of a flow process. The first strategy relies upon stabilization and half-life extension of an unstable organometallic species by applying cryogenic reaction conditions (typically −78 °C) before its electrophilic quench. Here, the focus lies on the maintenance of a narrowly predefined temperature and excellent heat dissipation, whereas mixing time, mixing efficiency, as well as residence time play only a subordinate role (Figure 3.2a) [10]. The second approach focuses on excellent mixing efficiency and very short mixing times in the range of seconds. In this case, the reaction time (=residence time) is defined as the mixing time, and therefore, the reaction occurs under quasi-adiabatic conditions (Figure 3.2b); the unstable organometallic intermediate is not stabilized over a certain period but is rather consumed instantaneously before its degradation – this concept is often referred to as flash chemistry.

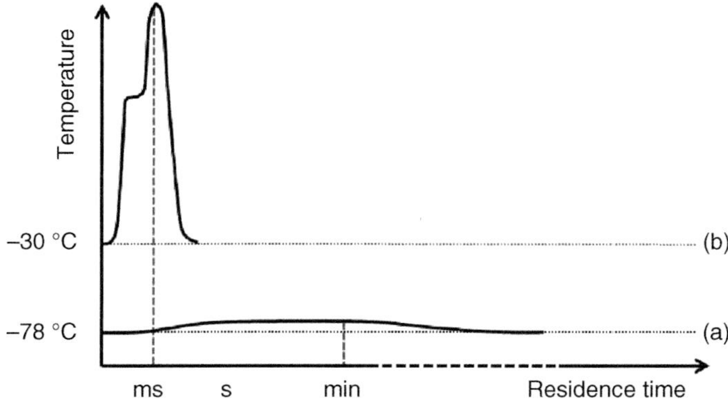

Figure 3.2 Possible temperature profiles: (a) accurate (quasi-isothermal) temperature control vs (b) quasi-adiabatic conditions.

The concept of *flash chemistry* was introduced by Yoshida et al. in 2008 and is defined as a field of chemical synthesis where extremely fast reactions (*milli*seconds to seconds) are conducted in a highly controlled manner to produce desired compounds with high selectivity. [11]

Organic transformations conducted in the range of (milli)-seconds under continuous flow conditions offer various advantages over classical organometallic batch reactions, which are discussed in the following sections.

3.2.3 Flash Chemistry: Functional Group Tolerance

Organometallic species are highly reactive carbanion equivalents and react spontaneously with any electrophilic moiety. Therefore, electrophilic functionalities such as carbonyls, carboxyls, nitriles, or nitro groups are in general not compatible with the generation of organometallic species in batch mode and synthetic routes have to be planned accordingly or strategies involving protecting groups have to be applied. Flow chemistry allows to tackle this issue technologically by improved mixing efficiency, reduced mixing times, and short residence times: undesired side reactions that are kinetically slower than the Br/Li exchange can thus be mitigated and controlled. In 2008, Yoshida et al. successfully highlighted the advanced functional group tolerance in the lithiation of ethyl-2-bromobenzoate using *s*-butyllithium at −58 °C and a mixing time (t^R) of 0.06 seconds (Table 3.1, entry 1). The synergetic effect of highly efficient mixing and short residence times allowed for the lithiation/quench sequence in the presence of "incompatible" electrophilic functionality yielding the desired product in high selectivity and good yield [12]. This work impressively demonstrates the capability of flow chemistry for very fast reactions (e.g. metal/halide exchange) and the scope was expanded to other halogenated arenes bearing electrophilic moieties such as nitro ($t^R = 0.01$ seconds) [13], nitrile ($t^R = 0.01$ seconds) [14], isocyanate ($t^R = 0.016$ seconds) [15], additional bromide ($t^R = 0.8$ seconds) [16], or even ketone carbonyl groups ($t^R = 0.003$ seconds) [17] (Table 3.1).

Table 3.1 Flash chemistry: functional group tolerance.

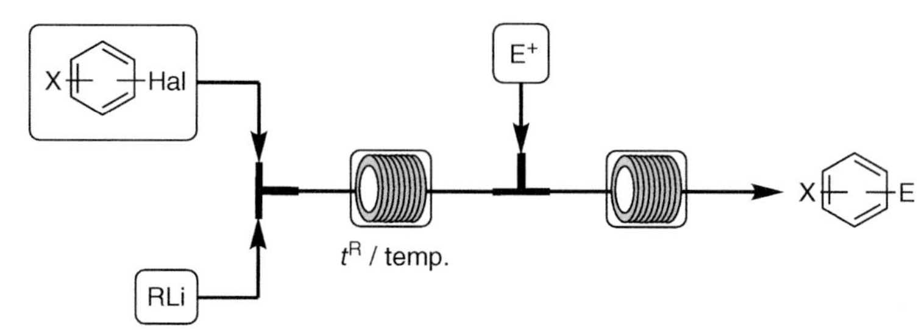

Entry	X	Hal	R	t^R (s)	Temperature (°C)	Yield (%)
1	COOEt	Br	sBu	0.06	−58 to (−48)	61–90
2	NO_2	I	Ph	0.01	−28 to 0	36–93
3	CN	Br	nBu	0.01	0 to 20	81–96
4	NCS	Br	nBu	0.016	25	75–99
6	COMe	I	Mes	0.003	−70	59–77
7	*o*-Br	Br	nBu	0.8	−78	68–81

In addition, the concept of flash chemistry can also be applied for the synthesis of labile aliphatic or alkenyl organometallic reagents, which are not easily accessible by classical batch processes. Metalated organoalkyl-[18] and organoalkenyl halides [19] (e.g. perfluoroalkyl lithium, dichloromethyl lithium, dichlorovinyl lithium, and (trifluoromethyl)vinyllithium) belong to the most labile organometallic species as they decompose rapidly under generation of inorganic salts. However, by using flow technology, these versatile compounds can be synthesized in a controlled manner and directly used for the synthesis of functionalized building blocks.

3.2.4 Flash Chemistry: Selectivity

The selectivity of metalation reactions depends on various factors: most importantly, the temperature, concentration, and residence time (respectively half-life). Especially, the residence time can have a significant impact on the selectivity of a metalation reaction as it is in the range of hours in a classical batch environment compared to seconds in continuous flow mode. In general, metal–halide exchange reactions as well as metalation by deprotonation selectively yield the kinetic reaction product. However, the kinetic reaction product can intramolecularly isomerize or rearrange to more stable thermodynamic reaction products when extended reaction times are applied before the electrophilic quench. Thus, the selectivity of the desired reaction decreases, which does not only reduce the yield but can also complicate downstream processing. Flow technology can solve these selectivity issues as the reaction time for isomerization can be accurately

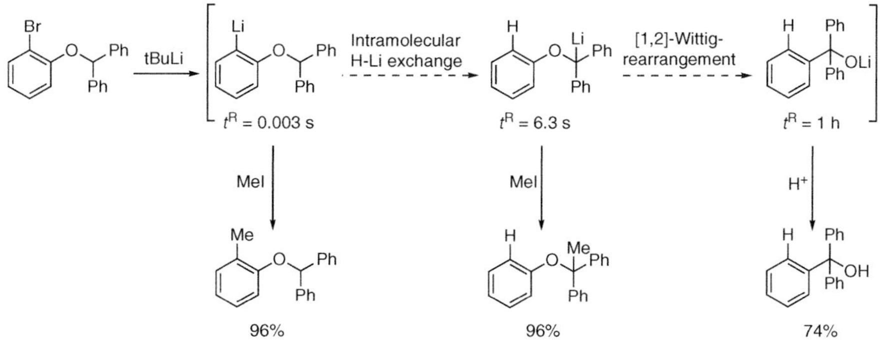

Scheme 3.2 Metalation of *ortho*-bromo-aryl benzyl ethers under flow conditions: residence time and selectivity.

controlled allowing to trap organometallic intermediates before isomerization [20].

A typical side reaction in halide/metal exchange reactions is the H–Li exchange reaction, which occur in the presence of acidic protons. *Ortho*-lithiated aryl benzyl ethers are prone for intramolecular H–Li exchange as the acidity of its benzylic proton is higher compared to the protons at the aromatic core. In 2018, Lee et al. demonstrated the selective functionalization of *ortho*-lithiated aryl benzyl ethers under flow conditions [21]. By applying varying residence times, isomerization and functionalization were accurately controlled, giving selective access to different substituted arenes (Scheme 3.2). Equally challenging are metalations of aromatics bearing multiple halogen substituents because of a phenomenon called "halogen dance" [22].

Another outstanding example of how reactivity and selectivity can be controlled under flow conditions is the outpacing of the anionic Fries rearrangement recently demonstrated by Kim et al. [23]. Once metalated in *ortho*-position, *ortho*-halogenated O-aryl carbamates rearrange quickly to *ortho*-metalated salicylamides. This rearrangement was outpaced when the metalation of *ortho*-iodophenyl diethylcarbamate occurred in 0.33 ms reaction time followed by the immediate quench with an electrophile. Following this protocol, various *ortho*-functionalized carbamates are accessible in good to excellent yields and high selectivity. In contrast, isomerization took place when the residence time was extended to 628 ms, facilitating the deliberate synthesis of the rearranged products in good yield and selectivity (Scheme 3.3).

3.2.5 Flash Chemistry: Stoichiometry and Chemoselectivity

Flow technology allows to accurately control the stoichiometry and concentration profile of a desired transformation across all reaction steps, which becomes a remarkable benefit when the desired reaction product is prone for consecutive over reactions [24]. Consecutive transformations such as multiple additions of an organometallic species to an ester group limit the scope of organometallic transformations in classical batch processes. In batch mode, the formation

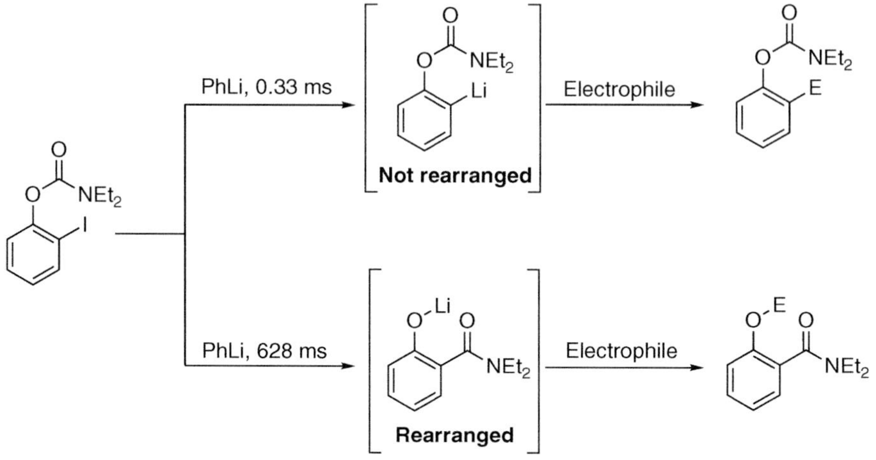

Scheme 3.3 Selectivity control gained by different residence times: outpacing the anionic *Fries*-rearrangement.

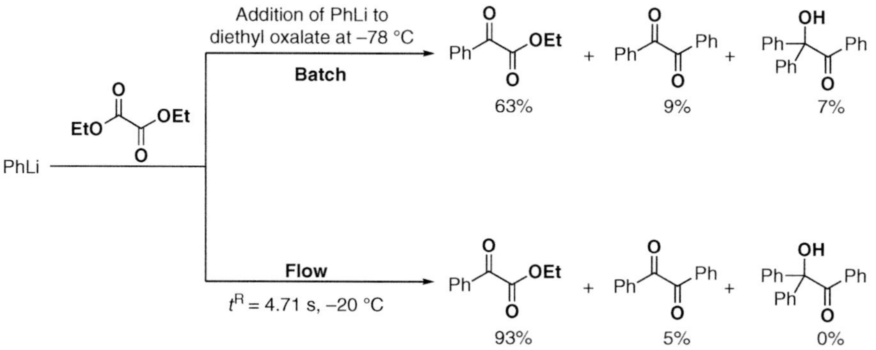

Scheme 3.4 Selectivity of the addition of phenyl lithium to diethyl oxalates in batch and flow mode.

of the undesired tertiary alcohol is unavoidable because the second addition to the ketone is faster than the desired first addition reaction to the ester. In 2013, the Yoshida group demonstrated the selective addition of aryl lithium moieties to dialkyl oxalates using a flow microreactor [25]. Although in batch mode the yield of the desired monoaddition product did not exceed 63% because of the formation of multiple addition by-products, the yield was increased to >90% when the reaction was realized under flow conditions (Scheme 3.4). This concept was extended to difunctionalized electrophiles bearing a ketone and an aldehyde functionality [26]. The key to success was high mixing efficiency and accurate control of a 1 : 1 reaction stoichiometry. Thus, organolithium reagents were carefully trapped by the electrophile preventing consecutive additions.

3.3 Continuous Flow Technology

Although continuous manufacturing offers numerous benefits for controlling chemical processes, there are also areas where certain chemistries or technical operations remain challenging. For a reliable chemical performance and robust operation in continuous flow mode, the selection of the appropriate equipment parts is therefore essential. A typical continuous flow unit for low-temperature chemistry comprises well-characterized hardware modules, which facilitate a smooth scale-up of chemical processes without necessity to perform additional laborious process development work (Figure 3.3).

Every chemical flow process starts with the preparation of feedstock solutions and the selection of appropriate storage vessels (Block A). Pump modules equipped with prefilters (Block B) provide a reagent stream at a desired flow rate (stoichiometry), followed by heat exchanger units (Block C) to adjust the temperature before the mixing point. Block D is dedicated to appropriate mixing and required residence time for the chemistry to react to completion. The final module consists of vessels for product and waste collection (Block E). The entire setup can be monitored by online and PAT tools at multiple positions to stay in control of the process and to initiate emergency actions if needed.

When preparing feedstock solutions (Block A) for a continuous operation, certain criteria must be considered and evaluated to ensure a robust, safe, and smooth operation: Solubility of starting materials in a favorable solvent at a desired temperature must be known to ensure that one is not storing a feedstock solution in a metastable zone and the dissolved compounds may spontaneously precipitate and freeze. Furthermore, the stability of a feedstock solution over extended periods must be assessed to avoid potential fouling or encrustation of the feedstock that causes issues in practical operation and product quality. Here, material compatibility may play a major role and selection of the feed vessel must be wisely chosen. Even though a reactor volume in continuous flow mode may be smaller compared to a traditional batch, the volume of potentially hazardous or dangerous feedstock might pose a severe potential for incidents and must be accounted for. A major hurdle for the application of flow chemistry in

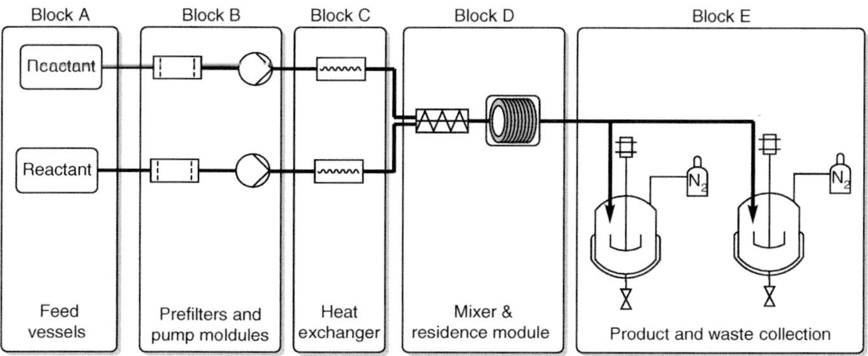

Figure 3.3 Schematic for a low temperature hardware setup.

organometallic reactions is clogging of mixers and flow reactors by solids arising during manufacturing. Thorough drying and inertization of the feedstock vessels and the entire flow reactor setup are crucial when operating with organometallic species because traces of protic solvents or water may rapidly lead to clogging of the system. An additional feature to prevent clogging issues is to use in-line filters for each feedstock solution, thus preventing any suspended particles to enter the flow reactor setup.

For providing the feedstock solutions, various pump modules (Block B) are suitable and commercially available. It is essential to ensure material compatibility of all equipment parts (O-rings, seals, syringes, connectors, etc.), which get in touch with chemistry. Furthermore, minimal pulsation is of importance to ensure an exact stoichiometry, residence time, and residence time distribution. Here, continuous syringe pumps or pressurized head tanks in combination with a Coriolis mass flow controller are recommended options. Before a feedstock solution entering the reactor zone, it passes through a heat exchanger unit (Block C) allowing to pretemper to a desired temperature before the mixing point.

The reactor module (Block D) typically consists of a high-performance mixer for rapid mixing and good heat removal characteristics followed by a modular residence time module. The selection of an appropriate mixer for a given chemical transformation is one of the most important and delicate tasks. The degree of influence of mixing on yield or quality of the reaction dictates whether a sophisticated mixer or a more simplistic device must be used. It is the chemists' and engineers' job to identify a device with maximal mixing efficiency and lowest susceptibility to clogging, which is always going to be a trade-off. Numerous mixer designs are described [27–30] and a general recommendation is impossible to give – every case is individual and requires careful observation and evaluation in the early development phases. As residence time modules, empty tubing or tubing filled with static mixer elements [31] are often used. Tube-based systems are commonly made of stainless steel, hastelloy, or perfluorinated polymers. Reactors manufactured from metals possess robustness and good heat dissipation capabilities and are therefore most suitable for pilot scale but might be less handy to use in a laboratory setup during process development phases. On lab scale, perfluorinated tubes have the advantage of broad chemical compatibility and low cost allowing for single use – if clogging of a reactor occurs, it can be easily discarded and replaced. An alternative to tube-based and structured reactors is continuously stirred tank reactors (CSTRs), which are of particular interest for reactions involving slurries or for chemical transformations requiring longer residence times [32, 33].

Independent from the equipment selection, inherent safety of a flow process can only be guaranteed with appropriate process understanding, monitoring of process parameters, and the ability to react to deviations promptly. A software-controlled platform enables for data acquisition and online monitoring of temperatures, flow rates, mass flow, system pressures, conductivity, as well as emergency shutdown actions based on predefined parameters, thereby mitigating the risk of incidents.

3.3.1 Clogging as a Major Hurdle in Flow Chemistry

Fouling, encrustation, and clogging are well-known problems when operating over an extended period of time [34]. Precipitants and fouled layers initially cause pressure fluctuation, impact heat and mass transfer, and gradually lead to clogging of the entire setup [10d, 35, 36]. Blockage of a flow system still embodies the largest obstacle in continuous manufacturing and hamper a reliable and robust continuous operation [34, 35, 37]. There are numerous factors that may contribute to clogging, namely solid formation because of moisture in the feedstocks or transfer lines, generation of impurities and degradants, precipitation of reaction intermediates and salts, or formation of polymer-like deposits. There are no "magic" countermeasures in place; however, in-depth process understanding and precise control of optimal process parameters can reduce the likelihood of interruptions. Figure 3.4 highlights prominent hot spots for clogging including the prefilters (I) connected to the organometallic feedstock, pump module (II) for organometallic feedstock, precooling module (III) for the organometallic feedstock, mixing element (IV) for metalation chemistry, and finally the residence time module (V). From the second mixing element onward, lesser issues are detected, which can be rationalized by higher dilution of the reaction mixture and increased overall flow rates.

The identification of the initial trigger for clogging is difficult as many process deviations are interlinked and each can cause knock-on effects. A key factor for successful operation is a reliable, accurate, and most importantly pulsation-free mass flow. As soon as a pump module deviates in mass flow, factors such as stoichiometries, solvent composition, solubility, or residence time are jeopardized. As a consequence, undesired by-products are formed and the system begins to run out of control.

Precautionary measures include usage of dry solvents for the preparation of high-quality feedstock solutions. Feedstocks may have to be diluted or alternative solvent compositions are considered to avoid solubility issues [11a,c], and its stability has to be ensured over the anticipated storage time and visually inspected for potential suspended matter. Nitrogen gas used to build up an inert atmosphere in the feed tanks may partially dissolve in the solution and degas later on.

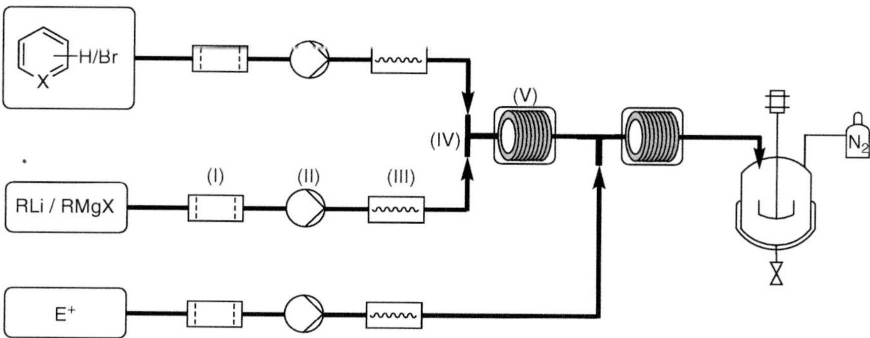

Figure 3.4 Hot spots for clogging.

This cavitation issue can consequently cause fluctuation in mass flow and system pressure and eventually result in malfunction of the whole pump module. Spring-based dampers or a bubble trap can counteract this phenomenon [38].

A common fouling problem occurs when organometallic reagents are used in THF at noncryogenic temperatures. The decomposition of THF [39] results in the formation of a yellowish, jelly-like polymer, which can lead to pressure increase and blocking of mixers and reactors. Selection of inline filters with appropriate mesh size and sufficient surface area mitigates the risk for interruptions.

When organometallic species are involved, clogging issues occur because of the moisture in feedstocks or decomposition of unstable intermediates releasing colorless, inorganic, water-soluble salts and polymeric degradants. For alkyllithium reagents in particular, the presence of LiOH is evident. In addition, the solubility of certain metalizing agents and the freezing temperature of alkane impurities within the organometallic feedstock (e.g. dodecane) must be considered to avoid blockages in the precooling elements and mixers. The precipitation of inorganic salts is almost inherent and scheduled flushing cycles of the flow system are recommended. Continual deplugging of mixing elements or residence time modules is possible by applying an ultrasonic device [40].

Overall, the most reliable handle to counteract fouling and clogging is the installation of redundant equipment parts where required. The blocked unit can be bypassed, dismantled, and cleaned while the redundant system maintains the continuous operation. Alternatively, recurrent and predefined cleaning cycles might help to increase the robustness of the flow process. Here, it is of paramount importance to ensure dryness of the entire equipment after rinsing with protic solvents or water to avoid instantaneous clogging once continuous operation is restarted.

3.3.2 Start-up and Shutdown Operation

The major focus during process development of a flow process lies on the "steady state" – the point at which the flow setup operates as planned: pumps provide an accurate flow rate and thus steady reagent concentrations at a predefined temperature. Heat management is stable, and product and by-product formation remain unchanged during operation. However, as the reagent stoichiometry and solvent composition are not ideal during start-up and shutdown, these phases must be carefully investigated to avoid any unforeseen or uncontrolled events.

3.3.3 Material of Construction

Today, flow reactors and mixers can be manufactured from various materials including aluminum alloys [41], silicon carbide [42], stainless steel [43], or polymers such as the widespread polyfluoroalkoxy (PFA), polytetrafluoroethylene (PTFE), and fluorinated ethylene propylene (FEP), among others. For selection of an appropriate material of construction, it is essential to consider factors such as thermal conductivity, chemical compatibility, surface characteristics, and material cost. Stainless steel is a prominent candidate because of its availability, low material cost, high pressure and temperature rating, and good heat conductivity, limiting its incompatibility with acidic media because of corrosion risks. Silicon-based hardware excels through its thermal conductivity, ease of manufacture, high temperatures, and pressures rating and its chemical resistance toward a wide range of chemicals, limiting its use in strong alkaline reaction mixtures. Usage of polymeric materials such as PFA, PFTE, and FEP shows low thermal conductivity and lower pressure and temperature rating, which are thus less suited for strongly exothermic chemistries. Polymer-based hardware, however, is still frequently used in small-scaled flow chemistry reactions because of the ease of manufacture, low material cost, and remarkable chemical stability.

3.3.4 Safety Concept and Emergency Strategies

Even though continuous manufacturing offers various safety advantages compared to classical batch processes, e.g. smaller reactor volumes or better heat transfer, flow processes are not inherently safe. Unexpected blockages, thermal events, or undesired leakages can lead to potential incidents. Precipitation, encrustation, or fouling are still major problems in continuous manufacturing and can lead to sudden, unexpected clogging and pressure build-up. Such an event can have dramatic health and safety consequences if it comes to rupture of equipment parts and exposure to operators. Furthermore, even though flow chemistry benefits from favorable surface-to-volume ratios and fast heat transfer, isothermal operation is not guaranteed [10d] and the entire energy balance must be fully understood. If the generated reaction heat is greater than the heat dissipation capacity of the equipment part, there will be a steady temperature increase, which in worst case could result in a runaway scenario. The time to realize such an event and to react accordingly is short and therefore a suitable safety measure must be installed. Undesired leakages are detected by liquid sensors placed at strategic positions around the flow setup notifying the operators. In addition, mass flow controllers also function as detectors for leakages. All the abovementioned points must be discussed in a process-hazard-analysis (PHA) to define adequate safety controls, emergency shutdown actions, and cleaning procedures.

3.4 Development of a Flow Process

In order to develop robust and efficient chemical processes, development laboratories have gone through major transformations. Although it was common to use a round-bottom flask a few decades ago, today's state-of-the-art laboratories are mostly equipped with automated batch reactor systems mimicking the desired plant setup on lab scale [44]. These sophisticated systems facilitate data-rich experimentation, elevate data quality, and are equally useful for the development of batch and flow processes. The following sections will discuss the different phases during process development of a flow synthesis and intends to provide a straightforward guide about how flow processes can be realized.

3.4.1 Screening Phase: Feasibility Study

Although the development of a flow process is the final goal, it does not implicate that all experimental investigations have to be performed purely in flow mode. Automated batch reactor systems allow to perform multiple experiments in parallel, generating valuable information in a short period of time. Thus, this equipment is the ideal starting point to assess the solubility of all starting materials, intermediates, and reaction products at various temperatures: data which is important to confirm the flowability of the desired chemistry. Although it is straightforward to evaluate the solubility of stable and isolatable compounds, it is more difficult to assess the likelihood of precipitation of unstable intermediates (e.g. metalated intermediates) – especially, if these intermediates tend to decompose rapidly even at very low, cryogenic temperatures. In these cases, an in-depth solubility assessment is not possible; however, two strategies can be used to get an idea of potential precipitations during a flow process. As the solubility of moderately stable organometallic intermediate reduces with lower temperatures, batch experiments at very low cryogenic temperature can be performed as a worst-case scenario to assess the solubility of all intermediates. If no precipitation at very low temperature is observed, then the likelihood for precipitation at elevated flow conditions can be deemed as low risk. If this is not a feasible approach because of the decomposition of the intermediate, simultaneous addition of all reagents into a vigorous stirred batch reactor at different temperatures allows to safely assess precipitation at the mixing point of a flow process. These experiments provide no guarantee for trouble-free operation but increase the chance to identify potential clogging issues in early development stages. Based on various reports in the literature, 0.3 mol/l for the aryl halide (in THF) and 1.6 mol/l for the organometallic reagent (e.g. *n*-butyllithium in hexane) are reasonable concentrations as a starting point [45].

Although solubility and solid formation are the most important parameters that have to be addressed before performing any flow experiments, additional considerations are crucial to ensure safe and seamless investigations under flow conditions. Start-up and shutdown operations as well as cleaning procedures have to be defined and established in flow mode to ensure smooth operation. For start-up of a flow operation, it would be advisable to switch on the feed pump for the organometallic reagent at last and for shutdown to switch off the feed pump for the organometallic reagent at first. Notably, as the reagent stoichiometry is not ideal during the start-up and shutdown phase, the impact of this process deviation on the flowability of the process must also be taken into account.

An additional prerequisite for running a first flow experiment is process safety and at least a basic assessment in batch mode should be available upfront. DSCs of starting materials and product give insight into the risk of thermal decomposition. In addition, expected reaction energy and adiabatic temperature rise can be initially assessed by single-shot addition of the organometallic reagent to the solution of the aryl halide.

Based on our own experience in the field of organometallic chemistries in flow mode, simple plug flow reactors consisting of PTFE tubings and T-pieces (as mixers) are often sufficient to demonstrate the initial feasibility of a process. For organolithium chemistries, residence times in the range of seconds at around −30 °C are reasonable under efficient mixing conditions [45]. These conditions can be regarded as a good starting point for a first trial run. In contrast, residence times in the range of minutes have to be considered when *Grignard* reagents are used, even at elevated (e.g. ambient temperature) temperature. Once the flowability is demonstrated and a safe and reliable process under steady-state conditions is ensured, the reaction conditions can be optimized by varying residence times, flow rates (correlates with mixing efficiency), reaction temperature, and stoichiometry. It is important to wait between changes and sampling for the flow system to reach steady state; a flow system requires approximately three residence times to enter steady-state operation.

In general, it is quite useful if the utilized flow equipment allows for a straightforward scale-up; however, various concepts are known in the literature, which can be used for further scaling-up the process once its feasibility and flowability have been demonstrated (see Section 3.4.2) [2c, 46].

3.4.2 Process Development Phase: Extended Evaluations Including Technical Feasibility

The next development phase starts once the feasibility of the flow process has been demonstrated and its flowability assessed over a short run time (several minutes to an hour). Similar to the development of a traditional batch process, now the in-depth assessment of the flow process starts and all important parameters are fine-tuned and examined based on their criticality on the whole process. Temperature profiles, flow rates, residence times, heat exchange, concentrations, solvent composition, mixing efficiency/sensitivity, and reagent stoichiometry are carefully evaluated to facilitate a seamless scale-up at optimum reaction conditions. Yield losses are investigated, critical by-products are identified, and process know-how is generated in flow mode supported by batch experiments. In general, two critical by-products can be quickly identified: a dehalogenated starting material (obtained in the presence of acidic protons) and the Wurtz coupling product (resulting from the reaction between the metalation reagent and the starting material). Although the amount of dehalogenated starting material often correlates with the amount of water in the solvents used, Wurtz coupling is preferred at high temperature and extended residence times (before an electrophilic quench) and can be suppressed during development phase (Scheme 3.5). Besides yield loss, these side reactions also form inorganic salts, which dramatically increase the risk of clogging and make its suppression a priority during process development.

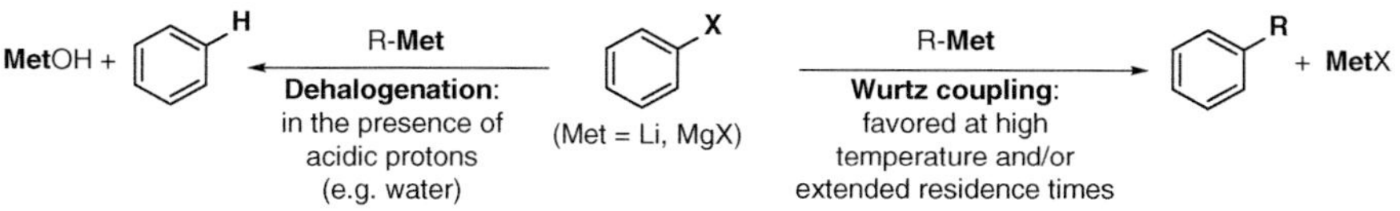

MetOH + Ph–H ← R-Met — **Dehalogenation**: in the presence of acidic protons (e.g. water) — Ph–X (Met = Li, MgX) — R-Met → **Wurtz coupling**: favored at high temperature and/or extended residence times — Ph–R + MetX

Scheme 3.5 Major side reactions when organometallic reagents are used.

Safety data have to be collected from all reaction mixtures and the reaction energy as well as decomposition temperatures determined. The entire energy balance of the reaction must be fully understood and heat dissipation capacities are assigned accordingly to avoid runaway scenarios.

It is a practical problem to assess the temperature profile of fast reactions inside narrow-bored, tubular reactors. As a first measure, temperatures are recorded at the mixing point and at the exit of the flow reactor allowing for a first indication of the reaction enthalpy and $\Delta T_{adiabatic}$. However, the real temperature profile within the reactor and potential local hot spots are not detected and may lead to decomposition or even thermal runaways if the heat exchange is not sufficient. Additional valuable data points can be generated using inline PT100 probes or thermocouples at predefined positions (residence times) along the flow reactor. Understanding of the kinetics and thermodynamics of a reaction becomes crucially important on scale-up and determines if a process can be run robustly in a selected flow reactor.

Strongly exothermic reactions and temperature-sensitive reaction mixtures can be controlled by splitting one reagent stream into multi-injections instead of the generally used single-injection principle [47].

Storage conditions of all feeds as well as the final product have to be evaluated and the feedstocks used accordingly. In addition, an assessment of waste streams has to take place and a concept for process deviations must be developed. For example, two different storage tanks can be used: storage tank-1 for product collection and storage tank-2 for waste collection (e.g. during start-up/shutdown or cleaning phases). Organometallic flow reactions tend to clog at one point during operation; thus, it is of great importance to have a reliable cleaning concept in place and appropriate infrastructure installed to prevent any impact on the final product quality. PAT tools allowing for in-line analytics can help to increase process robustness and should be considered to ensure the product quality over an extended period of time. To support the mixer and reactor design for the final plant process, material tests have to be performed at critical points (e.g. mixers) during which the reaction medium has to be exchanged as often as possible to gain reliable information about corrosion during continuous operation.

After all necessary information has been collected, a final HAZOP analysis [48] has to be performed to address safety concerns, develop countermeasures, and ensure the safety of the whole process. To evaluate the robustness of the process and its feasibility for large-scale production, an assessment over an extended period of time is necessary. Thus, the setup should be run for multiple hours/days to demonstrate its technical feasibility, test its back-up concepts (e.g. mixer cleaning), and ensure a constant product quality.

3.5 Literature Examples: Flow Processes on Multi 100 g Scale

With the advancement of continuous flow technology and a growing population of researchers educated in the field, the advantages of flow technology are providing increased momentum for implementing continuous processing into small-molecule manufacture. The following chemical examples highlight selected organometallic chemistries performed on quantities greater than 100 g scale, including feasibly studies, technical proof of concepts, and pilot campaigns.

3.5.1 Manufacture of Verubecestat (MK-8931)

Process researchers from Merck Research Laboratories reported on the manufacture of verubecestat (MK-8931), a drug candidate for the treatment of Alzheimer's disease [49]. A key step in the synthesis involves an organolithium addition to a chiral ketimine. A competitive proton transfer resulted in only moderate yield when performed in classical batch mode. However, when the reaction was conducted in continuous flow mode, the undesired proton transfer was outperformed and the yield was significantly improved compared to batch. The flow process was developed by implementing flow chemistry principles and delivered the desired product on kilogram scale (Scheme 3.6).

The deprotonation of the methyl sulfonamide and the electrophilic quench revealed to be highly sensitive to mixing, whereas the reaction temperature was of subordinate importance, enabling the continuous process to be performed under noncryogenic conditions. During process development, clogging risks that were basically invisible in small-scale reactions and short run time were identified and rationalized by the decomposition event at the mixing point of methyl sulfonamide and *n*-hexyllithium. Fouling was addressed by careful selection of the mixer type moving from Upchurch Scientific static mixing T-pieces to Koflo Stratos static tube mixers. To mimic processing on plant scale, a T-piece attached to a downstream Koflo Stratos static tube mixer was evaluated. Naber et al. demonstrated on kilogram scale that flow technology enables the Mannich-type reaction to outpace fast deprotonation of the electrophile, a side reaction that is unavoidable in batch mode. The flow process offered substantial yield improvements over batch while eliminating the requirement for cryogenic cooling and delivered the desired product in 87–91% assay yield and in a 92 : 8 ratio of diastereomers at steady-state operation.

3.5.2 Manufacture of Edivoxetine

A key objective for the Eli Lilly organization is the development of inherently safer organometallic chemistry via a continuous processing approach [50]. Researchers elaborated on the manufacture of a common starting material for the synthesis of edivoxetine under flow conditions utilizing both a continuous Grignard and a lithiation procedure (Scheme 3.7). The entire Grignard route that includes the formation of the organometallic reagent, formylation, reduction, and workup steps were run in CSTRs. The lithiation approach was performed in a combination of plug flow reactors (PFRs) for the halogen/metal exchange and formylation steps and CSTRs for the reduction and workup steps. Both flow routes offered significant benefits compared to the traditional batch processing.

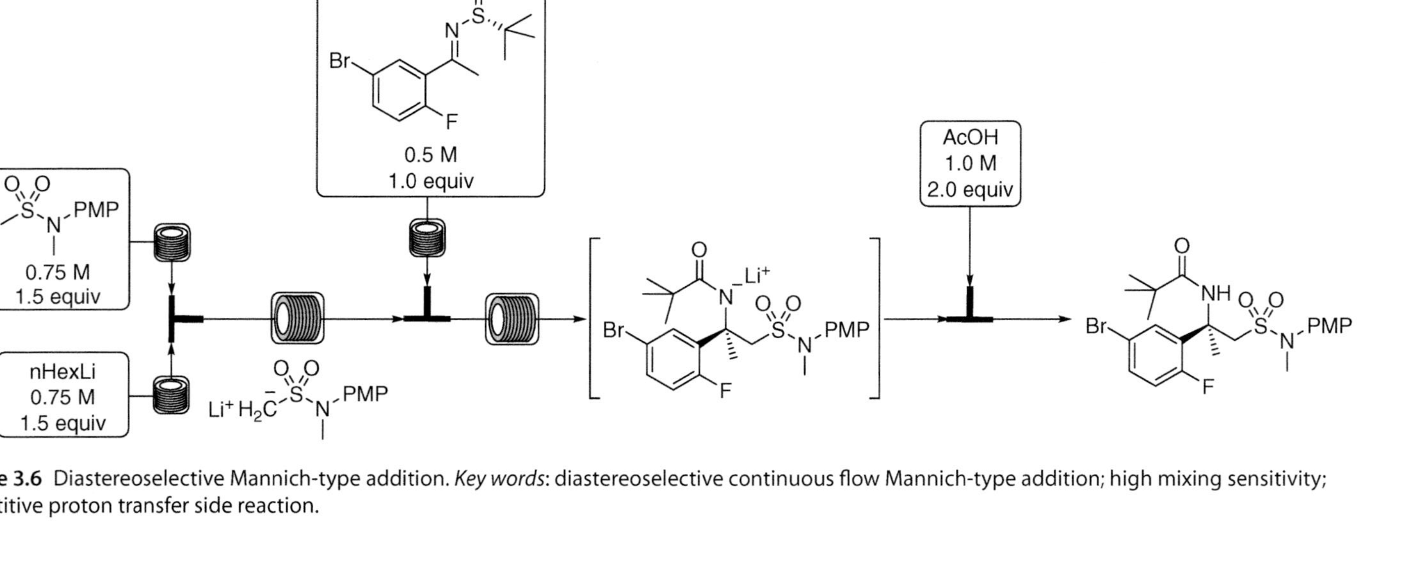

Scheme 3.6 Diastereoselective Mannich-type addition. *Key words*: diastereoselective continuous flow Mannich-type addition; high mixing sensitivity; competitive proton transfer side reaction.

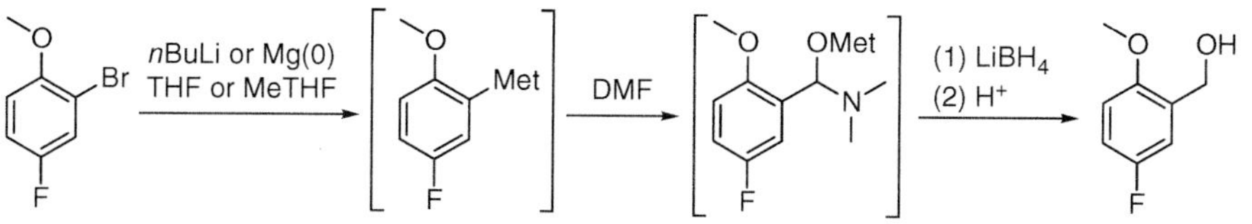

Scheme 3.7 Reaction sequence toward a regulatory starting material using CSTRs. *Key words*: Head-to-head comparison of batch vs Grignard and lithiation chemistry; PFR; CSTR; yield improvement; enhanced safety.

PFRs are of limited use for exothermic reactions involving solids or fines and CSTRs become significantly more practical. Consequently, a CSTR was chosen for the formation of the Grignard reagent using Mg(0). An operational challenge of magnesium fines exiting the CSTR was resolved by installing a settling pipe to keep magnesium solids inside the CSTR and an additional solid trap outside the CSTR. Mg particles accumulating in the outside trap were periodically transferred back into the CSTR. The kinetics for the Grignard formation are reported to be extremely fast and >99% conversion can be achieved in a single CSTR with $\tau \leq 1$ hour. The consecutive *Bouveault* formylation and reductive workup were conducted using four CSTRs in series. The formylation was performed using a 10% (w/w) dimethylformamide (DMF) solution in THF at 40–45 °C and a residence time of 30 minutes (CSTR1). A 4 M solution of $LiBH_4$ in THF was used for the reductive workup at 22 °C and a residence time of 30 minutes (CSTR2). The acidic quench with 25% (w/w) methanolic H_2SO_4 and addition of toluene occurred in CSTRs 3 and 4 at 22 °C. The final phase separation was performed in batch mode. The entire Grignard process was run over five days with overnight holds with all operations linked continuously for 32 hours of total processing time providing 3.93 kg (82.4% yield) of the desired product.

The concept for the corresponding organolithium chemistry in a PFR is to make and rapidly consume the unstable aryl lithium intermediate to afford a more stable solution of the DMF-quenched product that is then reduced with $LiBH_4$ to the desired benzyl alcohol. Down-scaling of flow process to low flow rates while achieving efficient and representative mixing performance poses a challenge. Static mixers may not be fully functional at reduced flow rates and the use of microreactors is of limited practicality because of the likely blockage of the narrow channel dimensions. Eli Lilly scientists addressed that challenge by usage of dynamic mixing, provided by magnetic stirrer bars sequestered within a steel tubing. The well-documented trouble of pumping chemically aggressive alkyl lithium reagents that often contain trace amount of inorganic precipitate was addressed by usage of Teledyne Isco 1000D syringe pumps in combination with an injection coil for the *n*-butyllithium reagent. The Br/Li exchange and the electrophilic quench with DMF were performed in PFRs in series (37 ml/min aryl bromide reagent feed) at a jacket temperature of −78 °C. Each PFR comprised four static mixer elements. Periodical flushing of the reactors with THF counteracted potential fouling and clogging issues. This reaction solution was collected in an agitated vessel and treated with a continuous stream of 4 M $LiBH_4$ in THF provided by a peristaltic pump. In analogy to the Grignard route, the acidic quench with methanolic H_2SO_4 and addition of toluene for extractive workup occurred in separate CSTRs; the final phase separation was performed in batch mode. Overall, 529 g of aryl halide was processed over a total run time of 74.3 minutes affording the desired product with an overall yield across the reaction chain of 81%.

A comparison of the initial batch process with the two flow procedures revealed a significant advantage of flow chemistry over batch mode. Given the more challenging handling of the *n*-butyllithium reagent and its associated hurdles, the Grignard approach was selected as superior; alkyl lithium reagents should be reserved for particular cases such as involvement of unstable aryl lithiums that must be produced at low temperature and consumed immediately.

3.5.3 Scale-up of Highly Reactive Aryl Lithium Chemistry

Mleczko and coworkers from Bayer Technology shared experiences obtained during commissioning and operation of a flow setup with a capacity of approximately 1 kg/h [10d]. The study describes the deprotonation of difluorobenzene by *n*-hexyllithium and consecutive quench with DMF and dimethyl sulfate (DMS) (Scheme 3.8). The 2,3-difluorophenyl lithium intermediate is highly reactive, raising safety concerns when accumulated in the batch reactor. A two-step flow process operating at only small hold-up of the critical aryl lithium moiety was envisioned to mitigate the safety risk. Continuous operation was demonstrated over a campaign length of up to 300 hours.

The utilization of flow technology was superior relative to batch with respect to chemical performance and space–time yield. In continuous mode, 40 kg of the product was produced over 24 hours within a total reactor volume of only 3 l, whereas in batch mode, only 20 kg of the product was synthesized over 24 hours within a 400 l batch vessel. Additionally, the reaction sequence was performed at higher temperature (−45 °C) compared to batch mode (−70 °C).

The pilot unit consisted of IKSM mixers connected to the residence time module with thermocouples positioned after each reactor module. Experimentation revealed extremely strong exothermicity for the reaction of difluorobenzene with *n*-hexyllithium as well as for the electrophilic quench and the temperature increase inside the IKSM modules could not be dissipated quickly enough for

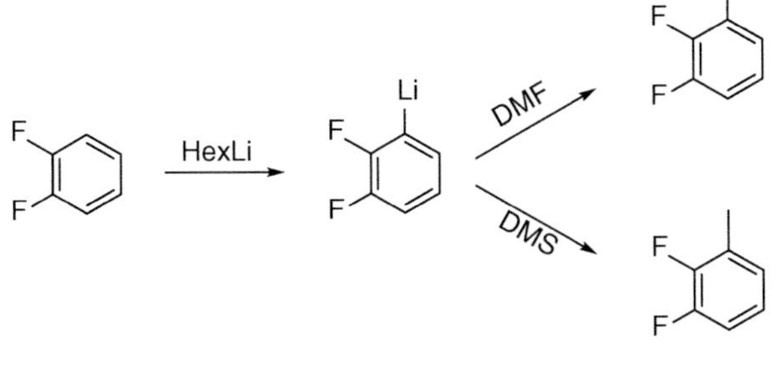

Scheme 3.8 Synthesis of difluorobenzaldehyde and difluorotoluene in continuous flow mode. *Key words*: higher space–time yields; yield improvements; scale-up of organolithium chemistry; temperature profile; isothermal operation; safety considerations.

continuous operation under isothermal conditions. To avoid unsafe reaction conditions, e.g. local hot spots or runaways, the throughput had to be reduced in order to ensure stable and safe operation; even though flow chemistry benefits from high heat exchange coefficients and large heat exchange surface area, the processing is not inherently safe by default and careful investigation is essential. For achieving stable operation and high yields, it was crucial to know the internal temperature profile and to adjust the throughput to match the heat dissipation capabilities of the flow setup.

3.5.4 Synthesis of Bromomethyltrifluoroborates in Continuous Flow Mode

Scientists from GlaxoSmithKline (GSK) developed a flow process toward potassium bromomethyl-trifluoroborate, a key precursor for a Suzuki–Miyaura coupling (Scheme 3.9) [51]. Certain factors supported the investigation into a continuous flow operation: a rapid and exothermic Br/Li exchange, handling of an unstable carbenoid intermediate, and cooling of narrow-bored tubing is less power intensive than cooling a conventional batch vessel. The process was finally scaled from grams to kilograms and only a small crew of operators produced approximately 100 kg of the desired product in just four weeks.

The initial synthesis of "methylene boronate" was performed at −60° on a plate reactor mixing a preprepared solution of 1.2 M dibromomethane and 1.05 M tri-isopropylborate in THF with a feed of 2.5 M *n*-butyllithium in hexanes. The exiting stream was collected and the transformation of boronate ester into trifluoroborate salt was performed in batch mode using an aqueous KHF_2 slurry.

In an attempt to reduce the reaction volume, KHF_2 was replaced by a 30% aqueous solution of hydrofluoric (HF) acid. To protect the operators from the new safety hazard, the dosage was automated, thereby mitigating the open handling and exposure risk. Also, usage of HF reduced the raw material costs for the process. Precipitation of insoluble lithium fluoride (LiF) rated the in-line quench impractical and the stream containing the methylene borate ester was collected, and simultaneously aqueous HF was dosed to the batch reactor, driving the conversion to the lithium trifluoroborate salt in semibatch mode.

During preparation of the first pilot plant campaign, an unforeseen issue arose: reaction calorimetry using a Mettler Toledo RC1 measured an adiabatic temperature rise for the transformation toward methylene borate ester of 200 °C for a 1 molar reaction mass. On lab scale using a highly sophisticated plate reactor with excellent heat dissipation capabilities, the reaction temperature could be maintained below −45 °C. However, in the pilot plant, the use of less efficient

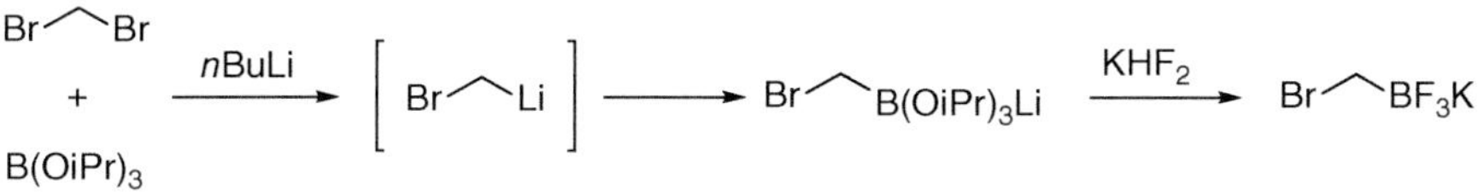

Br⌒Br + $B(OiPr)_3$ —nBuLi→ [Br⌒Li] ⟶ Br⌒$B(OiPr)_3Li$ —KHF_2→ Br⌒BF_3K

Scheme 3.9 Metalation sequence toward bromomethyltrifluoroborate. *Key words*: Br/Li exchange; unstable intermediate; batch limitations; handling of HF.

but more economical "shell and tube" reactors failed in dissipating the reaction heat that led to decomposition reactions and clogging of the flow setup. Three corrective actions were taken: dilution of the triisopropylborate solution (0.3 M instead of 1.0 M) to slow the rate of the exothermic addition reaction, operation at higher flow rate to dissipate the evolved heat more efficiently upon *n*-butyllithium addition, and multi-injection of *n*-butyllithium to further improve on the heat management. Following these adaptations, the pilot plant operated for 240 consecutive hours and produced 10 kg/d of lithium trifluoroborate salt for a total of 100 kg.

3.5.5 Two-Step Synthesis Toward Boronic Acids

Novartis scientists developed a concept "from feasibility to kilogram quantities" for organolithium chemistry focusing on the synthesis of aryl boronic acids (Scheme 3.10) [45, 52]. A standardized flow setup on laboratory scale – referred to as "boronic acid toolbox" – was established utilizing commercially available T-pieces and PFA tubing allowing for flexibility and modularity. Operating this standardized lab setup at ≥20ml/min total flow rate afforded excellent yields across a wide range of aryl boronic acids, ensured robust processing conditions, and achieved a throughput of approximately 300 mmol/h. Scale-up of this toolbox using Swagelok T-pieces and PTFE-static mixer elements fitted within PFA tubing gave a throughput of 1800 mmol/h.

For the "boronic acid toolbox," process parameters including concentrations of feedstock, mixers, reactors, jacket temperature, and residence times were predefined: (i) 0.3 M aryl bromide in THF, (ii) 1.6 M *n*-butyllithium in hexanes, (iii) 1.0 M trimethyl borate in THF, (iv) T-piece mixers made of PTFE with ID = 0.5 mm, (v) PFA tubing as reactors with ID = 0.78 mm, (vi) jacket temperature −30 °C, and τ_1 = 500 ms and τ_2 = 500 ms. The lab-based flow equipment suitable for early process development consists of precooling loops for all reagent streams, PFA tubular reactors, continuous Syrdos syringe pumps, and pressure sensors for each reagent stream. The entire platform is software controlled, enabling for online monitoring of temperatures, flow rates, and system pressures. Instead of optimizing the residence time of the Br/Li exchange reaction to achieve high conversion of the starting material, the yield for the halogen/metal exchange was optimized as a function of the total flow rate (=mixing efficiency).

The scale-up to kilogram quantities was performed in a linear, straightforward manner without the requirement of any redevelopment. The reactor setup for higher productivity consists of ¼″ Swagelok T-pieces and commercially available PTFE-static mixer elements (length = 4.8 mm) fitted within PFA tubing (ID = 5.0 mm; OD = 6.0 mm) as mixers. The three feeds are connected in 90 °C angle to the static mixers to minimize back-mixing effects and a PT100 thermocouple positioned at the exit of the second reactor allows for measuring the adiabatic temperature increase ($\Delta T_{\text{adiabatic}}$) of the metalation/borylation sequence and functions also as a basic PAT (process analytical technology) tool to monitor the steady state of the flow system. With a total reactor volume of just 2.4 ml, a throughput of 1800 mmol/h (~370 g/h or ~8.9 kg/d) is reported.

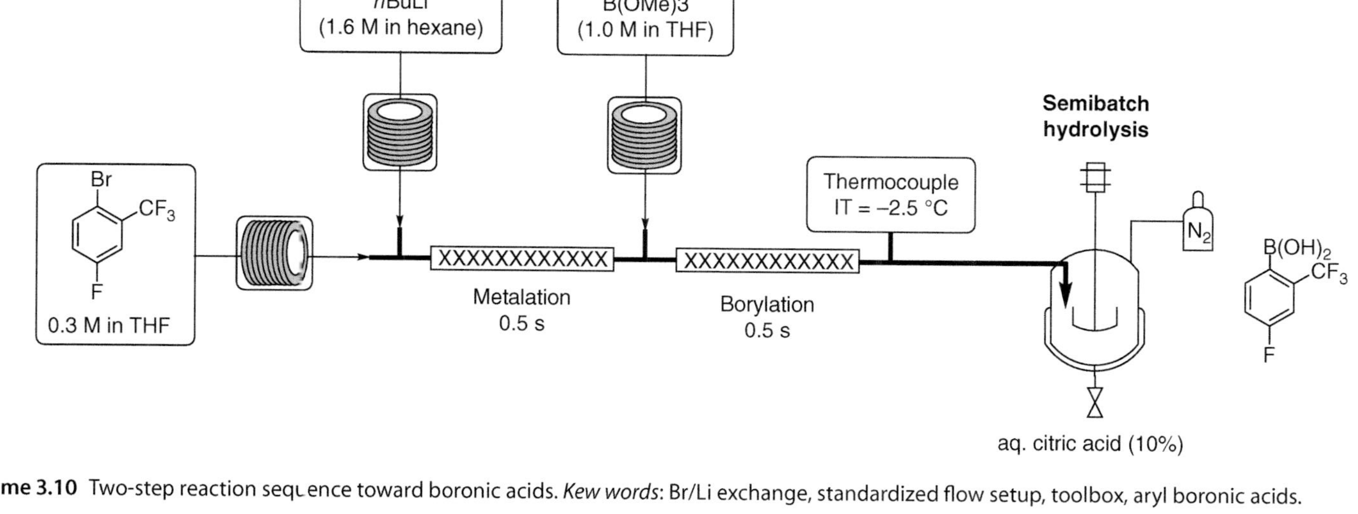

Scheme 3.10 Two-step reaction seqւence toward boronic acids. *Kew words*: Br/Li exchange, standardized flow setup, toolbox, aryl boronic acids.

3.5.6 Reaction Sequence Toward a Highly Substituted Benzoxazole Building Block

Novartis scientists developed a multistep reaction sequence toward a benzoxazole starting material required for clinical manufacturing of a lead candidate (Scheme 3.11) [34]. This compound significantly contributed to the final cost-of-goods of the drug substance. The supply chain for this building block revealed itself unreliable and an external contract manufacturing organization (CMO) entirely failed to meet the delivery request because of the challenging chemistry. CSTRs were constructed to address (i) the highly exothermic nature of the chemistry, (ii) unstable intermediates, and (iii) solid formation along the reaction chain. The selection of the reactor design was the key to a successful and robust upscaling of the reaction sequence and delivery of 17 kg of the desired benzoxazole moiety.

The reaction sequence involved both deprotonation of the pivaloyl-amide and regioselective *ortho*-lithiation of the aromatic core using *n*-butyl lithium. The lithiated intermediates and the aryne moiety are reported to be highly unstable and timely quench with a sulfur dioxide (SO_2) solution in THF to generate the stable aryl sulfinate was crucial. Oxidation and chlorination with sulfuryl chloride (SO_2Cl_2) afforded the aryl sulfonyl chloride that was consecutively transformed with N,O-dimethylhydroxylamine to the desired sulfonamide-substituted benzoxazole. On lab scale, using PTFE T-pieces and PFA tubular reactors, various obstacles were identified, which include (i) clogging issues because of partial insolubility of the aryl sulfinate salt, butyl sulfinate by-product, and LiCl; (ii) instantaneous and mixing-sensitive lithiation chemistry, as well as (iii) very high adiabatic temperature rise during *n*-butyl lithium addition. Despite these hurdles, an *in situ* yield of the aryl sulfinate is reported with 94% compared to moderate 57% when operated in conventional batch mode. In preparation of a mini-pilot campaign, process development activities were intensified using RC1 calorimetry, FT-IR, and ^{1}H NMR analysis for reaction optimization and reactor design. These studies revealed that in order to obtain high purity of aryl lithium intermediate, the ideal process would require a long dosing period of *n*-butyl lithium as well as efficient mixing to avoid a local-zone effect and to mitigate hot spots. For the lithiation chemistry toward the aryl sulfinate intermediate, where strong exotherm, mixing efficiency, hold-up time, and clogging revealed critical, a cascade of CSTRs was selected as the reactor type of choice. The highly exothermic *n*-butyllithium addition ($\Delta T_{ad} = 110\,°C$) was addressed by a multi-injection concept splitting the *n*-butyllithium stream across two reactor vessels. A CSTR consisting five vessels (5 × 10 ml) constructed of stainless steel was selected in order to control the inner temperature. The mixing time in the stainless steel cascade was approximately four seconds and the required residence time defined to be 1.4 minutes. Because of the acidic nature of the SO_2 quench solution, a second five-vessel CSTR (5 × 10 ml) was manufactured from borosilicate glass with a residence time of 1.1 minutes. Each CSTR vessel was specifically designed based on feedback from a simulation software, allowing vessel dimensions, impeller geometry, and stirring rate to be adjusted to perfectly match the requirements. Because the aryl sulfinate and aryl sulfonyl chloride

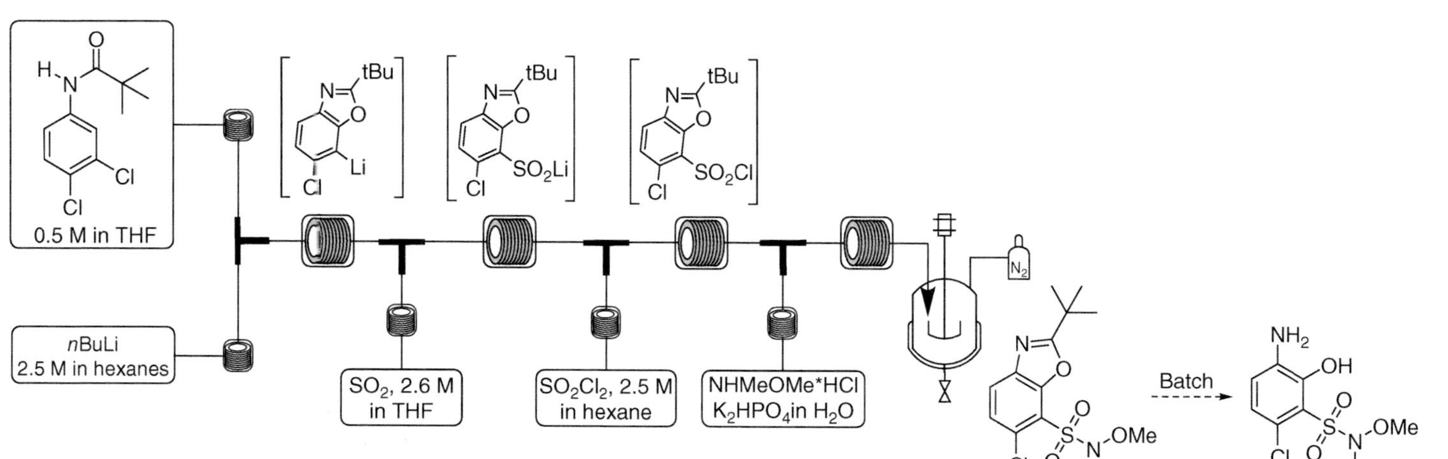

Scheme 3.11 Multi-step sequence toward a sulfonamide-substituted benzoxazole. *Key words: Ortho*-lithiation; aryne formation; multi-step; solid handling; PAT; kinetic modeling and simulation tools; high exothermy; CSTRs; multi-injection points; mini-plant.

were found to be sufficiently stable, the oxidation/chlorination and sulfonamide formation were performed in batch mode. A continuous process ran over 72 hours, produced 17 kg of the desired benzoxazole building block in 65% yield, and above 99.5 area% purity, which is notably higher compared to 34% for the optimized batch process.

3.6 Conclusion and Future Prospects

The development of continuous flow processes is no longer solely an academic exercise. Novel processing and production concepts based on flow technology are increasingly pursued and elaborated in the pharmaceutical industry. A key driver for this shift is modular chemical manufacturing facilities, enabling for accelerated process development and production [53, 54]. Even though flow technology offers various advantages over traditional batch operation, it is not a "magic bullet" and certain selection criteria must be fulfilled. The application of flow chemistry is beneficial for very fast and strongly exothermic reactions, hazardous reactions, mixing-controlled transformations, and reactions with selectivity issues. For a prompt and reliable development of a flow process, several resources from diverse disciplines including synthetic chemistry, chemical engineering, reaction kinetics, solubility, mixing performance, heat management, and process safety must be considered and merged. In this chapter, we provided an overview of the most important and relevant principles of flow chemistry, highlighted the importance of appropriate selection criteria, and gave a practical guide, accompanied by chemical examples, to scientists to perform organometallic chemistry in continuous flow mode.

Biography

Joerg Sedelmeier, born in 1977 in Freiburg (Germany), studied synthetic organic chemistry at the Albert-Ludwig University in Freiburg and conducted his Ph.D. thesis in the group of Professor Bolm at RWTH Aachen (Germany) in the field of sulfoximines and asymmetric metal catalysis. After a postdoctoral stay in Professor Ley's group at the University of Cambridge (UK) focusing on organic syntheses in continuous flow mode, he started his professional career as a process chemist at Novartis (Switzerland) in the Chemistry & Analytical Department in 2010. There, he was involved in process development, scale-up to pilot and plant scale, laboratory automation, and continuous flow technologies. In 2017, Joerg became a Novartis Fellow. In 2019, he moved to Roche (Switzerland) and is now leading a technology platform for continuous manufacturing. Jörg (co)-authored more than 24 publications and patents and won several company-internal science awards.

Andreas Hafner studied chemistry at the Karlsruhe Institute of Technology, received his diploma in 2010, and finished his Ph.D. in the field of organofluorine chemistry in 2013, both under the supervision of Prof. Stefan Bräse. After research with Prof. Steven Ley at the University of Cambridge as an Alexander-von-Humboldt fellow, he joined the chemical development unit of Novartis (Switzerland) as a postdoctoral fellow in 2015. Since 2017, Andreas has been working as a process chemist for Bayer CropScience in Switzerland.

References

1 (a) Plutschack, M.B., Pieber, B., Gilmore, K., and Seeberger, P.H. (2017). *Chem. Rev.* 117: 11796–11893. (b) Porta, R., Benaglia, M., and Puglisi, A. (2016). *Org. Process Res. Dev.* 20: 2–25. (c) Ley, S.V., Fitzpatrick, D.E., Myers, R.M. et al. (2015). *Angew. Chem. Int. Ed.* 54: 2–17. (d) Ley, S.V., Fitzpatrick, D.E., Ingham, R.J., and Myers, R.M. (2015). *Angew. Chem. Int. Ed.* 54: 2–18. (e) Pastre, J.C., Browne, D.L., and Ley, S.V. (2013). *Chem. Soc. Rev.* 42: 8849–8869. (f) Akwi, F.M. and Watts, P. (2018). *Chem. Commun.* 54: 13894–13928. (g) Jamison, T.F. and Koch, G. (2018). *Science of Synthesis: Flow Chemistry in Organic Synthesis*. Stuttgart: Thieme.

2 (a) Hartman, R.L., McMullen, J.P., and Jensen, K.F. (2011). *Angew. Chem. Int. Ed.* 50: 7502–7519. (b) Newman, S.G. and Jensen, K.F. (2013). *Green Chem.* 15: 1456–1472. (c) Teoh, S.K., Rathi, C., and Sharrat, P. (2016). *Org. Process Res. Dev.* 20: 414–431. (d) Hessel, V., Hofmann, C., Meudt, A. et al. (2004). *Org. Process Dev.* 8: 511–523. (e) Roberge, D.M., Ducry, L., Bieler, N. et al. (2005). *Chem. Eng. Technol.* 28: 318–322. (f) Mason, B.P., Price, K.E., Steinbacher, J.L. et al. (2007). *Chem. Rev.* 107: 2300–2318.

3 (a) Deadman, B.J., Collins, S.G., and Maguire, A.R. (2015). *Chem. Eur. J.* 21: 2298–2308. (b) Gutmann, B., Cantillo, B., and Kappe, C.O. (2015). *Angew. Chem. Int. Ed.* 54: 6688–6728. c Movsisyan, M., Delbeke, E.I.P., Berton, J.K.E.T. et al. (2016). *Chem. Soc. Rev.* 45: 4892–4928.

4 (a) Yoshida, J.-i., Takahashi, Y., and Nagaki, A. (2013). *Chem. Commun.* 49: 9896–9904. (b) Degennaro, L., Carlucci, C., De Angelis, S., and Luisi, R. (2016). *J. Flow Chem.* 6: 136–166.

5 (a) Ott, D., Kralisch, D., Dencic, I. et al. (2014). *ChemSusChem.* 7: 3521–3533. (b) Malet-Sanz, L. and Susanne, F. (2012). *J. Med. Chem.* 55: 4062–4098. (c) Mascia, S., Heider, P.L., Zhang, H. et al. (2013). *Angew. Chem., Int. Ed.* 52: 12359–12363. (d) Bogdan, A.R. and Dombrowski, A.W. (2019). *J. Med. Chem.*: asap. https://doi.org/10.1021/acs.jmedchem.8b01760.

6 Schlosser, M. (2002). *Organometallics in Synthesis: A Manual*. Chichester: Wiley.

7 Westermann, T. and Mleczko, L. (2016). *Org. Process Res. Dev.* 20: 487–494.

8 (a) Desai, A.A. (2012). *Angew. Chem. Int. Ed.* 51: 9223–9225. (b) Anderson, N.G. (2012). *Org. Process Res. Dev.* 16: 852–869.

9 (a) Plouffe, P. and Macchi, A. (2014). *Org. Process Res. Dev.* 18: 1286–1294. (b) Noël, T. (2016). *Topics in Organometallic Chemistry: Organometallic Flow Chemistry*. Springer.

10 (a) Browne, D.L., Baumann, M., Harji, B.H. et al. (2011). *Org. Lett.* 13: 3312–3315. (b) Newby, J.A., Blaylock, D.W., Witt, P.M. et al. (2014). *Org. Process Res. Dev.* 18: 1211–1220. (c) Newby, J.A., Blaylock, D.W., Witt, D.P.M. et al. (2014). *Org. Process Res. Dev.* 18: 1221–1228. (d) Laue, S., Haverkamp, V., and Mleczko, L. (2016). *Org. Process Res. Dev.* 20: 480–486. (e) Sleveland, D. and Bjørsvik, H.-R. (2012). *Org. Process Res. Dev.* 16: 1121–1130. (f) Kupracz, L. and Kirschning, A. (2013). *Adv. Synth. Catal.* 355: 3375–3380.

11 (a) Yoshida, J.-i. (2008). *Flash Chemistry: Fast Organic Synthesis in Microsystems*. Hoboken: Wiley-Blackwell. (b) Yoshida, J.-i. (2005). *Chem. Commun.*: 4509–4516. (c) Yoshida, J.-i., Nagaki, A., and Yamada, T. (2008). *Chem. Eur. J.* 14: 7450–7459.

12 Nagaki, A., Kim, H., and Yoshida, J.-i. (2008). *Angew. Chem. Int. Ed.* 47: 7833–7836.

13 Nagaki, A., Kim, H., and Yoshida, J.-i. (2009). *Angew. Chem. Int. Ed.* 48: 8063–8065.

14 Nagaki, A., Kim, H., Usutani, H. et al. (2010). *Org. Biomol. Chem.* 8: 1212–1217.

15 Kim, H., Lee, H.-J., and Kim, D.-P. (2015). *Angew. Chem. Int. Ed.* 127: 1897–1900.

16 Usutani, H., Tomida, Y., Nagaki, A. et al. (2007). *J. Am. Chem. Soc.* 129: 3046–3047.

17 Kim, H., Nagaki, A., and Yoshida, J.-i. (2011). *Nat. Commun.* 2: 264–266.

18 (a) Hafner, A., Mancino, V., Meisenbach, M. et al. (2017). *Org. Lett.* 19: 786–789. (b) Nagaki, A., Tokuoka, S., Yamada, S. et al. (2011). *Org. Biomol. Chem.* 9: 7559–7563. (c) Degennaro, L., Fanelli, F., Giovine, A., and Luisi, R. (2015). *Adv. Synth. Catal.* 357: 21–27.

19 (a) Nagaki, A., Matsuo, C., Kim, S. et al. (2012). *Angew. Chem. Int. Ed.* 51: 3245–3248. (b) Nagaki, A., Tokuoka, S., and Yoshida, J.-i. (2014). *Chem. Commun.* 50: 15079–15081.

20 (a) Nagaki, A., Takizawa, E., and Yoshida, J.-i. *Chem. Eur. J.* 16: 14149–14158. (b) Kim, H., Inoue, K., and Yoshida, J.-i. (2017). *Angew. Chem. Int. Ed.* 56: 7863–7866. (c) Becker, M.R. and Knochel, P. (2015). *Angew. Chem. Int. Ed.* 54: 12501–12505.

21 Lee, H.-J., Kim, H., and Yoshida, J.-i. (2018). *Chem. Commun.* 54: 547–550.

22 Schnürch, M., Spina, M., Khan, A.F. et al. (2007). *Chem. Soc. Rev.* 36: 1046–1057.

23 Kim, H., Min, K.-I., Inoue, K. et al. (2016). *Science* 352: 691–694.

24 (a) Moon, S.-Y., Jung, S.-H., Kim, U.B., and Kim, W.-S. (2015). *RSC Adv.* 5: 79385–79390. (b) Kim, H., Lee, H.-J., and Kim, D.-P. (2016). *Angew. Chem. Int. Ed.* 55: 1422–1426.

25 Nagaki, A., Ichinari, D., and Yoshida, J.-i. (2013). *Chem. Commun.* 49: 3242–3244.

26 Nagaki, A., Imai, K., Ishiuchi, S., and Yoshida, J.-i. (2015). *Angew. Chem. Int. Ed.* 54: 1914–1918.

27 (a) Reckamp, J.M., Bindels, A., Duffield, S. et al. (2017). *Org. Process Res. Dev.* 21: 816–820. (b) Schwolow, S., Hollmann, J., Schenkel, B., and Röder, T. (2012). *Org. Process Res. Dev.* 16: 1513–1522.
28 Little Things Factory, https://www.ltf-gmbh.com (accessed September 16 2019).
29 Corning, https://www.corning.com/emea/en/innovation/corning-emerging-innovations/advanced-flow-reactors.html (accessed September 16 2019).
30 Ehrfeld Mikrotechnik, https://ehrfeld.com/en/home.html (accessed September 16 2019).
31 Koflo, https://www.koflo.com/static-mixers.html; Sulzer, https://www.sulzer.com/en/products/static-mixers (accessed September 16 2019).
32 (a) Jolley, K.E., Chapman, M.R., and Blacker, A.J. (2018). *Beilstein J. Org. Chem.* 14: 2220–2228. (b) Karadeolian, A., Patel, D., Bodhuri, P. et al. (2018). *Org. Process Res. Dev.* 22: 1022–1028. (c) Levenspiel, O. (1998). *Chemical Reaction Engineering*, 3e. Hoboken, NJ: Wiley.
33 Susanne, F., Martin, B., Aubry, M. et al. (2017). *Org. Process Res. Dev.* 21: 1779–1793.
34 Schoenitz, M., Grundemann, L., Augustin, W., and Scholl, S. (2015). *Chem. Commun.* 51: 8213–8228.
35 Hartman, R.L. (2012). *Org. Process Res. Dev.* 16: 870–887.
36 Wolf, A., Michele, V., Schlüter, O.F.-K. et al. (2015). *Chem. Eng. Technol.* 38: 2017–2024.
37 Flowers, B.S. and Hartman, R.L. (2012). *Challenges* 3: 194–211.
38 Stucki, J.D. and Guenat, O.T. (2015). *Lab. Chip.* 15: 4393–4397.
39 Bates, R.E., Kroposki, L.M., and Potter, D.E. (1972). *J. Org. Chem.* 37: 560–562.
40 (a) Shu, W., Pellegatti, L., Oberli, M.A., and Buchwald, S.L. (2011). *Angew. Chem., Int. Ed.* 50: 10665–10669. (b) Sedelmeier, J., Ley, S.V., Baxendale, I.R., and Baumann, M. (2010). *Org. Lett.* 12: 3618–3621.
41 Harris, C., Despa, M., and Kelly, K. (2000). *J Microelectromech. Syst.* 9: 502–508.
42 Newman, S.G., Gu, L., Lesniak, C. et al. (2014). *Green Chem.* 16: 176–180.
43 Brandner, J.J. (2008). Fabrication of microreactors made from metals and ceramics. In: *Microreactors in Organic Synthesis and Catalysis* (ed. T. Wirth), 1–17. Weinheim: Wiley.
44 Caron, S. and Thomson, N.M. (2015). *J. Org. Chem.* 80: 2943–2958.
45 Hafner, A., Meisenbach, M., and Sedelmeier, J. (2016). *Org. Lett.* 18: 3630–3633.
46 (a) Levesque, F., Rogus, N.J., Spencer, G. et al. (2018). *Org. Process Res. Dev.* 22: 1015–1021. (b) Wakami, H. and Yoshida, J.-i. (2005). *Org. Process Res. Dev.* 9: 787–791. (c) Nagaki, A., Hirose, K., Tonomura, O. et al. (2016). *Org. Process Res. Dev.* 20: 687–691. (d) Hunter, S.M., Susanne, F., Whitten, R. et al. (2018). *Tetrahedron*: 3176–3182. (e) Roberge, D.M., Gottsponer, M., Eyholzer, M., and Kockmann, N. (2009). *Chem. Today* 27: 8–11. (f) Kockmann, N. and Roberge, D.M. (2009). *Chem. Eng. Technol.* 11: 1682–1694. (g) Thaisrivongs, D.A., Naber, J.R., Rogus, N.J., and Spencer, G. (2018). *Org. Process Dev.* 22: 403–408.

47 Haber, J., Kashid, M.N., Renken, A., and Kiwi-Minsker, L. (2012). *Ind. Eng. Chem. Res.* 51: 1474–1489.
48 Kockmann, N. (2012). *Chem. Ing. Tech.* 84: 715–726.
49 Thaisrivongs, D.A., Naber, J.R., and McMullen, J.P. (2016). *Org. Process Res. Dev.* 20: 1997–2004.
50 Kopach, M.E., Cole, K.P., Pollock, P.M. et al. (2016). *Org. Process Res. Dev.* 20: 1581–1592.
51 Broom, T., Hughes, M., Szczepankiewicz, B.G. et al. (2014). *Org. Process Res. Dev.* 18: 1354–1359.
52 Hafner, A., Filipponi, P., Piccioni, L. et al. (2016). *Org. Process Res. Dev.* 20: 1833–1837.
53 Bieringer, T., Buchholz, S., and Kockmann, N. (2013). *Chem. Eng. Technol.* 36: 900–910.
54 Brodhagen, A., Grünewald, M., Kleiner, M., and Lier, S. (2012). *Chem. Ing. Tech.* 84: 624–632.

4

Development of a Nickel-Catalyzed Enantioselective Mizoroki–Heck Coupling

Jean-Nicolas Desrosiers[1] *and Chris H. Senanayake*[2]

[1] *Pfizer, Inc., 558 Eastern Point Road, Groton, Connecticut 06340, USA*
[2] *Asta Tech (Chengdu) Biopharmaceutical Corp., 488 West Kelin Road, Chengdu, Sichuan 611130, P.R. China*

4.1 Introduction

4.1.1 Nonprecious Metal Catalysis Advantages for Industry

A survey of cross-coupling reactions realized on large scale indicated that the vast majority of these were catalyzed by palladium [1]. Nonprecious metals are rarely used for cross-coupling reactions on large scale in industry, and palladium clearly dominates this field [2]. The main reasons to explain this phenomenon are that chemists in the industry appreciate the level of predictability, tolerability, and reproducibility provided by palladium catalysts. The latter are compatible with a wide range of functional groups while being able to perform cross-couplings in water. The large amount of reports in literature also simplifies the selection of ideal ligands and precatalyst salts for a specific reaction. At the same time, the level of knowledge on related mechanisms help to predict the reactivity outcome and the related side products [3]. These characteristics make palladium an obvious choice of catalyst for a chemist who has to synthesize complex molecules containing various functional groups under the time pressures of industry [4]. The major downsides of using such catalysts are their low natural abundance, and consequently, small yearly production that does not provide a sustainable solution [5].

On the other hand, nonprecious metal catalysts, such as nickel, have a significantly higher natural abundance that translates to massive yearly productions and consequently very affordable costs [6]. These major advantages of using nickel catalysts are typically not convincing enough for an industrial chemist to employ them on large scale. The level of predictability of nickel is much lower than palladium's and the number of functional groups tolerated is often limited. Competing mechanisms involving Ni(I)/Ni(III), Ni(0)/Ni(II) catalytic cycles or radicals are commonly found [7]. Also, the choice of ligand for a specific transformation is less obvious. Overall, the large amount of nonprecious-metal-catalyzed cross-couplings in the literature has not directly translated to the world of industry. Considering the lower carbon dioxide

Organometallic Chemistry in Industry: A Practical Approach, First Edition.
Edited by Thomas J. Colacot and Carin C.C. Johansson Seechurn.

footprint, unique reactivity, and sustainability of nickel catalysts, the industry could greatly benefit from introducing those in large-scale processes [8].

4.1.2 Mizoroki–Heck Couplings in Industry with Palladium

Mizoroki and Heck independently discovered the vinylation of aryl halides in the presence of palladium catalysts (Scheme 4.1) [9, 10]. This Mizoroki–Heck coupling has had a tremendous impact on industry over the years because of its atom economical aspect, low cost of alkene coupling partner, level of complexity of product generated, broad substrate scope, and minimal waste produced [11].

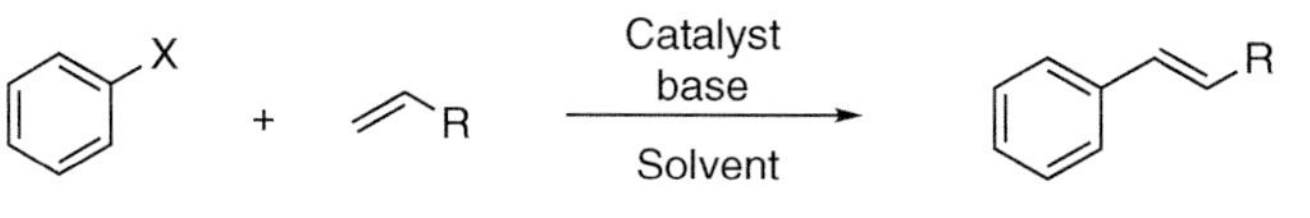

X
+
R
Catalyst
base
Solvent
R

Scheme 4.1 The Mizoroki–Heck coupling.

Numerous applications of this method can be found for the synthesis of natural products, active pharmaceutical ingredients, fine chemicals, and agrochemical preparation. For example, Albemarle developed a process to synthesize the nonsteroidal anti-inflammatory agents such as naproxen via a Mizoroki–Heck coupling (Scheme 4.2). This transformation was achieved on large scale under a pressurized atmosphere of ethylene to provide the desired 2-methoxy-6-vinylnaphthalene intermediate in 89% yield [12]. More recently, Pfizer utilized the Mizoroki–Heck coupling to incorporate a methyl ketone in the last steps of the commercial manufacturing process of palbociclib (Ibrance), which is a CDK4/6 inhibitor to treat breast cancer [13].

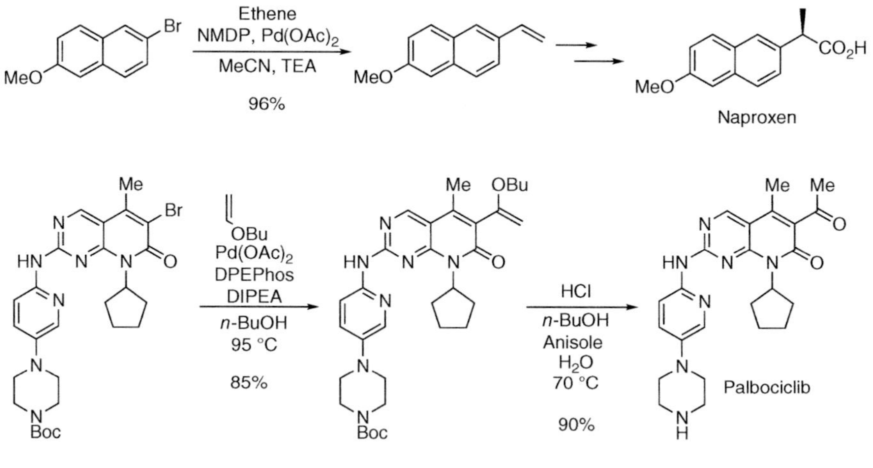

MeO
Br
Ethene
NMDP, Pd(OAc)2
MeCN, TEA
96%
MeO
MeO
CO2H
Naproxen
Me
Br
N
HN
N
N
O
N
N
N
Boc
OBu
Pd(OAc)2
DPEPhos
DIPEA
n-BuOH
95 °C
85%
Me
OBu
HCl
n-BuOH
Anisole
H2O
70 °C
90%
Me
Me
O
H
Palbociclib

Scheme 4.2 Selected applications of Mizoroki–Heck couplings at Albemarle and Pfizer (NMDP = neomenthyldiphenylphosphine; DPEPhos = ((oxydi-2,1-phenylene) bis(diphenylphosphine)).

4.1.3 Emergence of Nickel-Catalyzed Mizoroki–Heck Couplings

Despite the prevalence of palladium-catalyzed Mizoroki–Heck couplings in the industry, the application of the nickel variant has been limited [14]. This can probably be explained by multiple challenges to surpass in order to perform such reaction (Scheme 4.3) [15]. For instance, the nickel(II) hydride produced is not readily reduced back to nickel(0) as is the case for the palladium-catalyzed system [16]. Deleterious reactions, such as over-reductions and isomerizations, are known to occur in the presence of unreduced nickel(II) hydride species [17]. As opposed to palladium catalysts, the β-hydride elimination promoted by nickel requires more energy [16]. If the organo-nickel intermediate does not undergo a fast β-hydride elimination, homocoupling or protonation of that species can also occur [17]. In addition, the presence of coordinating heteroatoms, such as nitrogen, can complex the nickel catalyst and interfere with the rate level of conversion of the reaction by catalyst inhibition, which overall alters the reaction scope.

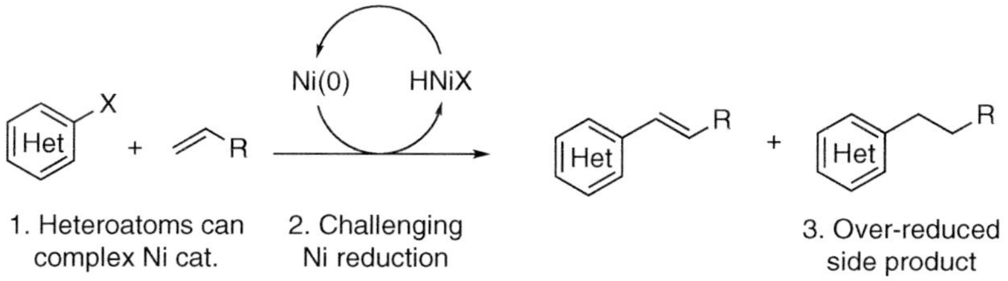

Ni(0) HNiX

Het X + R → Het R + Het R

1. Heteroatoms can complex Ni cat.

2. Challenging Ni reduction

3. Over-reduced side product

Scheme 4.3 Challenges of nickel-catalyzed Heck couplings.

In order to address these challenges, three different types of reaction conditions have been utilized to achieve the nickel-catalyzed Mizoroki–Heck coupling. The use of Ni(0) catalyst precursors in combination with aryl triflates or pivalates was reported by Srydstrup and coworkers [18], Watson and coworkers [19], and Jamison and coworkers [20] for the Mizoroki–Heck coupling under a cationic pathway (Scheme 4.4, type A). These conditions are somewhat limited because Ni(0) complexes are very sensitive to air and moisture, and the scope is limited to electron-rich alkenes. The second strategy consists of initiating the cross-coupling with a Ni(II) salt that is more amenable to large-scale processes because they are cheap and robust to air and oxygen (type B) [21]. In the presence of these air-stable precatalysts and aryl halides, the resulting cross-coupling typically occurs only at very high temperatures in polar aprotic solvents. Exogenous sources of halides, such as TBAI, are employed to ensure a neutral catalytic pathway. The limitations of such manifold are mainly the harsh reaction conditions and the scope of alkenes that is narrowed down to electron-deficient ones. The last type of reaction conditions is a combination of Ni(II) catalyst precursor with an external reductant (type C) [19, 22]. Many options of reductant are available, but the most commonly used is zinc dust, which leads to good yields for the Mizoroki–Heck couplings. This combination allows milder reaction conditions because the reaction temperature can be decreased down to about 65 °C. It also facilitates the cross-coupling of electron-rich and poor alkenes.

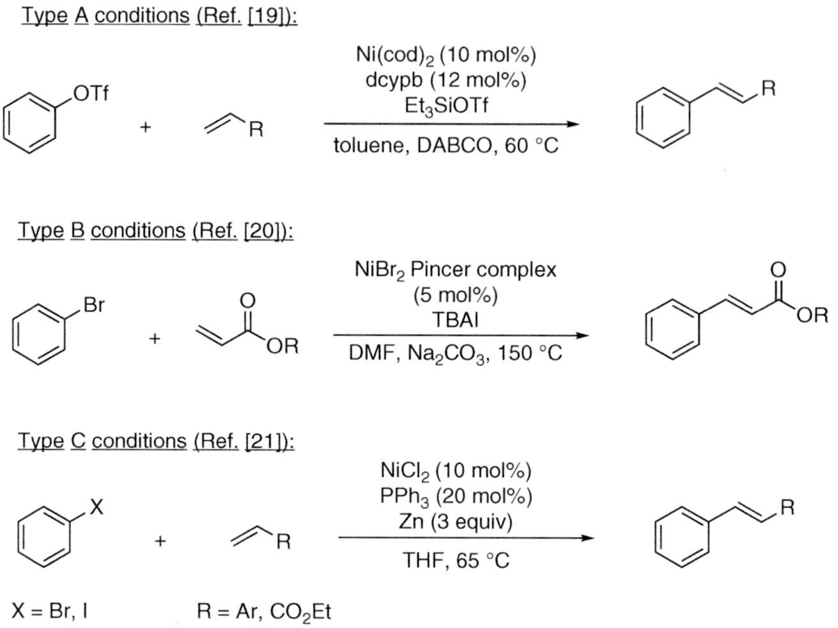

Type A conditions (Ref. [19]):

OTf + R

$Ni(cod)_2$ (10 mol%)
dcypb (12 mol%)
Et_3SiOTf
toluene, DABCO, 60 °C

R

Type B conditions (Ref. [20]):

Br + O OR

$NiBr_2$ Pincer complex
(5 mol%)
TBAI
DMF, Na_2CO_3, 150 °C

O OR

Type C conditions (Ref. [21]):

X + R

$NiCl_2$ (10 mol%)
PPh_3 (20 mol%)
Zn (3 equiv)
THF, 65 °C

R

X = Br, I   R = Ar, CO_2Et

Scheme 4.4 Types of reaction conditions for nickel-catalyzed Heck couplings.

4.1.4 Enantioselective Nickel-Catalyzed Couplings

When the numerous advantages of nickel catalysis are combined with asymmetric transformations, the resulting processes are cost competitive, shorter in step count, and lead to a substantial reduction of waste [23]. In other words, the sustainability of nonprecious metals with the absence of chiral auxiliary or resolution to synthesize enantioenriched molecules can provide an overall greener process.

The development of enantioselective nickel-catalyzed variants of the classic carbon–carbon cross-couplings has started in 1973, when Consiglio and Botteghi [24] reported the asymmetric Kumada coupling (Scheme 4.5). Shortly after, Kumada and coworkers [25] himself published a similar transformation, but using a phosphine-amine ferrocenyl ligand that provided much higher enantioselectivities. The research group of Fu has been very active in the development of enantioselective nickel-catalyzed cross-couplings. In 2005, they reported the asymmetric Negishi coupling [26] followed by the Suzuki–Miyaura coupling in 2008 [27]. Despite the vast knowledge built over the years in this field, there was no known method to achieve the enantioselective Mizoroki–Heck coupling with nickel catalysts [28, 29]. The closest examples we could find were the nickel-catalyzed enantiospecific Mizoroki–Heck coupling [30], the nickel-catalyzed tandem cyclization/cross-coupling with arylboron derivatives [31], and finally a nickel-catalyzed photo-redox system to build indolines [32]. In this chapter, we shall describe efforts in the Catalysis Group of Boehringer Ingelheim directed toward the development of the first enantioselective nickel-catalyzed Mizoroki–Heck coupling. This work was accomplished via collaborations with the research groups of Garg and coworker [33], Kozlowski and coworkers [34], and Zhang et al. [35] who are academic experts in this field.

Scheme 4.5 Enantioselective nickel-catalyzed classical couplings.

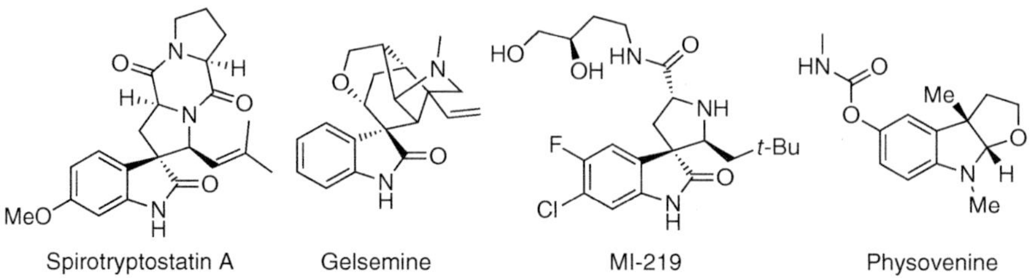

Spirotryptostatin A Gelsemine MI-219 Physovenine

Figure 4.1 3,3-Disubstituted oxindoles in natural products and synthetic biologically active molecules.

4.1.5 Synthesis of Oxindoles via Mizoroki–Heck Cyclizations

To reach this goal, we decided to study the synthesis of oxindoles. 3,3′-Disubstituted oxindoles are a prevalent scaffold found in multiple biologically active natural products [36]. Spirotryprostatins A and B (Figure 4.1) were isolated from Aspergillus fumigatus and found to inhibit the G2/M progression of cell division in tsFT210 cells [37]. Nature-inspired oxindoles containing an all-carbon quaternary center have been designed for the development of active pharmaceutical ingredients for a wide range of therapeutic areas. Oncology research programs, for example, have focused their efforts on 3,3-disubstituted oxindole-core-related candidates for the development of MDM2-p53 antagonists [38], such as MI-219 [39]. The elevated sp^3 character of these oxindoles that offers the possibility to further explore the chemical space combined with its amide functionality that acts as a hydrogen bond acceptor makes it an obvious motif of choice for the elaboration of a bioactive compound with high binding affinity.

One plausible route to access 3,3-disubstituted oxindoles is the intramolecular Mizoroki–Heck coupling between an activated aryl electrophile and an α,β-unsaturated amide moiety. A seminal work on this approach where the C3—C3a bond is formed was mostly accomplished using palladium catalysts in combination with phosphine-based ligands. The first examples were described by Ban and coworkers [40] and Heck and coworker [41], and these systems were later studied in depth by the Overman group [42] and others [43] (Scheme 4.6). This strategy was successfully applied for the total synthesis of a wide number of natural products such as gelsemine, spirotryptostatins A and B, physovenine, physostigmine, esermethole, and chimonanthine [44].

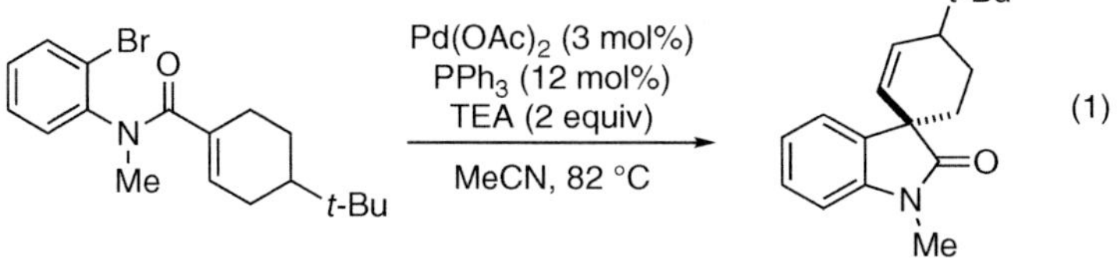

$Pd(OAc)_2$ (3 mol%), PPh_3 (12 mol%), TEA (2 equiv), MeCN, 82 °C (1)

Scheme 4.6 Palladium-catalyzed intramolecular Mizoroki–Heck reaction.

4.2 Development of a Nickel-Catalyzed Heck Cyclization to Generate Oxindoles with Quaternary Stereogenic Centers

4.2.1 Precedents and Challenges

The ability to generate oxindoles with quaternary centers via an intramolecular Mizoroki–Heck cyclization using nonprecious metal catalyst precursors, such as nickel catalysts, is even more appealing considering that they are inexpensive, sustainable because of their natural abundance, air-stable, while offering unique reactivity. Ban and coworkers [45] and Durandetti and coworkers [46] attempted such transformation for the synthesis of oxindoles and indoles (Scheme 4.7). Conclusions drawn from these systems were a limited substrate scope, modest yields, complex mixtures of regioisomers, and side products with reduced alkenes.

In order to develop the enantioselective nickel-catalyzed Mizoroki–Heck coupling to build 3,3-disubstituted oxindoles, multiple challenges would have to be addressed (Scheme 4.8). First, a judicious choice of reductant is required in order to obtain mild reaction conditions and relatively low reaction temperatures to allow for enantioselectivity. The reductant has to be compatible with the chiral ligand and the organo-nickel intermediate species to avoid the formation of reduced product. This reductant has to be involved in each catalytic cycle to regenerate Ni(0) from the Ni(II) catalyst. The chiral ligand has to be picked with the intention to control the enantioselectivity while allowing the β-hydride elimination, which is challenging for nickel. The subsequent generation of a quaternary stereogenic center is not known under those conditions and might be challenging. Finally, all of these steps have to occur via one type of mechanism while suppressing competing ones that can reduce the enantioselectivity, such as the formation of radicals.

In order to develop the enantioselective system, we first explored the nonasymmetric analogous method. This would allow us to (i) simplify the initial task to achieve, (ii) develop a platform with mild reaction conditions compatible with asymmetric catalysis, and (iii) understand the critical parameters of the nickel catalytic cycle. This would also be an accomplishment in itself considering that there are no known nickel-catalyzed Mizoroki–Heck coupling, which leads to quaternary centers [47, 48].

4.2.2 Optimization of Reducing Agent and Base

Based on our previous conditions developed for the nickel-catalyzed direct arylation of pyridinium ions [49], we initially treated **1** as a template substrate with a combination of phenanthroline and $NiCl_2$(dme) (Table 4.1). In the absence of external reducing agents, nonreproducible results were obtained (entry 1) for the formation of **2**. With the objective to obtain reproducible results under milder reaction condition, various reductants were explored. During the optimization

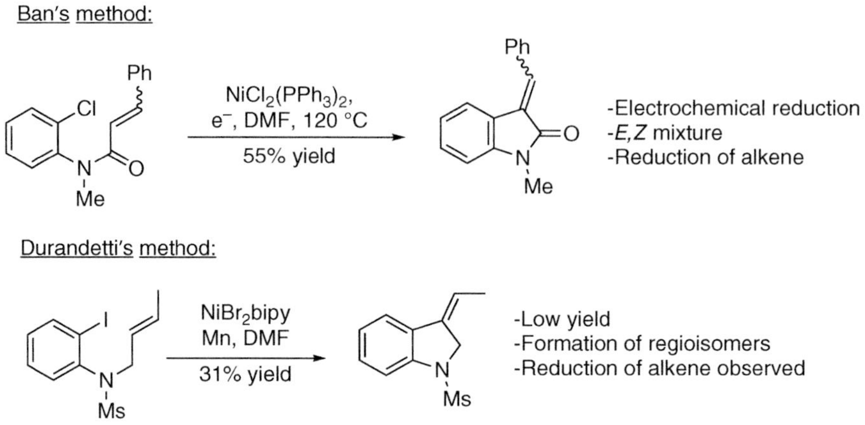

Ban's method:

$NiCl_2(PPh_3)_2$, e^-, DMF, 120 °C

55% yield

-Electrochemical reduction
-*E*,*Z* mixture
-Reduction of alkene

Durandetti's method:

$NiBr_2$bipy
Mn, DMF

31% yield

-Low yield
-Formation of regioisomers
-Reduction of alkene observed

Scheme 4.7 Precedents on nickel-catalyzed Mizoroki–Heck cyclizations.

phase of this reaction, all the reagents were weighed out in a glove box. The reactions were then carried out in an Amigochem workstation. It was found that this platform was ideal to minimize the moisture and oxygen while being able to perform multiple reactions in parallel and sample the reactions at various intervals. Notably, the reaction mixtures were stirred at room temperature for one hour to induce the Ni(I) reduction before heating them up to the desired temperature. Initially, reaction mixtures were analyzed directly for conversions by UPLC.

The reducing agent has to be judiciously selected as the redox potential will impact the reduction of nickel(II) species to turn over the catalyst and generate higher yield [50]. On the other hand, a too strong reductant will lead to the problematic formation of over-reduced product. For example, triethylsilane (entry 5) gave high overall conversion of the starting material, but with a substantial amount of **3**. Boronic acid derivatives (entries 2 and 3) were too weak reductants to lead to any product formation. Chromium afforded a negligible amount of oxindole, while zinc and manganese were the optimal reagents found. Manganese provided the best selectivity, especially at slightly lower temperature. As opposed to previous reports on nickel-catalyzed Mizoroki–Heck couplings [46], it was found that the presence of both reductant and base was crucial to obtain reactivity.

Considering that the starting amide is also base-sensitive, a careful study of the impact of the base was executed. Carbonate bases were ideal to induce high conversions while providing mild conditions to prevent substrate decomposition. Organic bases such as pyridine, PMP, DABCO, and Cy_2NMe did not improve the yield of the reaction and stronger bases such as NaO*t*-Bu and DBU were simply too basic and led to decomposition.

4.2.3 Ligand Screening

Using the manganese/sodium carbonate combination, a wide range of ligands were explored via a screening in the Amigochem workstation as mentioned in the previous section and selected examples are presented in Scheme 4.9. For this

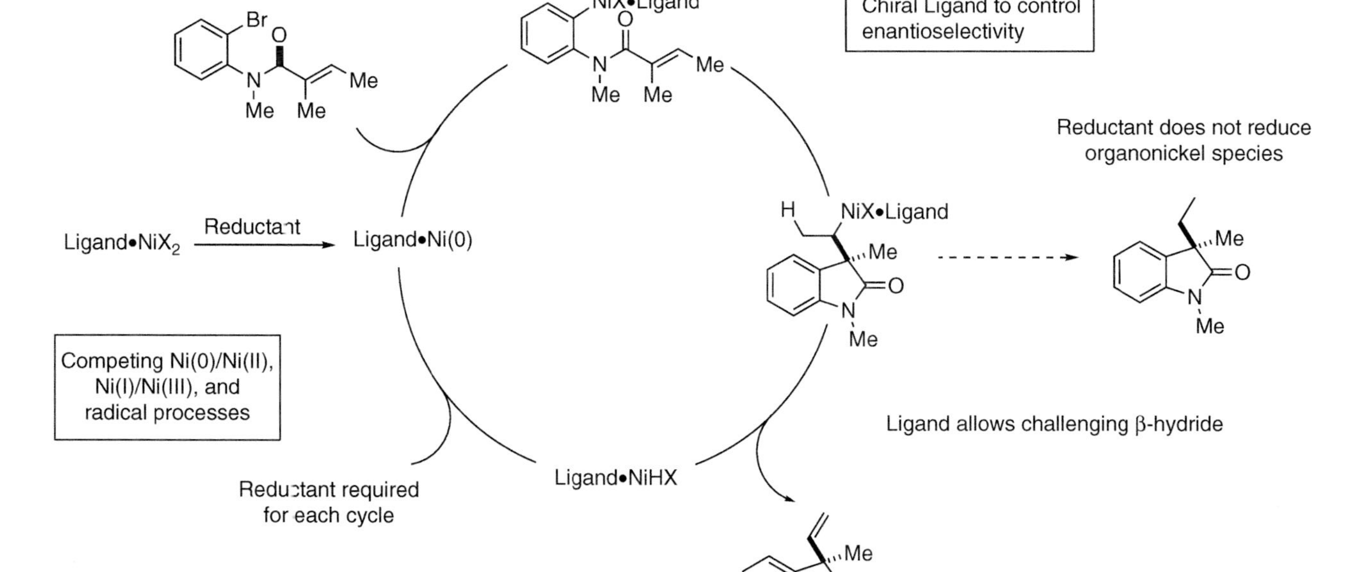

Scheme 4.8 Catalytic cycle and challenges associated with a nickel-catalyzed enantioselective Mizoroki–Heck coupling.

Table 4.1 Optimization of the reducing agent.

$NiCl_2(DME)$ (10 mol%), 1,10-phenanthroline, Reductant (3 equiv), DMF, K_2CO_3, 80 °C; **1** → **2** + **3**

Entry	Reductant	2 (%)[a]	3 (%)[a]
1	None	10–57	0–15
2	$PhB(OH)_2$	0	0
3	PhB(pin)	0	0
4	$PhSiH_3$	0	0
5	Et_3SiH	36	64
6	Cr	14	0
7	Zn	53	47
8	Mn	77	23
9[b]	Mn	79	12

a) Determined by UPLC area %.
b) Performed at 60 °C.
Source: Desrosiers et al. 2016 [49]. Reprint adapted with permission of John Wiley & Sons.

screening, reactions were worked-up and yields were determined by ^1H NMR by using dimethyl fumarate as an internal standard in order to ensure the accuracy of results generated. Various classes of ligands were tested, and neocuproine (L4) was first identified as almost equivalent to phenanthroline. All other phenanthroline derivatives (L1–L3 and L5–L7) led to significantly lower oxindole formation. Bisphosphines were then evaluated, and among them, dppe and dppp gave higher conversion than phenanthroline, albeit with a non-negligible proportion of reduced oxindole **3**. Electron-rich unhindered alkyl monophosphines turned out to be ideal for this methodology. For example, triethylphosphine and tributylphosphine led to 72% and 90% yield, respectively.

4.2.4 Impact of Aryl Electrophile and of Stereochemistry of Alkene Moiety

The impact of aryl-activating group on the Mizoroki–Heck cyclization was next examined (Scheme 4.10). Aryl iodide (**5**), methoxide (**6**), pivaloyl (**10**), and triflate (**9**) all gave poor conversions to the spirooxindole. The bromo (**1, 7, 11**) and chloro (**4** and **8**) compounds were the electrophiles that led to the highest yields for most substrates tested. The geometry of the alkene had no significant impact on the conversion as substrate 7 containing an E alkene with a bromo substituent led to 64% and **11**, which is composed of a *Z* alkene, afforded the corresponding oxindole with 66% yield.

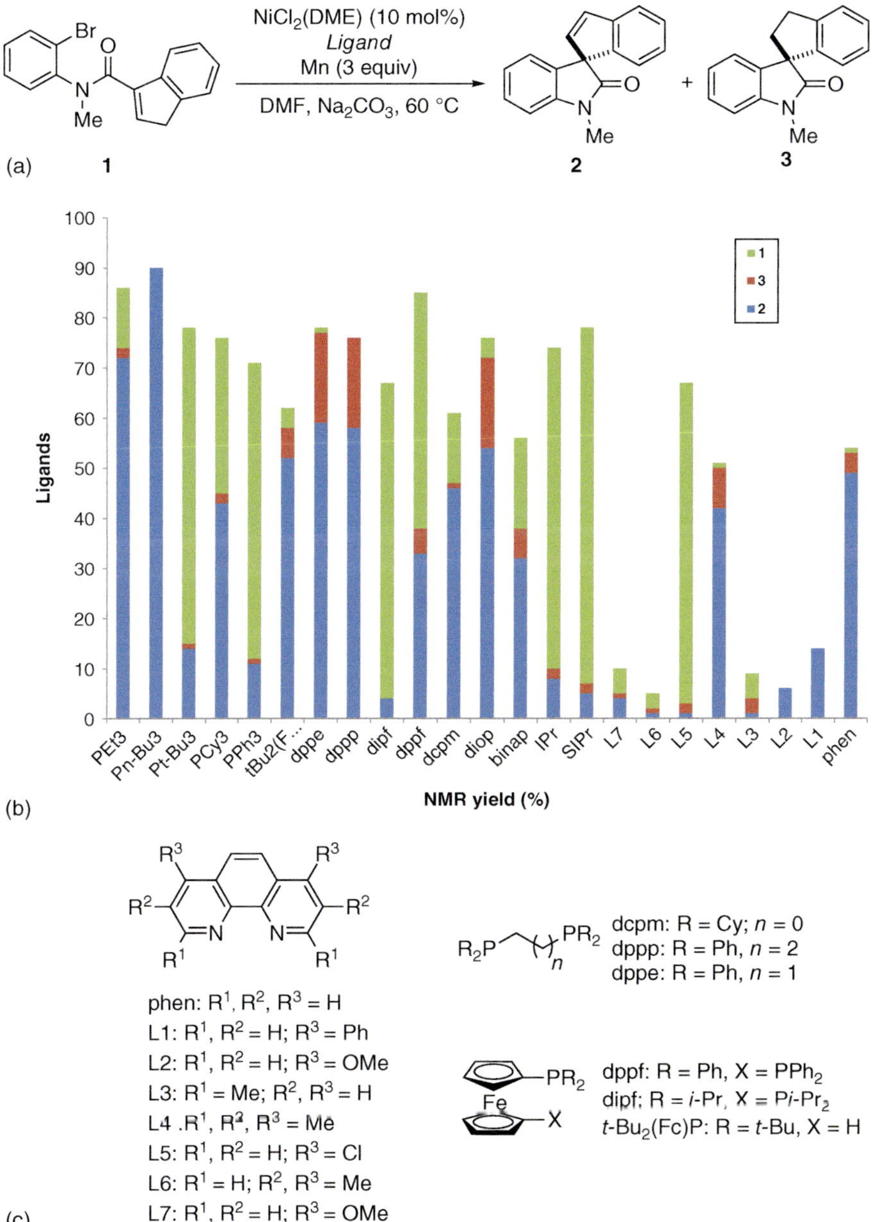

Scheme 4.9 Ligand screening. Source: Desrosiers et al. 2016 [49]. Reprint adapted with permission of John Wiley & Sons.

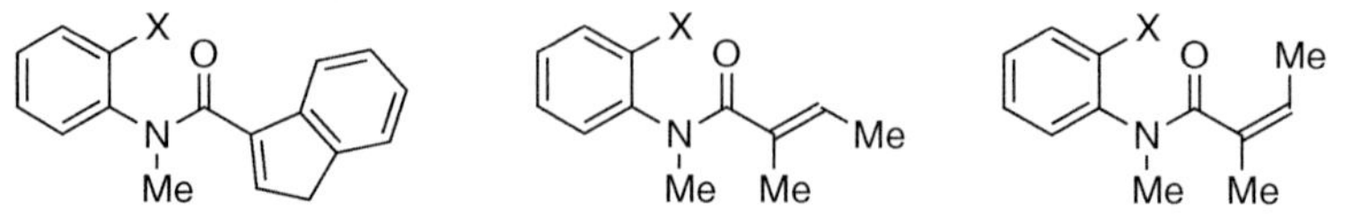

X=
Br (**1**): 85% yield
Cl (**4**): 80% yield
I (**5**): 7% yield
OMe (**6**): 7% yield

X =
Br (**7**): 64% yield
Cl (**8**): 65% yield
OTf (**9**): 21% yield
OPiv (**10**): 0% yield

X =
Br (**11**): 66% yield

Scheme 4.10 Impact of the aryl electrophile and alkene geometry.

Table 4.2 Robustness screening test.

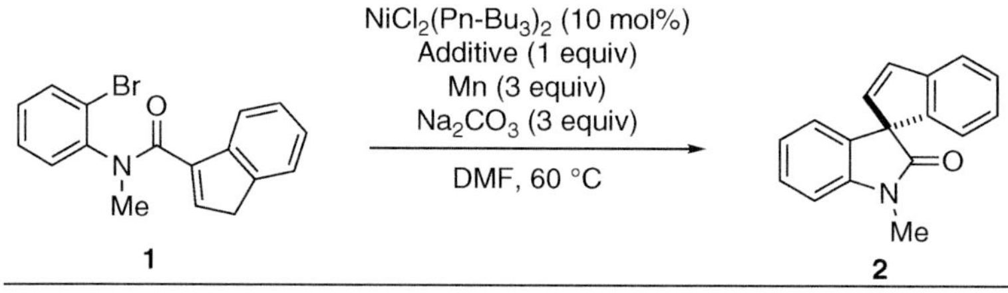

$NiCl_2(Pn\text{-}Bu_3)_2$ (10 mol%), Additive (1 equiv), Mn (3 equiv), Na_2CO_3 (3 equiv), DMF, 60 °C; **1** → **2**

Entry	Additive	Yield (%)[a]	Additive recovery (%)
1	Benzonitrile	84	72
2	*N,N*-dimethylbenzamide	62	85
3	*Tert*-butyl benzoate	56	99
4	2-Methyl-1-phenylpropan-1-one	72	99
5	*Tert*-butanol	78	67

a) Determined by ^{1}H NMR using dimethylfumarate as an internal standard.

4.2.5 Exploration of the Substrate Scope

In order to evaluate the tolerability of this nickel-catalyzed Mizoroki–Heck cyclization toward various functional groups, a robustness screening developed by Glorius and coworker [51] was performed by introducing 1 equiv of additive in the reaction mixture (Table 4.2). Both the amount of recovered additive and the yield of **2** were determined. From this study, it is fair to admit that the reaction is compatible with alcohols, nitriles, esters, ketones, and amides. To confirm these conclusions, a full substrate scope was submitted to the optimal conditions as well.

Different substituents on the arene moiety were tested (Scheme 4.11) under the standard reaction conditions (100.0 mg of the substrates were reacted with 10 mol% of $NiCl_2(Pn\text{-}Bu_3)_2$, 3.0 equiv of Mn, 3.0 equiv of Na_2CO_3 in DMF (0.3M) for 12 hours) [52]. Compounds with electron-donating groups, such as **12** and **16**, gave 71% and 62% yields, respectively. Electron-withdrawing substituents also provided the corresponding 3,3′-spirooxindoles with good

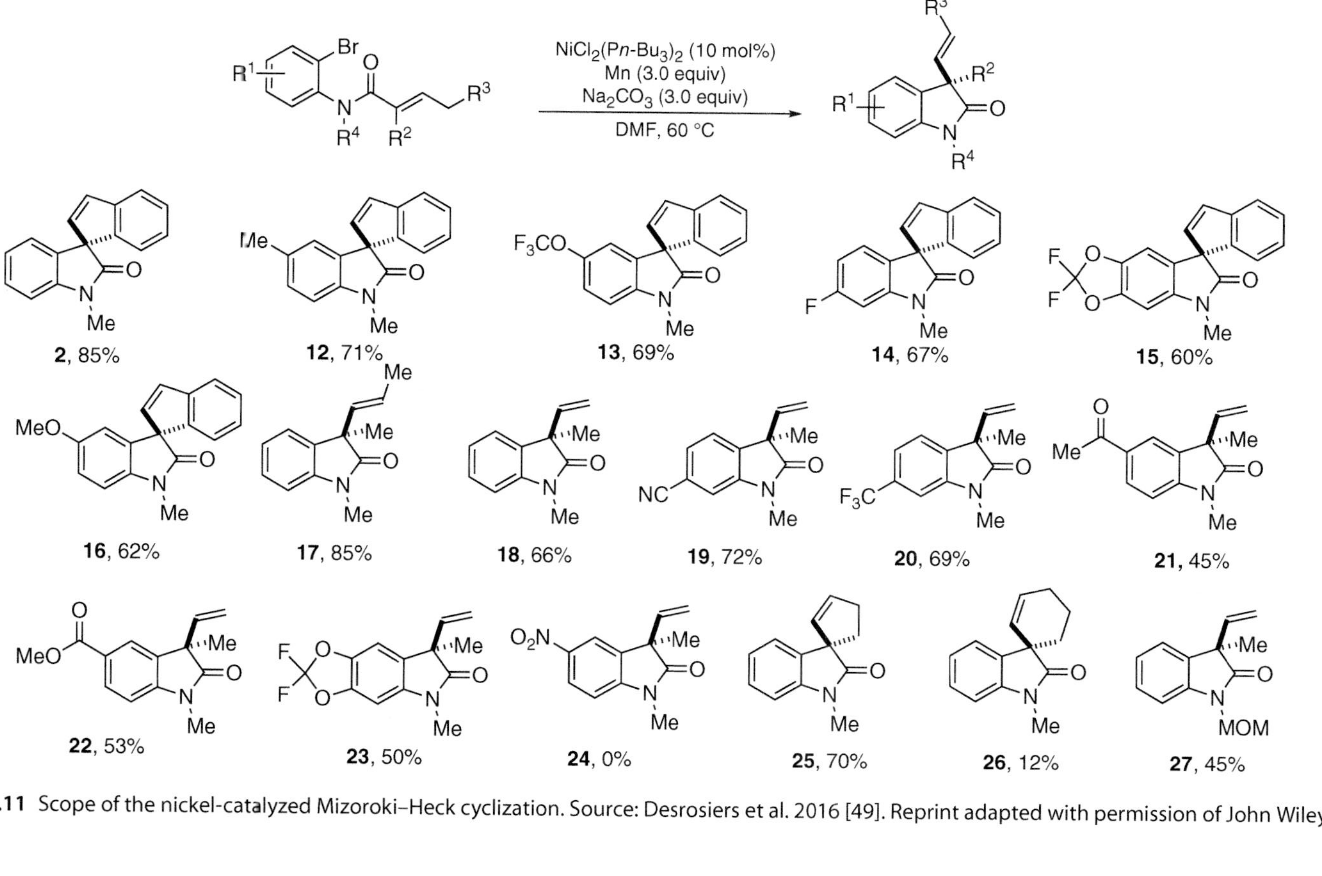

Scheme 4.11 Scope of the nickel-catalyzed Mizoroki–Heck cyclization. Source: Desrosiers et al. 2016 [49]. Reprint adapted with permission of John Wiley & Sons.

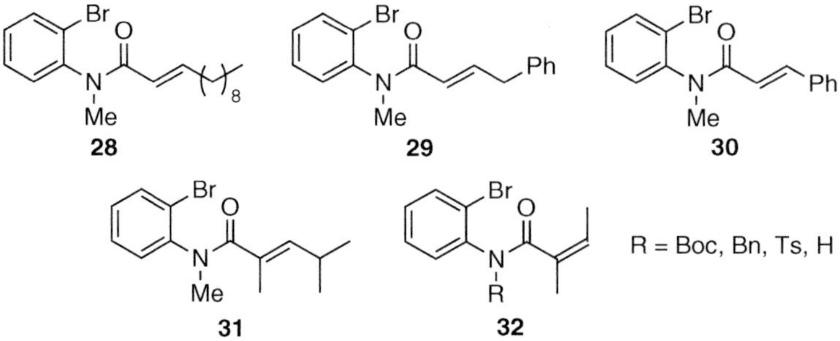

R = Boc, Bn, Ts, H

Figure 4.2 Limitations of the methodology.

yields. For instance, trifluoromethyl ether (**13**) led to 69% yields while the fluoro-spirooxindole (**14**) was obtained with 67%. Overall, electron-donating and electron-withdrawing groups provided the corresponding oxindoles with good yields. The only limitation found was the nitro substituent (**24**), which did not generate product. Cyclopentene-derived substrate **25** required an iodo-substituted arene to generate the oxindoles with 70% yield, while the cyclohexyl parent compound (**26**) only gave 12% yield. Various protecting groups were tested; methyl- and MOM (**18** and **27**)-protected anilines gave rise to the desired cyclized product.

4.2.6 Limitations of the Methodology

During the exploration of the substrate scope, notable limitations were observed in addition to the nitro aryl compounds mentioned in the section above. As opposed to trisubstituted alkene, disubstituted olefins such as **28** could not generate the corresponding oxindole (Figure 4.2). Addition of a benzyl (**29**) or phenyl (**30**) alkene substituent did not improve the reactivity of this class of substrate. When the gamma position was disubstituted by alkyl groups (**31**), the cyclization could not be promoted. Moreover, other N-substituted anilines (**32**) including carbamates, sulfonamides, and *N*-benzyl anilines provided only traces or none of the corresponding spirooxindoles.

4.2.7 Mechanistic Considerations

A series of control experiments were carried out with the objective to understand the nature of the catalytic cycle. In order to understand the role of Mn, $(n\text{-Bu}_3\text{P})_2\text{Ni(cod)}$ [53] was synthesized and used as a source of Ni(0) precatalyst. The preformed Ni(0) species with manganese (Table 4.3, entry 3) gave rise to similar results to standard conditions that were spiked with cod (entry 2). This indicates that, regardless of the origin of the catalytic precursor (Ni(II) or Ni(0)), the reaction follows the same catalytic pathway because Mn reduces the Ni(II) catalyst precursor down to Ni(0) under the standard conditions. In the absence of manganese (entry 4), the yield of the reaction dropped significantly. Those results

Table 4.3 Control experiments to probe the mechanism of the reaction.

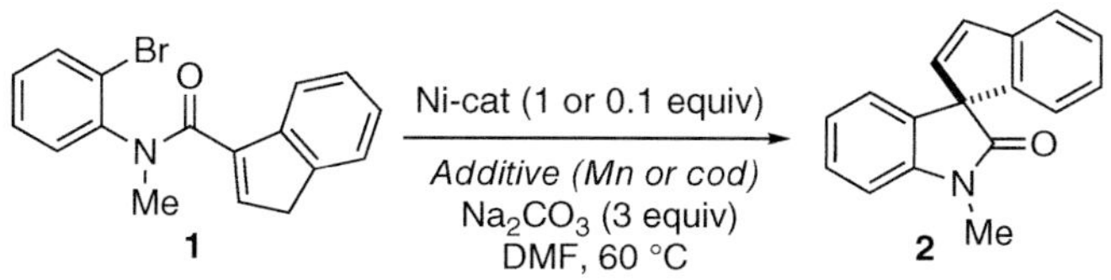

Entry	Ni source (mol%)	Additive (equiv)	Yield (%)	Conclusion/comment
1	$NiCl_2$(dme) (10)	Mn (3)	85	Standard conditions
2	$NiCl_2$(dme) (10)	Mn (3) + cod (0.1)	64	Detrimental effect of cod
3	$(n\text{-}Bu_3P)_2Ni(cod)$ (10)	Mn (3)	63	Mn used for initiation and in the catalytic cycle
4	$(n\text{-}Bu_3P)_2Ni(cod)$ (10)	None	39	
5	$(n\text{-}Bu_3P)_2Ni(cod)$ (100)	None	75	Ni(0)/Ni(II) cycle

suggest that the role of Mn is not only limited to the initial reduction of Ni(II), but Mn is also involved in the catalytic cycle.

In order to distinguish the oxidation state of nickel in the catalytic cycle (Ni(0)/Ni(II) vs Ni(I)/Ni(III)), an experiment was conducted with a stoichiometric amount of $(n\text{-}Bu_3P)_2Ni(cod)$ without manganese (entry 5). Hence, we avoided the possibility of accessing Ni(I) intermediates. Under these conditions, a yield comparable to the standard conditions was obtained, which strongly suggests a Ni(0)/Ni(II) catalytic cycle as the major pathway.

A radical clock experiment was then conducted to confirm that the catalytic process operates under a Ni(0)/Ni(II) regime (Scheme 4.12). A cyclopropyl-derived substrate (**33**) was synthesized to execute this control reaction. In the case of a radical-induced cyclization (path A), the diene oxindole (**34**) would be generated with a high yield [54]. If the reaction proceeds via a nickel-catalyzed oxidative addition (path B), the β-hydride elimination from organonickel intermediate (**35**) would be challenging. Therefore, a catalyst deactivation or a slow β-carbon elimination [55], which would lead to low yields of the same diene product, would be observed. When this experiment was performed, only 7% of the ring-opened product was isolated. Briefly, the outcome of this radical clock experiment suggests that a radical-induced cyclization is not responsible for the major reaction pathway.

The above mechanistic investigations point toward a classical Ni(0)/Ni(II) cycle as the major pathway (Scheme 4.13). The catalyst precursor $LNi^{II}X_2$ is initially reduced by Mn to LNi^0. The oxidative addition then occurs followed by migratory insertion to the olefin, which results in a LNi^{II}(alkyl)(X) intermediate. β-hydride elimination from LNi^{II}(alkyl)(X) generates the product with a quaternary center and LNi^{II}(H)(X). The resulting nickel hydride species can then be reduced by Mn to regenerate the active LNi^0 catalyst, which starts a new cycle.

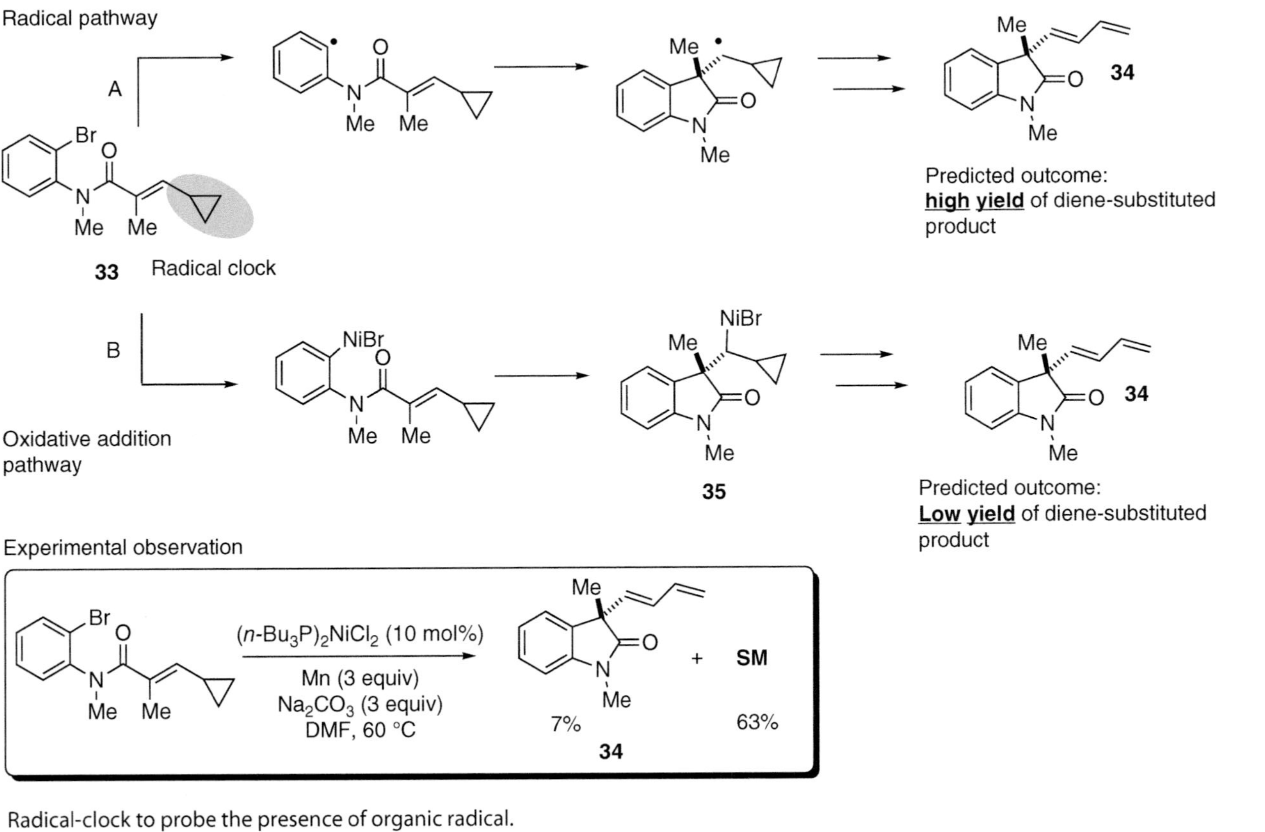

Scheme 4.12 Radical-clock to probe the presence of organic radical.

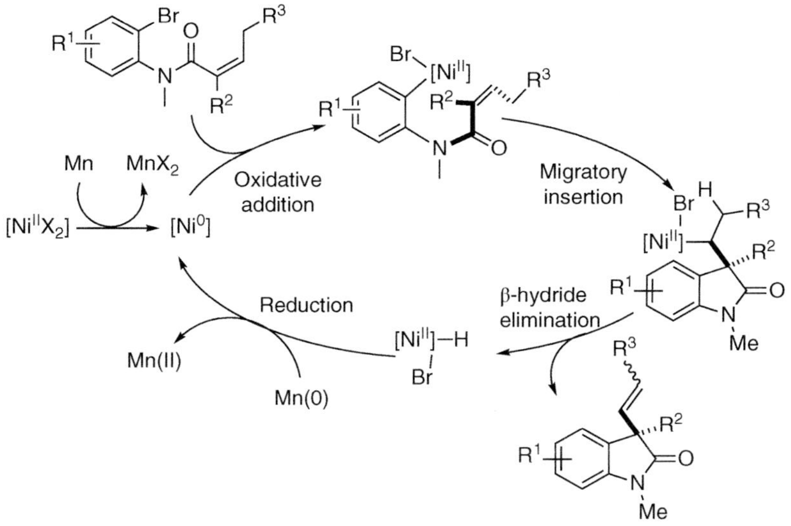

Scheme 4.13 The proposed catalytic cycle for Ni-catalyzed intramolecular Mizoroki–Heck coupling.

4.3 Development of First Enantioselective Nickel-Catalyzed Heck Coupling

4.3.1 Ligand Screening

This nonasymmetric method that operates under a relatively low reaction temperature, a neutral pathway, and a Ni(I)/Ni(II) mechanism in the absence of radicals is an ideal platform to develop the enantioselective nickel-catalyzed Mizoroki–Heck coupling. As mentioned above in Section 4.1.4, the enantioselective Ni-catalyzed Mizoroki–Heck coupling had not been reported.

The previously developed conditions were then directly used to explore a wide library of ligands to induce enantioselectivity. We first selected the indene amide **1** as the template substrate to perform this study (Scheme 4.14). This screening was once again performed in an Amigochem workstation, which allowed sampling of the reaction mixture under inert atmosphere robotically. The samples were then directly analyzed by UPLC for conversions and HPLC for ee values. Among 150 ligands screened, bis- or tris-alkylphosphine provided best conversions and selectivity. QuinoxP* [56] was the only ligand that generated spirooxindole **2** with more than 40% ee.

4.3.2 Impact of Alkene Stereochemistry

After looking at an extensive survey of ligands, the impact of the alkene geometry of the substrate was then investigated. Initially, indene-derived amide **1** with an *E* alkene was employed as the template substrate for optimization (Scheme 4.15). The indene fragment was then replaced by another *E* alkene prepared from tiglic

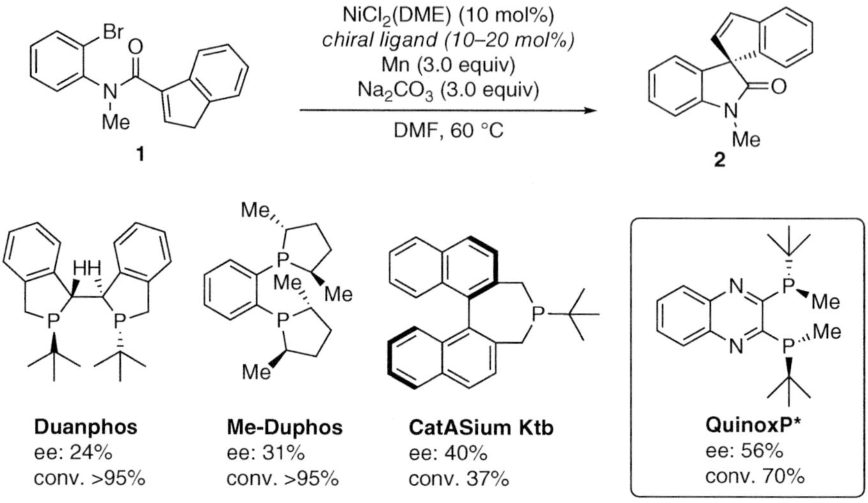

$NiCl_2(DME)$ (10 mol%)
chiral ligand (10–20 mol%)
Mn (3.0 equiv)
Na_2CO_3 (3.0 equiv)
DMF, 60 °C

1 → **2**

Duanphos	**Me-Duphos**	**CatASium Ktb**	**QuinoxP***
ee: 24%	ee: 31%	ee: 40%	ee: 56%
conv. >95%	conv. >95%	conv. 37%	conv. 70%

Scheme 4.14 Optimal chiral ligands identified after library screening.

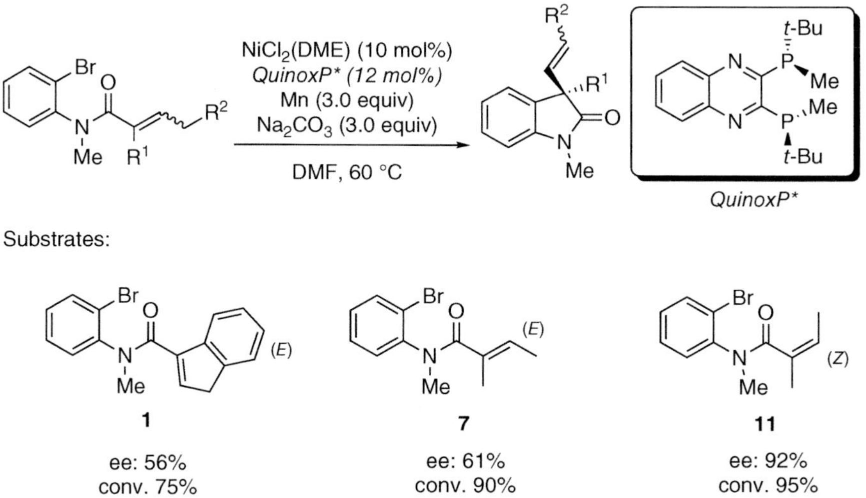

$NiCl_2(DME)$ (10 mol%)
QuinoxP (12 mol%)*
Mn (3.0 equiv)
Na_2CO_3 (3.0 equiv)
DMF, 60 °C

*QuinoxP**

Substrates:

1 (*E*)	**7** (*E*)	**11** (*Z*)
ee: 56%	ee: 61%	ee: 92%
conv. 75%	conv. 90%	conv. 95%

Scheme 4.15 Impact of alkene geometry.

acid (**7**). This modification gave rise to a slightly better conversion and enantioselectivity. A significant improvement was then observed when a substrate with a *Z* alkene synthesized from angelic acid (**11**) was submitted to the nickel-catalyzed conditions. A much cleaner reaction profile (95% conversion) and higher enantioselectivity (92% ee) was obtained. This phenomenon is also in line with experimental results obtained from the parent Pd system [44a].

4.3.3 Neutral vs Cationic Pathways

Control experiments were then performed to confirm the best pathway for this methodology. Conditions that are known to promote the cationic pathway of

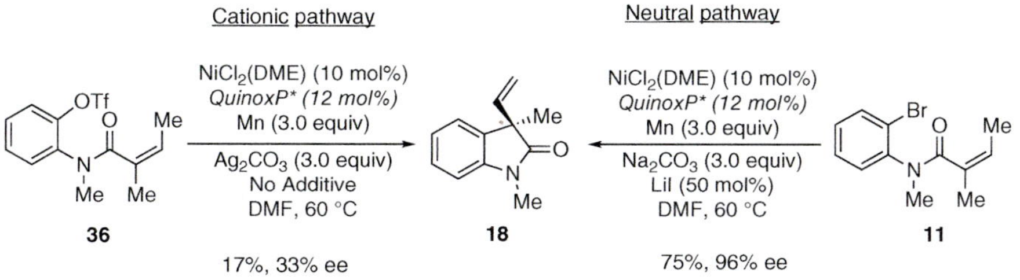

Scheme 4.16 Reaction conditions promoting the neutral or cationic pathway.

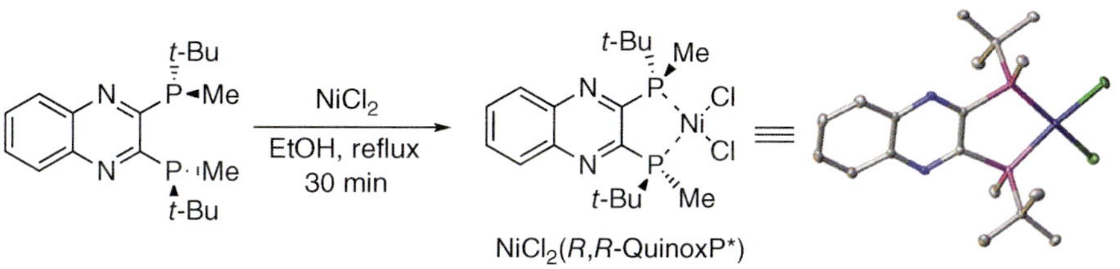

Scheme 4.17 Synthesis of the QuinoxP*•$NiCl_2$ complex. Source: Desrosiers et al. 2017 [58]. Reprint adapted with permission of American Chemical Society.

Mizoroki–Heck couplings were first tested. An aryl triflate substrate (**36**) was treated with silver carbonate as a halide scavenger (Scheme 4.16). Under these conditions, very low desired oxindole **18** was observed with low selectivity as well. Reaction conditions that favor the neutral pathway were then investigated [10]. The starting aryl bromide (**11**) was submitted to sodium carbonate and an exogenous source of halide. It was found that 50 mol% of LiI kept the high level of conversion while giving higher enantioselectivity at 60 °C. TBAI was also a competent additive to improve the level of enantioinduction, but purification was more challenging because of the TBA cation carrying over in the crude mixture. A similar observation, where addition of halide salts improved the selectivity by promoting the neutral pathway, was previously reported by Overman and coworkers [44c].

4.3.4 Nickel Precatalyst Complex Synthesis

At this stage of the methodology development, we started to experience some reproducibility issues. We quickly realized that the complexation of the bisphosphine ligand to $NiCl_2$ precursors required long time periods at high temperature [57]. Therefore, any variation of that precomplexation time induced various loadings of active precatalyst. In order to solve this issue, the QuinoxP*•$NiCl_2$ complex was prepared by mixing both $NiCl_2$ and QuinoxP* in ethanol under reflux, followed by crystallization. The resulting precatalyst is a dark red crystalline solid that is air-stable for several months without noticeable decrease in reactivity. X-ray analysis of this complex revealed a slightly distorted square-planar configuration (Scheme 4.17).

4.3.5 Exploration of the Substrate Scope

By using the QuinoxP*•$NiCl_2$ complex in combination with the exogenous source of halide LI, reproducible results were obtained, and these optimal conditions were utilized to study the scope of this methodology (Scheme 4.18) [58]. As for the nonasymmetric variant developed initially, this system tolerates a wide array of functional group such as esters (**22**), nitriles (**19**), ethers (**13** and **16**), and ketal (**15**) as well. The trend observed from the substrates tested is that the electron-donating groups on the aniline moiety lead to high enantioselectivities ranging from 90% to 96% ee. On the other hand, electron-withdrawing functionalities generate oxindoles with lower ee values (77–84%). The longer alkyl side chain (**17**) provided the corresponding oxindole with 92% ee and the MOM-protected amide gave rise to compound (**27**) with 61% yield and 89% ee. In the case of the spirooxindole composed of an indene moiety (**2**), lower selectivity was observed because of the *E* alkene geometry of the starting material (**1**).

4.3.6 Mechanistic Studies

The mechanistic studies that were performed on the nonasymmetric system provided a relatively clear picture of the catalytic cycle of this transformation. In order to gain additional knowledge on the origin of the enantioselectivity, we initiated a collaboration with Professor Marisa Kozlowski from the University of Pennsylvania to perform DFT studies. The initial calculations were executed on the migratory insertion step, which is the most likely to be responsible for the enantioinduction (Scheme 4.19). The $\Delta\Delta G^{\ddagger}$ of the diastereomeric transition states of the cyclization from the *Z* or *E* alkene did not match the experimental results. The calculated values predicted a much lower enantioselecivity (15% ee from *Z* alkene and 36% ee from the *E* alkene in selectivity of insertion box in Scheme 4.19) than the experimental outcomes (>90% ee). In addition, the predicted absolute configuration was opposite to the results obtained in the laboratory. This inconsistency was indicative that the enantioinduction event occurs before the migratory insertion step. Such concept would be in line with reports from Stephenson and coworkers [59] and Curran and coworkers [60]. They proposed a selective oxidative insertion, for the Pd system, which then generates an axially chiral intermediate that transfers chirality during the cyclization event.

The energetic profile of the transition states of the oxidative addition was then computed to confirm the origin of the enantio-determining step. The resulting $\Delta\Delta G^{\ddagger}$ value (88% ee) was in line with the results obtained from the *Z* alkene (96% ee). In the case of *E* alkene, the lab results (61% ee) were lower than the predicted calculated value (85% ee). To explain this phenomenon, the 3D structure of the transition states from the *E* and *Z* alkenes was analyzed (Figure 4.3). In both cases, the alkene moiety is pointing away from the reaction center and therefore should have minimal impact on the selectivity outcome. This discrepancy between the experimental and theoretical ee values could then arise from the energetics of the oxidative addition organo-nickel intermediates.

The *E* alkene oxidative addition organo-nickel intermediate seems to favor by 7.5 kcal an equilibration toward another which is more stable preinsertion

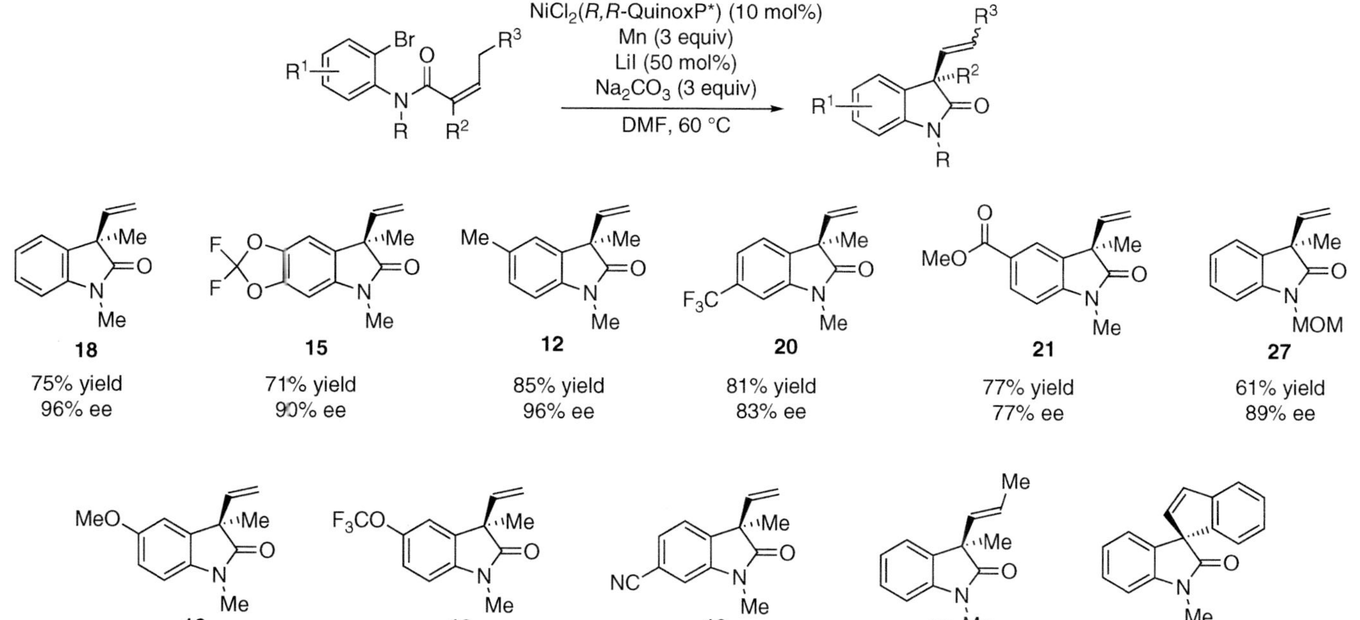

Scheme 4.18 Study of the scope of the enantioselective nickel-catalyzed Mizoroki–Heck coupling. Source: Desrosiers et al. 2017 [58]. Reprint adapted with permission of American Chemical Society.

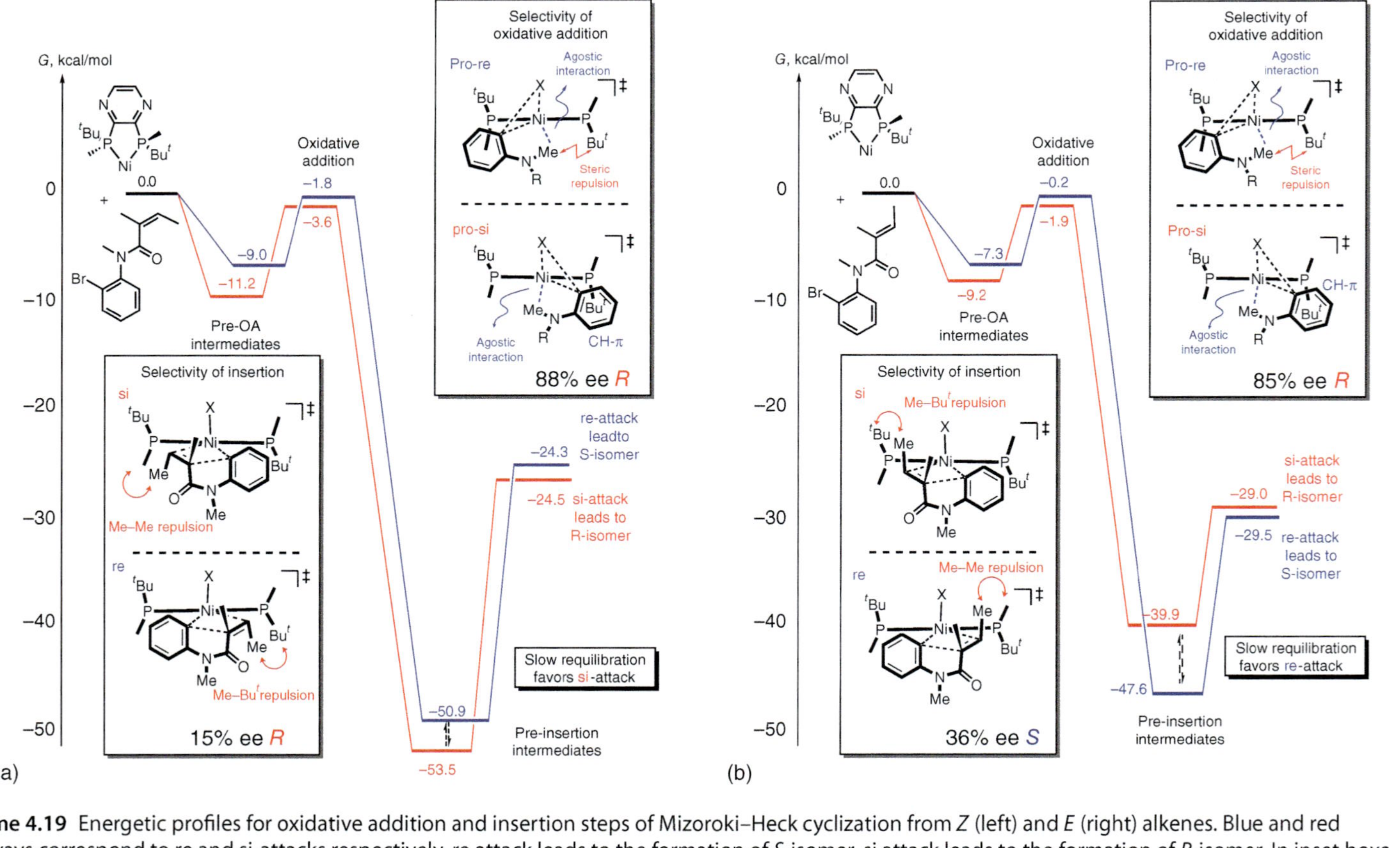

Scheme 4.19 Energetic profiles for oxidative addition and insertion steps of Mizoroki–Heck cyclization from *Z* (left) and *E* (right) alkenes. Blue and red pathways correspond to re and si-attacks respectively. re attack leads to the formation of *S*-isomer, si attack leads to the formation of *R*-isomer. In inset boxes, blue and red indicate stabilizing and destabilizing interactions that lead to selectively, respectively. Free energies were computed using M06/6-311 + G(d,p)-SMD-DMF//B3LYP/6-31G(d); values are in kcal/mol. Source: Desrosiers et al. 2017 [58]. Reprint adapted with permission of American Chemical Society.

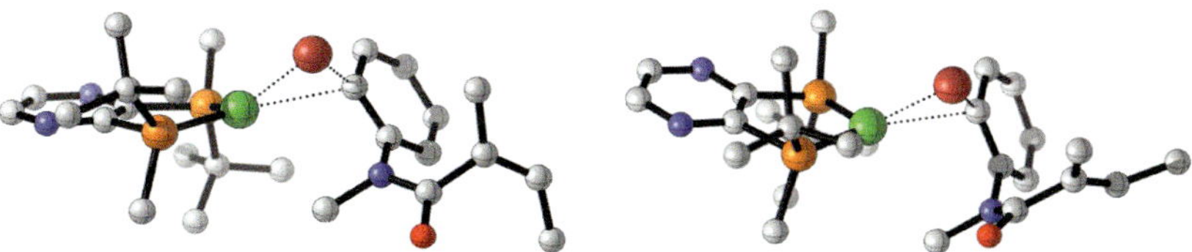

Figure 4.3 Geometries of the stereodetermining oxidative addition transition state (left = from *Z* alkene, right = from *E* alkene). Source: Desrosiers et al. 2017 [58]. Reprint adapted with permission of American Chemical Society.

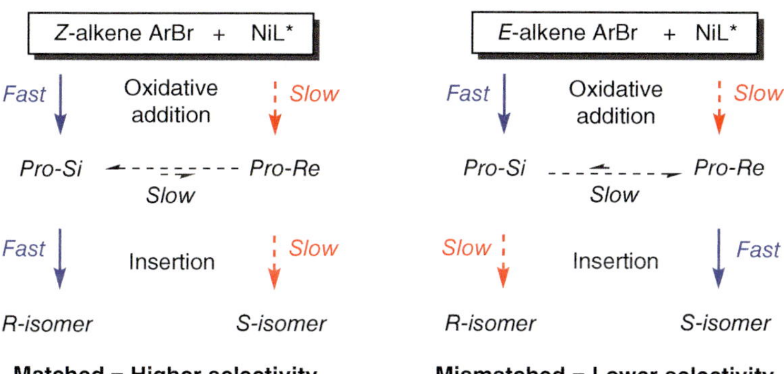

Scheme 4.20 Proposed stereoselection mechanism for the *E* and *Z* isomers. Source: Desrosiers et al. 2017 [58]. Reprint adapted with permission of American Chemical Society.

intermediate (Scheme 4.20). The latter is more responsible for the lower ee values obtained, as it leads to the minor enantiomer *S*. In the case of the *Z* alkene, this disfavored equilibration at the preinsertion stage is less impactful as only 2.6 kcal differentiates both intermediates. In other terms, there is a match case with the *Z* alkene while the *E* alkene leads to a mismatch effect.

4.4 Conclusions

In conclusion, the intramolecular Mizoroki–Heck coupling via nonprecious metal catalysis is strenuous because of multiple side reaction pathways. After an extensive optimization of reaction conditions, the nickel-catalyzed Mizoroki–Heck cyclization was achieved via a combination of manganese with nonhindered electron-rich monophosphine ligand P*n*-Bu_3 to afford 3,3′-spirooxindoles.

This transformation was originally designed to support the portfolio of Boehringer Ingelheim, which consisted of a growing number of candidates with 3,3-disubstituted oxindoles. Those valuable pharmacophores could be generated from various aryl halides, aniline derivatives, and tri-substituted cyclic or linear alkenes. The fact that an air-stable nickel catalyst precursor can be utilized is another aspect that brings cost efficiency to this reaction system. In order to facilitate the execution of this method, we developed a glove-box-free procedure that provided similar results on gram scale to reactions that were executed in a glove box on milligram scale. Based on these results, we believe that this reaction could be scaled-up further with adequate development. That being said, a large-scale and late-stage nickel-catalyzed coupling of a highly functionalized API precursor is a major challenge, and current reaction conditions would have to be further optimized to achieve such goal.

This methodology was the first nickel-catalyzed intramolecular Mizoroki–Heck reaction that forges quaternary stereocenters. Using this reaction platform as an initial achievement, we then focused our efforts toward the development of the first enantioselective Mizoroki–Heck coupling. This objective was achieved after screening multiple ligands and finding that QuinoxP* provided the best enantioselectivities with a Z alkene moiety under neutral conditions. The reaction conditions are applicable to a variety of substrates containing a range of different functionalities. Experimental mechanistic investigations combined with computational studies led to a proposed Ni(0)/Ni(II) catalytic cycle where the enantiodetermining step is the oxidative addition. This body of work has had an impact in this field [61], considering follow-up methods derived from this system [62], other examples on nickel-catalyzed Mizoroki–Heck couplings which led to quaternary centers [63], and the reports of nickel-catalyzed reductive asymmetric Mizoroki–Heck couplings [64].

Biography

Jean-Nicolas Desrosiers received his bachelor degree at the University of Montreal in 2003. He obtained his Ph.D. degree in 2008 under the supervision of Prof. André Charette, where he performed studies on enantioselective processes catalyzed by chiral bisphosphine monoxide copper complexes. In 2008, he joined the research group of Prof. Eric N. Jacobsen at Harvard University as a NSERC postdoctoral fellow. His postdoctoral research mainly focused on silyl ketene acetal acylation through anion-binding catalysis.

In 2010, Nick started is industry career in the Chemical Development Department of Boehringer Ingelheim in Ridgefield, CT, where he worked in their process chemistry and catalysis groups. He then joined the process chemistry group of Pfizer in Groton, CT, in 2017, where he is a senior principal scientist.

Dr. Chris H. Senanayake obtained his Ph.D. degree under the guidance of Professor James H. Rigby at Wayne State University in 1987, where he worked on the total synthesis of complex natural products. He then undertook a postdoctoral fellow with Professor Carl R. Johnson and worked on the total synthesis of polyol systems. In 1989, he joined the Department of Process Development at Dow Chemical Co. In 1990, he joined the Merck's process research group. In 1996, he joined Sepracor, Inc., where he was an Executive Director of Chemical Process Research. In 2002, he joined Boehringer Ingelheim Pharmaceuticals, as the vice president of Chemical Development. In 2018, he joined AstaTech (Chengdu) Biopharmaceutical Corporation as the partner and CEO, where he oversees all subsidiaries including cofounder of AstaGreen Chem Inc., in Richmond VA as the CEO. Dr. Senanayake is the coauthor of more than 500 papers, patents, book chapters, and review articles in many areas of synthetic organic chemistry, drug development, and design of improved chemical entities. In 2010, he received the prestigious Siegfried Gold Medal Award for development of practical processes for APIs and Process Chemistry. He is a board member of several journals and professional bodies.

References

1 (a) Magano, J. and Dunetz, J.R. (2011). *Chem. Rev.* 111: 2177–2250. (b) Colacot, T.J. (ed.). *New Trends in Cross-Coupling :Theory and Applications*, vol. 2015. Cambridge: RSC Publishing. (c) Magano, J. and Dunetz, J.R. (eds.) (2013). *Transition Metal-Catalyzed Couplings in Process Chemistry*. Wiley-VCH.

2 (a) Ritter, S. (2017). *C&E News* 95: 7. (b) Colacot, T.J. (ed.) (2015). *New Trends in Cross-Coupling : Theory and Applications*, VII–X. Cambridge: RSC Publishing.

3 (a) Johansson Seechurn, C.C., Kitching, M.O., Colacot, T.J., and Snieckus, V. (2012). *Angew. Chem. Int. Ed.* 51: 5062–5085. (b) Schlummer, B. and Scholz, U. (2004). *Adv. Synth. Catal.* 346: 1599–1626. (c) Hanley, R. (2014). *Platinum Met. Rev.* 58: 93–98. (d) Ruiz-Castillo, P. and Buchwald, S.L. (2016). *Chem. Rev.* 116: 12564–12649.

4 (a) Torborga, C. and Bellera, M. (2009). *Adv. Synth. Catal.* 351: 3027–3043. (b) Balasubramanian, M. Industrial scale palladium chemistry. In: *Palladium in Heterocyclic Chemistry*, 2e, 587–618. Elsevier Science. (c) Biajoli, A.F.P., Schwalm, C.S., Limberger, J. et al. (2014). *J. Braz. Chem. Soc.* 25: 2186–2214. (d) Buchwald, S.L., Mauger, C., Mignani, G., and Scholzc, U. (2006). *Adv. Synth. Catal.* 348: 23–39.

5 (a) Bihani, M., Ansari, T.N., Smith, J.D., and Handa, S. (2018). *Curr. Opin. Green Sust. Chem.* 11: 45–53. (b) Ludwig, J.R. and Schindler, C.S. (2017). *Chem* 2: 313–316.

6 (a) Tasker, S.Z., Standley, E.A., and Jamison, T.F. (2014). *Nature* 509: 299–309. (b) Rosen, B.M., Quasdorf, K.W., Wilson, D.A. et al. (2011). *Chem. Rev.* 111: 1346–1416.

7 Ananikov, V.P. (2015). *ACS Catal.* 5: 1964–1971.

8 (a) Mesganaw, T. and Garg, N.K. (2013). *Org. Process Res. Dev.* 17: 29–39. (b) Hazari, N., Melvin, P.R., and Mohadjer Beromi, M. (2017). *Nat. Rev.* 1: 1–16. (c) Shintani, T. and Hayashi, T. (2005). Asymmetric synthesis. In: *Modern Organonickel Chemistry* (ed. Y. Tamaru). Weinhem: Wiley-VCH.

9 Mizoroki, T., Mori, K., and Ozaki, A. (1971). *Bull. Chem. Soc. Jpn.* 44: 581.

10 Heck, R.F. and Nolley, J.P. Jr., (1972). *J. Org. Chem.* 37: 2320.

11 (a) Jagtap, S. (2017). *Catalysts* 7 (267): 1–53. (b) Beletskaya, I.P. and Cheprakov, A.V. (2000). *Chem. Rev.* 100: 3009–3066. (c) Zafar, M.N., Mohsin, M.A., Danish, M. et al. (2014). *Russ. J. Coord. Chem.* 40: 781–800. (d) Whitecombe, N.I., Hii, K.K.M., and Gibson, S.E. (2001). *Tetrahedron*

57: 7449–7476. (e) Franzén, R. (2000). *Can. J. Chem.* 78: 957–962. (f) Biffis, A., Zecca, M., and Basato, M. (2001). *J. Mol. Catal. A Chem.* 173: 249–274. (g) Jutand, A. (2009). Mechanisms of the Mizoroki-Heck reaction. In: *The Mizoroki–Heck Reaction* (ed. M. Oestreich), 1–42. Wiley. (h) Dounay, A.B. and Overman, L.E. (2003). *Chem. Rev.* 103: 2945–2963.

12 (a) Wu, T.-C. (1994). Process for preparing aryl-substituted aliphatic carboxylic acids and their esters using cyclic phosphine catalysts. US Patent 5, 315,026, filed 29 September 1993 and issued 24 May 1994. (b) Wu, T.-C. (1996). Process for preparing olefins. US Patent 5, 536,870, filed 17 February 1995 and issued 16 July 1996. (c) Beller, M.; Tafesh, A.; Herrmann, W.A. (1996). A process for preparing aromatic olefins by catalysis of palladacycles. DE 19503119, filed 02 February 1995 and issued 8 August 1996.

13 (a) Duan, S., Place, D., Perfect, H.H. et al. (2016). *Org. Process Res. Dev.* 20: 1191–1202. (b) Toogood, P.L. and Ide, D. (2016). Palbociclib (Ibrance). In: *Innovative Drug Synthesis*, 1e (eds. J.J. Li and D.S. Johnson), 167–195. Wiley.

14 (a) Qian, Q., Zang, Z., Chen, Y. et al. (2013). *Mini Rev. Med. Chem.* 13: 802–813. (b) Wanga, S.-S. and Yang, G.-Y. (2016). *Catal. Sci. Technol.* 6: 2862–2876. (c) Narayanan, R. (2010). *Molecules* 15: 2124–2138.

15 For examples of non-asymmetric nickel catalyzed Heck couplings, see: (a) Iyer, S., Ramesh, C., and Ramani, A. (1997). *Tetrahedron Lett.* 38: 8533–8536. (b) Lin, P.-S., Jeganmohan, M., and Cheng, C.-H. (2007). *Chem. Asian J.* 2: 1409–1416. (c) Wang, Z.-X. and Chai, Z.-C. (2007). *Eur. J. Inorg. Chem.*: 4492–4499. (d) Ma, S., Wang, H., Gao, K., and Zhao, F. (2006). *J. Mol. Catal. A* 248: 17–20. (e) Bhanage, B.M., Zhao, F., Shirai, M., and Arai, M. (1998). *Catal. Lett.* 54: 195–198.

16 Lin, B.-L., Liu, L., Fu, Y. et al. (2004). *Organometallics* 23: 2114–2123.

17 (a) Chakraborty, S., Zhang, J., Krause, J.A., and Guan, H. (2010). *J. Am. Chem. Soc.* 132: 8872–8873. (b) Lin, S., Day, M.W., and Agapie, T. (2011). *J. Am. Chem. Soc.* 133: 3828–3831. (c) Tran, B.L., Pink, M., and Mindiola, D.J. (2009). *Organometallics* 28: 2234–2243. (d) Tolman, C.A. (1972). *J. Am. Chem. Soc.* 94: 2994–2999.

18 Gøgsig, T.M., Kleimark, J., Nilsson Lill, S.O. et al. (2012). *J. Am. Chem. Soc.* 134: 443–452.

19 Ehle, A.R., Zhou, Q., and Watson, M.P. (2012). *Org. Lett.* 14: 1202–1205.

20 Tasker, S.Z., Gutierrez, A.C., and Jamison, T.F. (2014). *Angew. Chem. Int. Ed.* 53: 1858–1861.

21 (a) Inamoto, K., Kuroda, J.-I., Hiroya, K. et al. (2006). *Organometallics* 25: 3095–3098. (b) Sun, L., Pei, W., and Shen, C. (2006). *J. Chem. Res.*: 388–389. (c) Kelkar, A.A., Hanaoka, T., Kubotam, Y., and Sugi, Y. (1994). *Cat. Lett.* 29: 69–75.

22 (a) Boldrini, G.P., Savoia, D., Tagliavini, E. et al. (1986). *J. Organomet. Chem.* 301: C62–C64. (b) Lebedev, S.A., Lopatina, V.S., Petrow, E.S., and Beletskaya, L.P. (1988). *J. Organomet. Chem.* 344: 253–259.

23 (a) Chirik, P. and Morris, R. (2015). *Acc. Chem. Res.* 48: 2495. (b) Bergin, E. (2012). *Annu. Rep. Prog. Chem., Sect. B: Org. Chem.* 108: 353–371.

24 Consiglio, G. and Botteghi, C. (1973). *Helv. Chim. Acta.* 56: 460–463.

25 (a) Hayashi, T., Tajika, M., Tamao, K., and Kumada, M. (1976). *J. Am. Chem. Soc.* 98: 3718–3719. (b) Hayashi, T., Fukushima, M., Konishi, M., and Kumada, M. (1980). *Tetrahedron Lett.* 21: 79–82. (c) Hayashi, T., Konishi, M., Fukushima, M. et al. (1983). *J. Org. Chem.* 48: 2195–2202.
26 Fischer, C. and Fu, G. (2005). *J. Am. Chem. Soc.* 127: 4594–4595.
27 Saito, B. and Fu, G. (2008). *J. Am. Chem. Soc.* 130: 6694–6695.
28 For diastereoselective Ni-catalyzed Mizoroki–Heck cyclizations, see: (a) Solé, D., Cancho, Y., Llebaria, A. et al. (1994). *J. Am. Chem. Soc.* 116: 12133–12134. (b) Kim, H. and Lee, C. (2011). *Org. Lett.* 13: 2050–2053.
29 For Ni-catalyzed arylcyanation cyclizations, see: (a) Watson, M.P. and Jacobsen, E.N. (2008). *J. Am. Chem. Soc.* 130: 12594–12595. (b) Nakao, Y., Ebata, S., Yada, A. et al. (2008). *J. Am. Chem. Soc.* 130: 12874–12875.
30 Harris, M.R., Konev, M.O., and Jarvo, E.R. (2014). *J. Am. Chem. Soc.* 136: 7825–7828.
31 Huan, C. and Fu, G.C. (2014). *J. Am. Chem. Soc.* 136: 3788–3791.
32 Tasker, S.Z. and Jamison, T.F. (2015). *J. Am. Chem. Soc.* 137: 9531–9534.
33 Dander, J.E. and Garg, N.K. (2017). *ACS Catal.* 7: 1413–1423.
34 Gutierrez, O., Tellis, J.C., Primer, D.N. et al. (2015). *J. Am. Chem. Soc.* 137: 4896–4899.
35 Zhang, W., Chi, Y., and Zhang, X. (2007). *Acc. Chem. Res.* 40: 1278–1290.
36 (a) Bin Yu, B., Yu, D.Q., and Liu, H.M. (2015). *Eur. J. Med. Chem.* 97: 673–698. (b) Cheng, D., Ishihara, Y., Tan, B., and Barbas, C.F. III, (2014). *ACS Catal.* 4: 743–762. (c) Ziarani, G.M. and Lashgari, N. (2012). *Arkivoc* 2012, Part (i): 277–320. (d) Zhou, F., Liu, Y.-L., and Zhou, J. (2010). *Adv. Synth. Catal.* 352: 1381–1407.
37 (a) Ding, K., Lu, Y., Nikolovska-Coleska, Z. et al. (2005). *J. Am. Chem. Soc.* 127: 10130–10131. (b) Cui, C.-B., Kakeya, H., and Osada, H. (1996). *Tetrahedron* 52: 12651–12666.
38 Gollner, A. (2015). *Synlett* 26: 426–431.
39 (a) Shangary, S., Qin, D., McEachern, D. et al. (2008). *Proc. Natl. Acad. Sci. U.S.A.* 105: 3933–3938.
40 Mori, M., Oda, I., and Ban, Y. (1982). *Tetrahedron Lett.* 23: 5315–5318.
41 Terpko, M.O. and Heck, R.F. (1979). *J. Am. Chem. Soc.* 101: 5281–5283.
42 (a) Abelman, M.M., Oh, T., and Overman, L.E. (1987). *J. Org. Chem.* 52: 4130–4133. (b) Ashimori, A. and Overman, L.E. (1992). *J. Org. Chem.* 57: 4571–4572.
43 (a) Amengual, R., Genin, E., Michelet, V. et al. (2002). *Adv. Synth. Catal.* 344: 393–398. (b) Kiely, D. and Guiry, P.J.J. (2003). *Organomet. Chem.* 687: 545–561. (c) Busacca, C.A., Grossbach, D., Campbell, S.J. et al. (2004). *J. Org. Chem.* 69: 5187–5195. (d) Busacca, C.A., Grossbach, D., So, R.C. et al. (2003). *Org. Lett.* 5: 595–598.
44 (a) Ashimori, A., Matsuura, T., Overman, L.E., and Poon, D.J. (1993). *J. Org. Chem.* 58: 6949–6951. (b) Madin, A., O'Donnell, C.J., Oh, T. et al. (1999). *Angew. Chem. Int. Ed.* 38: 2934–2936. (c) Ashimori, A., Bachand, B., Calter, M.C. et al. (1998). *J. Am. Chem. Soc.* 120: 6488–6499. (d) Overman, L.E. and Rosen, M.D. (2000). *Angew. Chem., Int. Ed* 39: 4596–4599.

45 (a) Mori, M. and Ban, Y. (1978). *Heterocycles* 9: 391–397. (b) Mori, M., Hashimoto, Y., and Ban, Y. (1980). *Tetrahedron Lett.* 21: 631–634. (c) Mori, M. and Ban, Y. (1976). *Tetrahedron Lett.* 21: 1807–1810.
46 Lhermet, R., Durandetti, M., and Maddaluno, J. (2013). *Beilstein J. Org. Chem.* 9: 710–716.
47 For an example of nickel-catalyzed intramolecular Mizoroki–Heck-type cyclization, see: (a) Sole, D., Cancho, Y., Llebaria, A. et al. (1996). *J. Org. Chem.* 61: 5895–5904. (b) For an example of reductive nickel-catalyzed intramolecular Mizoroki–Heck-type cyclization, see: Kim, H. and Lee, C. (2011). *Org. Lett.* 13: 2050–2053.
48 Few nickel-catalyzed methods to build quaternary stereocenters are available; examples include arylcyanation, C—O bond cleavage, Friedel–Crafts alkylation, and Michael addition. For selected examples, see the following: (a) Spielvogel, D.J. and Buchwald, S.L. (2002). *J. Am. Chem. Soc.* 124: 3500–3501. (b) Zhang, A. and Rajanbabu, T.V. (2006). *J. Am. Chem. Soc.* 128: 5620–5621. (c) Zultanski, S.L. and Fu, G.C. (2013). *J. Am. Chem. Soc.* 135: 624–627. (d) Hong, Y.C., Gandeepan, P., Mannathan, S. et al. (2014). *Org. Lett.* 16: 806–809. (e) Gao, J.-R., Wu, H., Xiang, B. et al. (2013). *J. Am. Chem. Soc.* 135: 2983–2986. (f) Cornella, J., Jackson, E.P., and Martin, R. (2015). *Angew. Chem. Int. Ed.* 54: 4075–4078. g Clariana, J., Galvez, N., Marchi, C. et al. (1999). *Tetrahedron* 55: 7331–7344.
49 Desrosiers, J.-N., Wei, X., Gutierrez, O. et al. (2016). *Chem. Sci.* 7: 5581–5586.
50 For examples discussing the use of external reductants for nickel-catalyzed systems, see: (a) Hie, L., Ramgren, S.D., Mesganaw, T., and Garg, N.K. (2012). *Org. Lett.* 14: 4782–4185. (b) Colon, I. and Kelsey, D.R. (1986). *J. Org. Chem.* 51: 2627–2637.
51 For the robustness screening concept, see: Collins, K.D. and Glorius, F. (2013). *Nat. Chem.* 5: 597–601.
52 Desrosiers, J.-N., Hie, L., Biswas, S. et al. (2016). *Angew. Chem., Int. Ed.* 55: 11921–11924.
53 Stanger, A. and Shazar, A. (1993). *J. Organomet. Chem.* 458: 233–236.
54 Newcomb, M. (2001). Kinetics of radical reactions: radical clocks. In: *Radicals in Organic Synthesis*, 1e, vol. 1 (eds. P. Renaud and M.P. Sibi), 317–336. Weinheim: Wiley-VCH.
55 Yu, C.C., Ng, K.P.D., Chen, B.-L., and Luh, T.-Y. (1994). *Organometallics* 13: 1487–1497.
56 (a) Nagata, K., Matsukawa, S., and Imamoto, T. (2000). *J. Org. Chem.* 65: 4185–4188. (b) Imamoto, T., Sugita, K., and Yoshida, Y. (2005). *J. Am. Chem. Soc.* 127: 11934–11935.
57 Magano, J. and Monfette, S. (2015). *ACS Catal.* 5: 3120–3123.
58 Desrosiers, J.-N., Wen, J., Tcyrulnikov, S. et al. (2017). *Org. Lett.* 19: 3338–3341.
59 (a) McDermott, M.C., Stephenson, G.R., and Walkington, A.J. (2007). *Synlett* 1: 51–54. (b) McDermott, M.C., Stephenson, G.R., Hughes, D.L., and Walkington, A.J. (2006). *Org. Lett.* 8: 2917–2920.
60 (a) Lapierre, A.J.B., Geib, S.J., and Curran, D.P. (2007). *J. Am. Chem. Soc.* 129: 494–495. (b) Bruch, A., Ambrosius, A., Frohlich, R. et al. (2010). *J. Am. Chem.*

Soc. 132: 11452–11454. (c) Guthrie, D.B., Geib, S.J., and Curran, D.P. (2011). *J. Am. Chem. Soc.* 133: 115–122.

61 Lautens, M. and Franzoni, I. (2017). *Synfacts* 13: 962.

62 (a) Li, Y., Wang, K., Ping, Y. et al. (2018). *Org. Lett.* 20: 921–924. (b) Yen, A. and Lautens, M. (2018). *Org. Lett.* 20: 4323–4327. (c) Yoon, H., Marchese, A.D., and Mark Lautens, M. (2018). *J. Am. Chem. Soc.* https://doi.org/10.1021/jacs.8b06966.

63 Medina, J.M., Moreno, J., Racine, S. et al. (2017). *Angew. Chem. Int. Ed.* 129: 6667–6671.

64 Qin, X., Wen, M., Lee, Y., and Zhou, J.S. (2017). *Angew. Chem. Int. Ed.* 56: 12723–12726.

5

Development of Iron-Catalyzed Kumada Cross-coupling for the Large-Scale Production of Aliskiren Intermediate

Srinivas Achanta[2], Debjit Basu[2], Uday K. Neelam[2], Rajeev R. Budhdev[2], Apurba Bhattacharya[1], and Rakeshwar Bandichhor[2]

[1] *Texas A&M University-Kingsville, Department of Chemistry, 920 W Santa Gurtrudes, Kingsville, TX 78363, USA*
[2] *Integrated Product Development, Innovation Plaza, Dr. Reddy's Laboratories Ltd, Bachupally, Qutubullapur, R.R. Dist., Hyderbad, Andhra Pradesh 700072, India*

5.1 Introduction

Transition-metal-catalyzed cross-coupling reactions in process research have gained significant momentum as a highly powerful synthetic tool to perform C—C and (or) C—heteroatom bond formation by involving a variety of organometallic reagents and organic electrophiles as coupling partners [1, 2]. The first report on the use of organomagnesium reagents for metal-catalyzed cross-coupling reactions dates back to 1940s. Kharasch in 1941 described the synthesis of biaryls using aryl Grignard reagents and aryl bromides catalyzed (3–5 mol%) by halides of iron, nickel, and cobalt (metals in the order of increased effectiveness) [3]. Pioneering work by Kochi in 1970s demonstrated that simple ferric salts can be used as effective catalysts for C–C cross-coupling reactions using Grignard reagents as coupling partners [4]. However, despite providing complementary reactivity, further development of iron-catalyzed cross-couplings attracted less attention in comparison to Pd and Ni [5].

Kumada–Corriu coupling, a type of transition-metal-catalyzed cross-coupling reaction, signifies the use of a Grignard reagent and organic halide to architect molecules of medicinal significance [6]. This reaction was reported independently by Kumada, Tamao group [7a], and Corriu, Masse group [7b] in 1972. Originally reported with nickel as a catalyst, subsequent development showed that Kumada–Corriu cross-coupling can also be achieved using palladium, manganese, and iron as catalysts [8]. Although this coupling reaction has the reputation of being among one of the earliest reported catalytic cross-coupling methods, [9] the obvious drawback of using a Grignard reagent intolerant of many functional groups for coupling led to the concurrent development of mechanistically related variants (Suzuki, Sonogashira, Stille, Hiyama, and Negishi) that found more widespread applications in the art of organic synthesis [10]. Over the past several years, with rapid access to functionalized Grignard reagents becoming readily available, the substrate scope has been broadened and

Organometallic Chemistry in Industry: A Practical Approach, First Edition.
Edited by Thomas J. Colacot and Carin C.C. Johansson Seechurn.

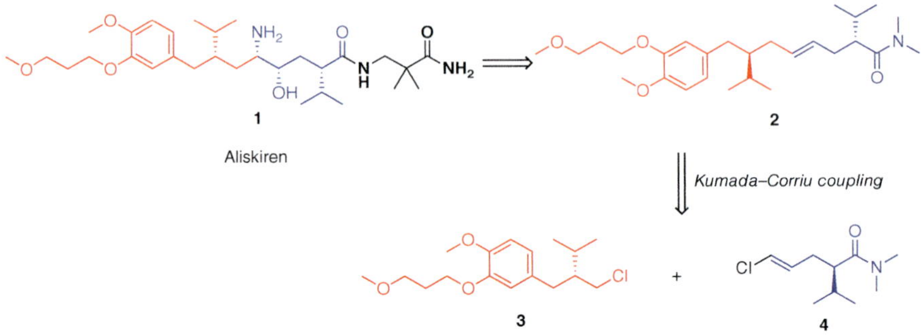

Figure 5.1 Retrosynthetic analysis of Aliskiren.

applications of iron-catalyzed Kumada–Corriu cross-coupling for the synthesis of pharmaceuticals have exponentially amplified [11, 12].

Herein, we discuss our efforts that led to the multikilogram synthesis of renin inhibitor, Aliskiren **1**, featuring an iron-catalyzed Kumada–Corriu coupling as the key step in the synthesis [13a]. Retrosynthetic analysis of Aliskiren **1** is shown in Figure 5.1. The *vicinal*-amino alcohol moiety present in Aliskiren **1** was envisioned to arise from the corresponding alkene **2**, which was envisioned to be assembled by the Kumada–Corriu cross-coupling with alkyl chloride **3** and alkenyl chloride **4** as coupling partners as shown in Figure 5.1. With access to commercial quantities of cross-coupling partners, we began our studies by identifying optimal conditions for the preparation of Grignard reagent of substrate **3** to affect the cross-coupling reaction.

Although C—C bond formation is largely dominated by the use of metals such as Pd and Ni catalysts, [14] we preferred the use of iron catalyst for the Kumada–Corriu coupling because of its environmentally benign character, low cost, and low toxicity. More importantly, the acceptable limit of heavy metals in active pharmaceutical ingredient (API) as per International Conference on Harmonisation (ICH) guidelines is very low (in the levels of ppm)[1] [15]. For this reason, the use of heavy metals is generally not recommended in the late-stage synthesis of APIs as it requires additional unit operations for scavenging the elemental impurities. This usually leads to material loss and thereby incurring additional cost [16]. Fortunately, because of the low inherent toxicity, iron does not come under this category as the acceptable limit of this is quite high as per current ICH guidelines [17].

Preparation of multikilogram quantities of intermediate **2** necessitated the need to develop a safe and robust process for the preparation of Grignard reagent of substrate **3**, which was positioned to be utilized for the Kumada–Corriu coupling. Surprisingly, literature reported synthetic methodologies to prepare Grignard reagent of substrate similar to **3** were found to be ineffective and all our attempts led to inconsistent formation of Grignard reagent that translated into poor chemical conversion during the Kumada–Corriu coupling step [13b]. This prompted us to investigate further with a goal to develop a reproducible and safer

1 http://www.ich.org

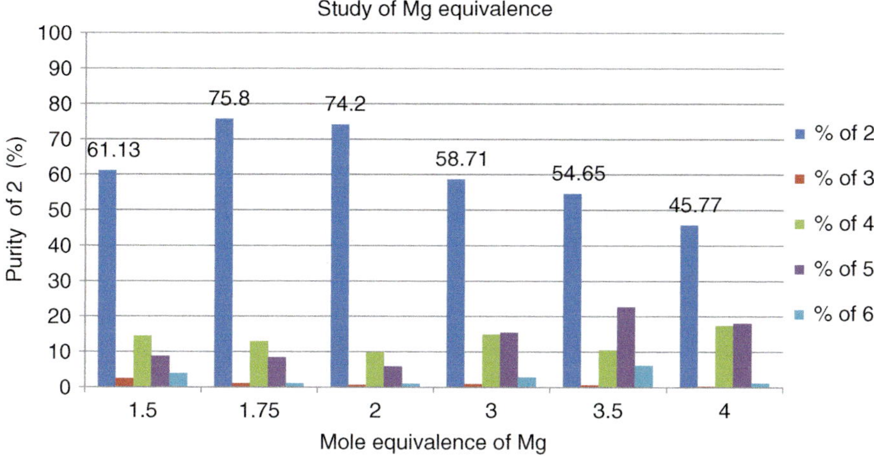

Figure 5.2 Effect of Mg equivalence on purity of 2 (%).

process amenable to large-scale production of Grignard reagent of substrate **3** that would take part in the ensuing Kumada–Corriu cross-coupling reaction to afford product **2** in high yields. Initial focus was on identification of optimal conditions required for the preparation of Grignard reagent by systematically evaluating the effect of grade and equivalents of magnesium, equivalents of 1,2-dibromoethane, solvent concentration, and the rate of addition of substrate **3** to magnesium before finally embarking on optimizing the conditions for the coupling reaction.

5.2 Optimization of Grade and Equivalents of Mg Metal

The effect of Mg turnings vs Mg powder on the Grignard reagent preparation was profound. In our hands, utilization of Mg turnings for preparation of Grignard reagent was unsuccessful, as Mg turnings required higher temperature (70–75 °C), maintained over a longer period of time, for activation. This led to the decomposition of the reagent. However, with Mg powder, initiation was achieved at 50–55 °C and required shorter time for activation. This can be attributed to the high surface area of Mg powder that could have complemented for the activation/initiation at low temperature [18]. For this reason, Mg powder was finalized and further optimization showed that 1.75 equiv of Mg powder is optimal for Grignard preparation (Figure 5.2).

5.3 Optimization of Equivalents of 1,2-Dibromoethane

Many methods are known for the activation of magnesium such as crushing the Mg pieces *in situ*, rapid stirring, sonication of the suspension, addition of iodine, [19] 1,2-dibromoethane, [20] and Vitride for Mg metal activation [21]. Most

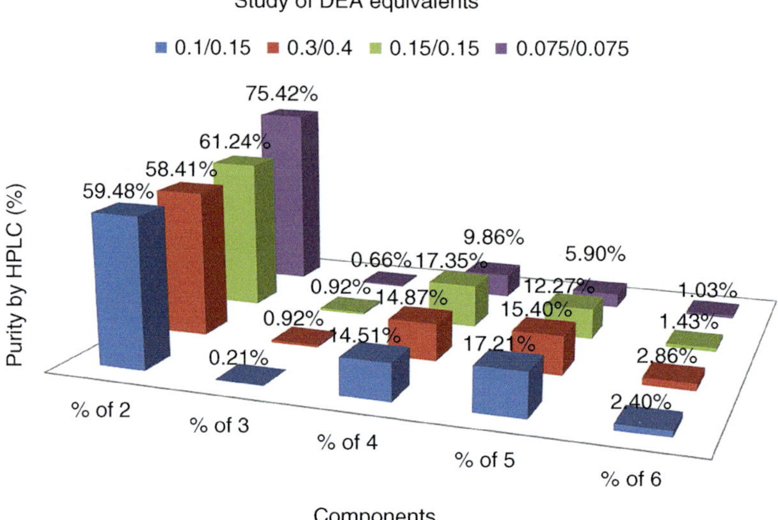

Figure 5.3 Effect of DBE equivalents on Grignard and coupling reaction.

commonly used initiator among these is 1,2-dibromoethane (DBE), and hence, the effect of DBE initiator on the Grignard preparation and Kumada–Corriu coupling was further studied. Initial experiments showed that addition of DBE (0.3 equiv) in one portion led to sudden exotherm up to 70 °C, leading to decomposition of the Grignard reagent and low yields in the cross-coupling reaction. For this reason, it was decided to add DBE lot-wise with the first lot added for Mg activation and the second lot added along with the substrate **3**. Further experimentation revealed that lowering the equivalents of DBE increases the purity of **2** as shown in Figure 5.3 and 0.15 equiv of DBE is optimal for the reaction.

As discussed earlier in this section, 0.15 equiv of DBE was added in two portions to the reaction mixture. The first lot of 0.075 equiv was added for Mg activation and subsequent initiation of Grignard reagent. The second lot of 0.075 mol was mixed with alkyl halide **3** and slowly added. The second lot of DBE ensured continuous Mg activation for an efficient preparation of Grignard reagent.

5.4 Effect of Solvent Concentration on Preparation of Grignard Reagent and Kumada–Corriu Coupling

The choice of solvent and its concentration played a major role in the preparation and stabilization of Grignard reagent. Solvent screening showed that ether-based solvents, notably diethylether and tetrahydrofuran (THF), are effective for preparation of the Grignard reagent and the subsequent Kumada–Corriu coupling. However, on industrial scale as the usage of diethyl ether is not safe because of the low ignition temperature and peroxide formation, we chose THF as the reaction solvent. First, we undertook a more detailed study to understand the effect of solvent concentration on the Grignard preparation. During the course of the optimization, two impurities, identified as alkane impurity **5** and hydroxy impurity **6**, were observed consistently albeit in varying proportions depending on the solvent concentration. The formation of alkane **5** can be explained as a direct consequence of quenching of Grignard reagent of substrate **3** with moisture present in the system. Hydroxy impurity **6** on the other hand is presumably formed via the hydrolytic quench of the peroxy Grignard reagent generated from the reaction of Grignard reagent of substrate **3** with oxygen as shown in Figure 5.4. It was evident from the mechanism of formation that a stringent control of moisture and oxygen content in the solvent and the reaction medium is essential for the consistent preparation of Grignard reagent.

The effect of THF volume on the Grignard reagent preparation and the subsequent coupling reaction were further studied. The results are summarized in Figure 5.5. The results showed that a minimum volume of THF is superior for preparation of Grignard and the subsequent coupling reaction.

Further understanding of the effect of THF volume on the coupling reaction showed that increase in the solvent volume led to decrease in purity of **2** (Figure 5.5) and also led to the increase in impurities, alkane **5** and hydroxy impurity **6** (Figure 5.5). Hence, an optimal total 6 vol of THF (2 vol for Mg initiation, 2 vol for dissolution of **3**, and 2 vol for dissolution of **4**) was finalized for Grignard reagent preparation and coupling reaction.

5.5 Effect of Alkyl Chloride 3 Addition Time on the Grignard Reagent Preparation

Another important parameter for Grignard reagent preparation is the rate of addition of the alkyl halide (**3**) solution. It is important to have a steady and continuous generation of Grignard reagent that depends on the addition rate of alkyl halide **3**. We observed that fast addition (~15 to 20 minutes) of alkyl chloride **3** solution in THF at 55–60 °C led to uncontrolled exotherm and decomposition of the Grignard reaction. The best result was obtained by slow addition (45–60 minutes) of alkyl chloride **3** solution at 55–60 °C to the Mg powder. After the completion of addition, the reaction mixture was maintained at reflux for 60–70 minutes to ensure complete formation of Grignard reagent of substrate **3**.

5.6 Stability of Grignard Reagent at 0–5 °C

After the formation of Grignard reagent, we evaluated the stability of the reagent at various temperatures before the solution was used for the coupling reaction. A quick screening showed promising results at lower temperature particularly at 0–5 °C. The stability of Grignard reagent with time was also analyzed. It was observed that holding at 0–5 °C for an extended time (>1.5 hours) has a detrimental effect on the stability of the Grignard reagent. As shown in Figure 5.6, when the Grignard reagent was stored at 0–5 °C for three hours, significant decomposition was observed. Alkane (**5**) was the major decomposition impurity observed. The optimal holding time was found to be 45–60 minutes at 0–5 °C.

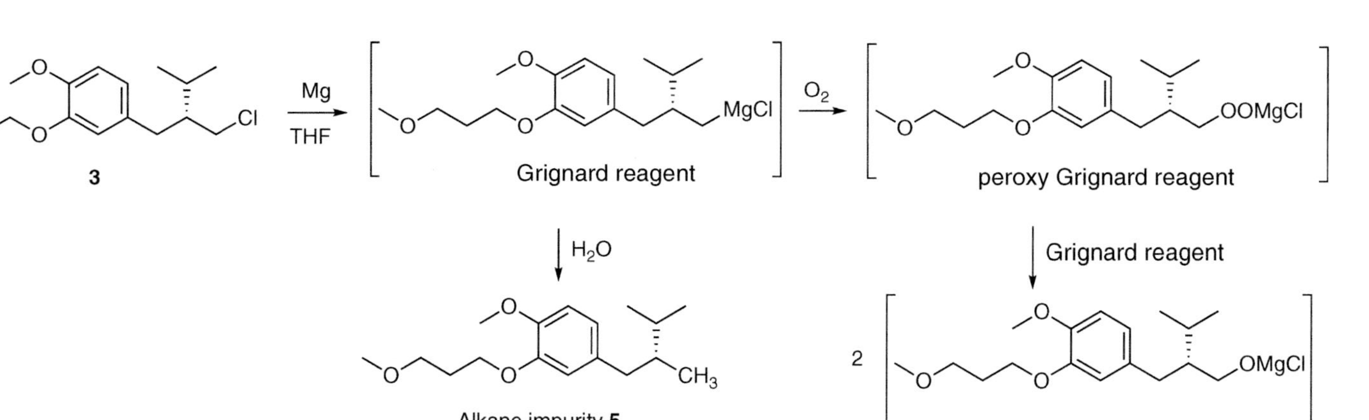

Figure 5.4 Proposed mechanism for formation of impurities.

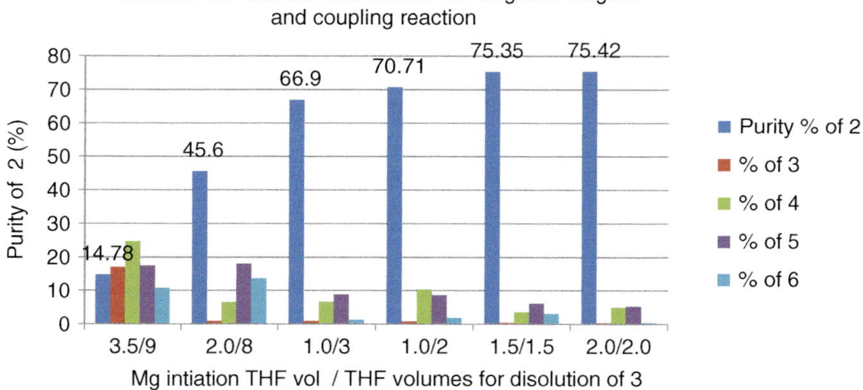

Figure 5.5 Effect of THF concentration on the Grignard reagent and coupling reaction.

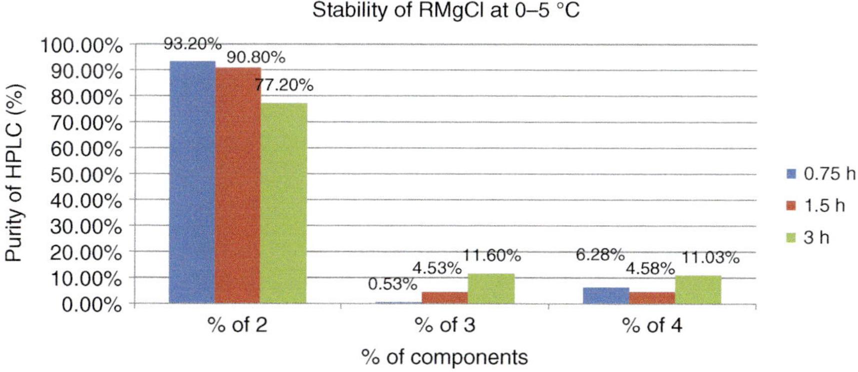

Figure 5.6 Effect of Grignard stirring time on conversion of **2**.

5.7 Iron-Catalyzed Cross-coupling Reaction

With the optimized process to prepare Grignard reagent in hand, we then decided to focus on the Kumada–Corriu cross-coupling reaction to prepare alkene intermediate **2**. As mentioned earlier, the field of C–C coupling reactions has been largely dominated by Pd and Ni catalysts, distinguished by their wide substrate scope encompassing aryl (alkenyl) iodides, bromides, and triflate. However, alkyl/alkenyl/aryl chlorides require the use of higher temperatures and ancillary ligands to render Pd and Ni catalysts sufficiently reactive. In this context, simple ferric salt-catalyzed cross-coupling reactions developed by Kochi showed significant promise because of their tolerance of nucleophiles containing β-hydrogens [4]. In 1998, Cahiez and colleagues discovered that addition of *N*-methylpyrrolidone (NMP) as a minor cosolvent allowed overcoming several limitations of the Kochi ferric salt system (excess alkenyl halide for coupling) and significantly broadened the substrate scope [22]. It is also worth noting that while NMP has been employed extensively in iron-catalyzed cross-coupling methodology, its role in enabling improved catalytic performance remains poorly understood or defined [23]. Cahiez attributed this to one of the effects of NMP stabilizing the iron organometallic species that are the real catalytic intermediates of this reaction by limiting the decomposition processes (i.e. β-hydride elimination) [22]. In striking contrast to the detailed insights available into the palladium and nickel catalysis, the lack of mechanistic understanding about the iron-catalyzed cross-coupling reactions has remained a formidable challenge in this field. Although a general consensus on the formal oxidation states of the operating catalytic species has never been reached, based on the empirical data, depending on the nature of the Grignard reagent (on the availability of β-hydrogens), two distinctly different mechanisms were proposed. Grignard reagents containing β-hydrogens (EtMgBr or higher analogs) proceed via the low-valent redox manifold, whereas Grignard reagents that do not possess β-hydrogens (MeMgBr, PhMgBr, etc.) proceed via organoferrate manifold [24]. For this reason, the catalytic cycle shown in Figure 5.7 is a formal representation and does not imply any detailed mechanistic or structural information about the iron-catalyzed Kumada–Corriu cross-coupling reaction.

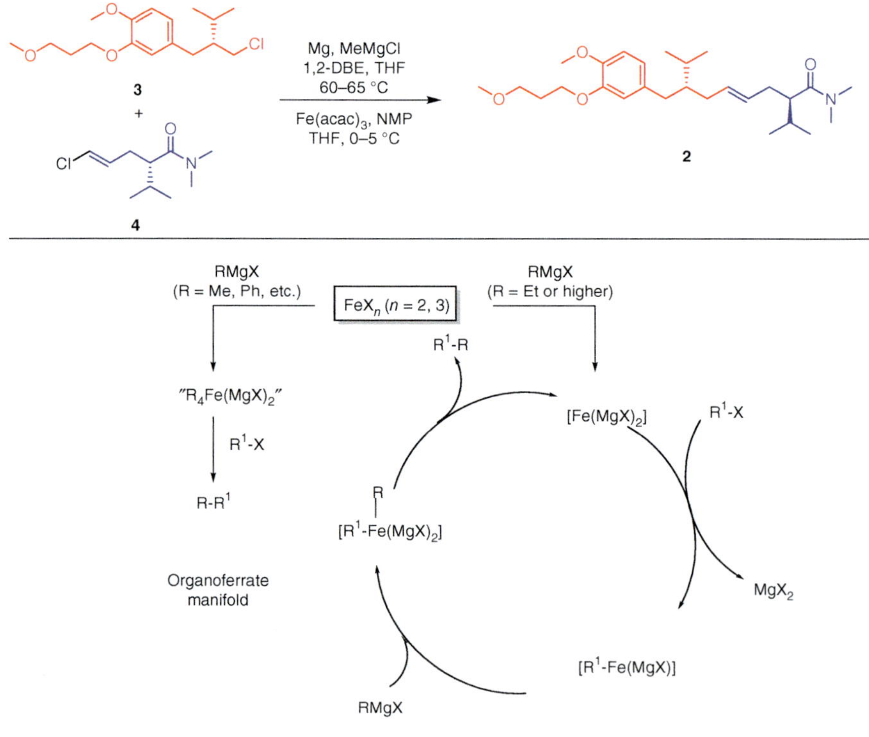

Figure 5.7 Kumada–Corriu cross-coupling reaction and mechanistic understanding.

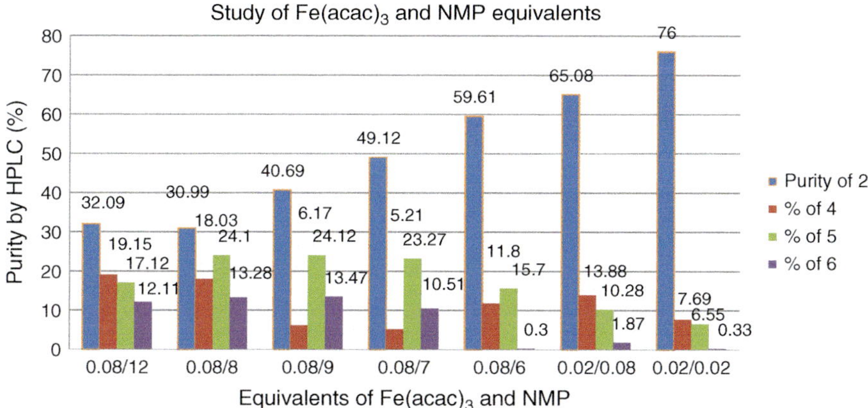

Figure 5.8 Study of $Fe(acac)_3$ and NMP equivalents.

5.8 Optimization of Equivalents of NMP and $Fe(acac)_3$

Based on the conditions reported by Cahiez, during the initial development, 0.08 equiv of $Fe(acac)_3$ and 9.0 equiv of NMP were used for evaluating the coupling reaction. Unfortunately, this resulted in poor conversion of the coupled product **2**. It can also be inferred from Figure 5.8 that lowering NMP equivalents with respect to $Fe(acac)_3$ led to improved purity of cross-coupled product **2**. Optimization of the ratio of NMP and $Fe(acac)_3$ with respect to the substrate **3** showed 0.02 equiv each of $Fe(acac)_3$ and NMP as part of the optimal conditions for the coupling reaction (Figure 5.8).

5.9 Optimization of Equivalents of Substrate 4 and Its Rate of Addition

The effect of equivalents of **4** with respect to substrate **3** was also studied. Although the purity of the coupled product **2** did not vary between 0.9 and 1.2 equiv of compound **4**, removal of unreacted **4** posed significant problem during downstream processing. For this reason, 1.05 equiv of **4** was finalized for the coupling reaction (Figure 5.9). The protocol for coupling required addition of feed solution (substrate **4**, $Fe(acac)_3$, and NMP) to the precooled solution of Grignard reagent at 0–5 °C. It was observed that the addition of **4** is exothermic and temperatures (>35 °C) proved detrimental for coupling. For this reason, the rate of addition of feed solution to the Grignard reagent was temperature controlled with the addition starting at 0–5 °C and addition rate adjusted to ensure that the reaction temperature does not increase beyond 35 °C.

5.10 Execution at Pilot Scale and Scale-up Issues

Although the optimized conditions were sufficient to reproduce the Grignard reagent synthesis and coupling reaction on lab scale (~500 g scale), we

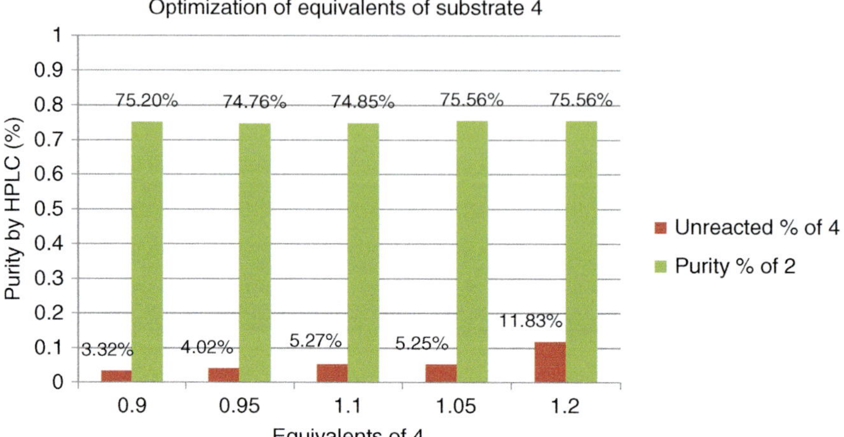

Figure 5.9 Study of four equivalence.

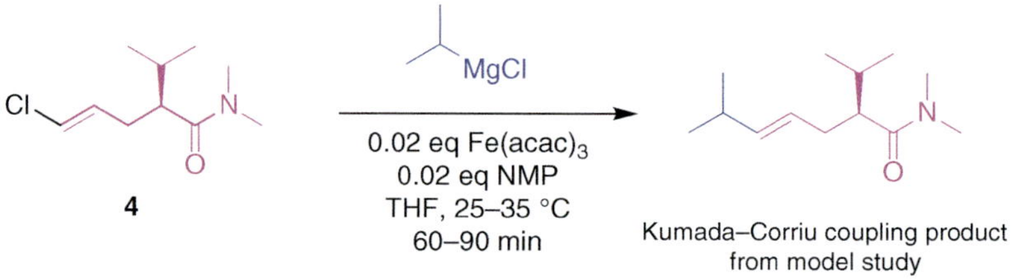

Cl
4
MgCl
0.02 eq Fe(acac)$_3$
0.02 eq NMP
THF, 25–35 °C
60–90 min
Kumada–Corriu coupling product from model study

Figure 5.10 Model study for coupling reaction.

encountered problems during the execution of pilot batch (5 kg scale). In the pilot batch, only 5–10% yield of coupling product was obtained. To ascertain the reason for low yield, isopropyl magnesium halide was used as a model Grignard reagent, and it was coupled with **4** on 5 kg scale. In this experiment, isopropyl-coupled product **8** was isolated in 80–85% yield (Figure 5.10), implying the possibility of low conversion during the formation of Grignard reagent, which could have led to low yield in the coupling reaction. This was corroborated by the HPLC analysis of the crude product from the pilot batch that showed unreacted **4** and impurity **5** as major by-products.

As discussed earlier, formation of the alkane **5** impurity suggested the possibility of premature quenching of the Grignard reagent with moisture. Unfortunately, ensuring completely dry conditions on commercial-scale equipment is difficult because the equipment is large and other supporting piping is fairly complex. For this reason, additives that scavenge adventitious moisture were investigated and MeMgCl showed promise. Addition of MeMgCl as a sacrificial reagent (0.2 equiv with respect to **3**) improved the consistency of Grignard reagent formation. We suspect that MeMgCl reacted with the adventitious moisture by liberation of gaseous methane. In order to render the process more robust, before starting the batch, anhydrous THF was refluxed in the reactor and recirculated through the transferring lines and receivers. Additionally, before charging Mg into the

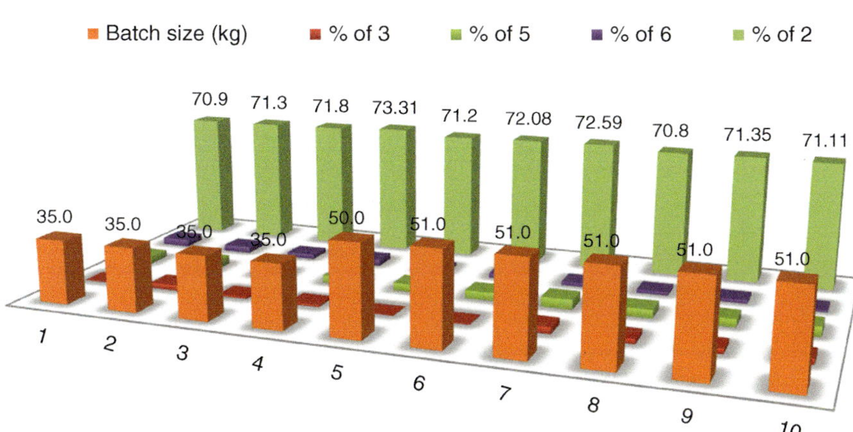

Figure 5.11 Purity trend for 10 large scale batches.

reactor, Mg was predried in a secondary reactor by carrying out vacuum/nitrogen gas cycles. The optimized process involved addition of compound **3** in THF solvent (2 vol) with DBE (0.075 equiv) and MeMgCl (0.2 equiv with respect to **3**) to Mg in THF for the preparation of Grignard reagent. These additional changes to the process ensured robustness as evident from successful execution of multiple batches on 50 kg scale (Figure 5.11).

5.11 Agitated Thin Film Evaporator (ATFE) for Purification of 2

After completion of reaction, it was quenched with aqueous HCl solution and the product was extracted with toluene. The organic layer was then concentrated to obtain crude product **2** with a purity of ~70% to 75%. The crude Kumada–Corriu product **2** was converted to Aliskiren **1** in a series of straightforward steps comprising bromolactonization **7**, azidation **8**, aminolysis of lactone **9**, and azidoreduction to obtain **1** as shown in Figure 5.12. Unfortunately, the downstream conversions from crude product **2** were irreproducible. At this juncture, it was decided to purify the crude product **2** before proceeding to conversion to Aliskiren **1**. Incidently, because of the liquid nature of the coupled product **2** and the corresponding impurities, purification required either resorting to chromatography or vacuum distillation. High-vacuum distillation (HVD) was finally opted for purification of the coupled product. Although the HVD (temperature: 240–250 °C, 1–2 Torr vacuum) was successful at lab scale, extensive decomposition was observed during the execution at pilot scale. We rationalized that the long hours of stirring during distillation in a stirred tank (Stainless steel) reactor presumably led to the degradation of compound **2** and resulted in lower assay (~60% to 65%).

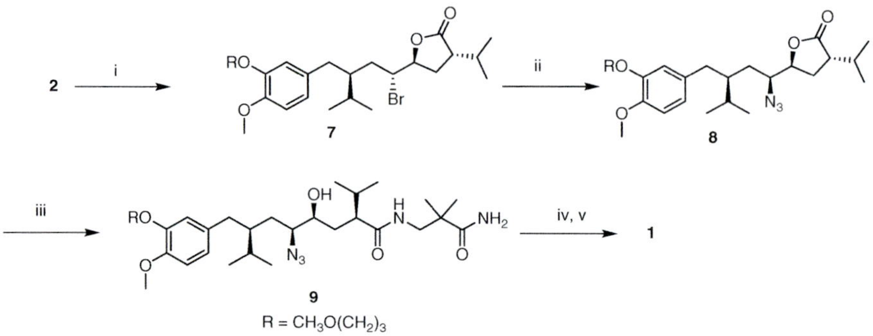

$R = CH_3O(CH_2)_3$

Figure 5.12 Reagents and conditions: (i) N-Bromosuccinimide (NBS), 42.5% H_3PO_4, THF, 0–5 °C, 90%; (ii) NaN_3, tripropylene glycol, H_2O, 78–83 °C, 24 hours, 85%; (iii) 3-amino-2,2-dimethylpropanamide (Synthon-C), Triethylamine (TEA), 2-hydroxy pyridine, 80–85 °C, 25 hours, 90%; (iv) 10% Pd/C, methanolic ammonia, MeOH, 25–35 °C, 8 hours 90%; and (v) fumaric acid, MeOH, CH_3CN, 25–35 °C, 12 hours, 85%.

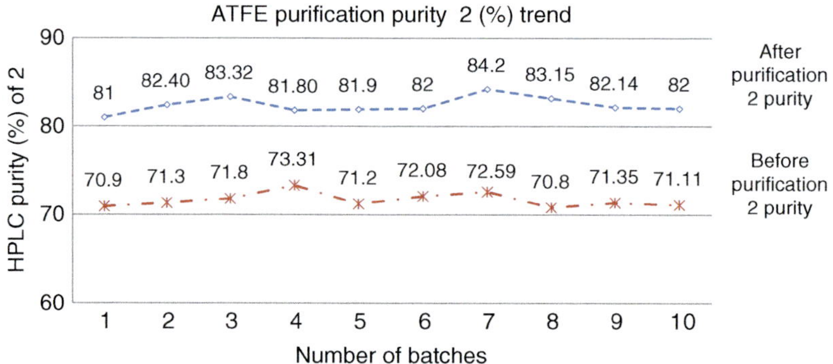

Figure 5.13 ATFE purification purity trend (before/after) of 10 large-scale batches.

This challenge was addressed at scale by using Agitated Thin Film Evaporator (ATFE) technology [25].[2] The short surface contact residence time of compound at higher temperature in ATFE ensured that the assay and purity of the compound **2** was not compromised and the desired product was isolated consistently with the required specification (Figure 5.13). Subsequent conversion to Aliskiren **1** proceeded [13b, 26] uneventfully on 50 kg scale.

5.12 Conclusion

An iron-catalyzed Kumada–Corriu cross-coupling reaction for the synthesis of Aliskiren on scale multikilogram was developed and executed, thereby demonstrating the use of earth-abundant environmentally friendly metal in cross-coupling vs the conventional use of Pd. The coupling product was purified by using ATFE technique.

2 https://lcicorp.com/assets/documents/CE_TFE_Scaleup.pdf

Acknowledgments

We thank the management of Dr. Reddy's Laboratories Ltd. for supporting this work. The AR&D and Process Engineering division of IPDO are also acknowledged.

Biography

Debjit Basu was born in India and obtained M.S. degree from Pune University in the year 2003. He joined Dr. S Chandrasekhar group at the Indian Institute of Chemical Technology – India, receiving his Ph.D. degree on research related to the organic synthesis of complex natural products. He came to France in 2009 and joined Dr. Prof. Jean Marie Beau's group in ICSN-Paris as a researcher to investigate novel neuraminidase inhibitors targeting Flu viruses. In 2010, he moved to Aurigene Discovery Technology – India and led multidiscipline teams engaged in the design and development of synthetic routes for new pharmaceuticals. In 2012, he joined Dr. Prof. Daniel Rauh's group in TU Dortmund, Germany, and was involved in various interdisciplinary areas of chemical biology. Currently, he is the technical lead of process development of various active pharmaceutical ingredients in Dr Reddy's Laboratory, India.

Srinivas Achanta was born in Nidadavole, India. He received his Bachelor of Technology degree in Chemical Engineering from Osmania University, India, and Ph.D. in Chemistry under the guidance of Prof. V. B. Birman at Washington University in St. Louis, USA. After completing the postdoctoral studies at York University, Canada, with Prof. M. G. Organ, he worked as a scientist in Jubilant Chemsys, TEVA API India Pvt Ltd., and is currently working in process development in Dr. Reddy's Laboratories, India.

Uday Kumar Neelam studied chemistry at the National Institute of Technology, Warangal, India, where he completed his M.Sc. degree. Later, he joined as a scientist at Dr. Reddy's laboratories Ltd., Hyderabad, India. Neelam completed Ph.D. thesis in 2014 at Osmania university, Hyderabad, India, in the group of Dr. Rakeshwar Bandichhor in collaboration with Dr. Reddy's laboratories Ltd. He then worked with Marisa Kozilowski at Upenn, USA, as a postdoctoral fellow from 2014 to 2015. Presently, he is working as a tech lead in API process R&D division at Dr. Reddy's laboratories Ltd., Hyderabad, India.

Shortened version

Dr. Apurba Bhattacharya (Apu) is a professor in the Department of Chemistry at TAMUK. Dr. Apu joined TAMUK in Fall, 1999. At the end of 2005, he took a leave of absence from TAMUK and served as the Global R&D Head and Senior Vice President in Dr. Reddys Laboratories in India before returning to TAMUK in Fall 2008.
Dr. Apu has over 20 years of experience in pharmaceutical industry in the areas of drug delivery and drug development (before TAMUK), including Merck (eight years), Hoechst (eight years), Bristol Myers Squibb (two years), and Dr. Reddy's Laboratories in India (two years). Dr. Apu has 7 book chapters, 123 referred publications, 50 patents, and 168 research presentations in national and international meetings. His students have made 60 research presentations. He is also a scientific advisor for several pharmaceutical companies. Dr. Apu's research interests include pharmaceutical process research and development and environmentally benign processes in organic synthesis.

Rajeev R. Budhdev is currently working as a senior vice president and global head – API R&D at Dr. Reddy's Laboratories Limited, Hyderabad. He holds a Masters and a Ph.D. degree in Organic Chemistry from Maharaja Sayajirao (M.S.) University of Baroda. He brings with him over 28 years of experience in the areas of academic research at M.S. University, followed by Process Chemistry of generic APIs at Sun Pharma's R&D center at Baroda, leading to numerous DMF/ANDA filings for a gamut of small molecules of varying complexities, including Peptides. Rajeev is very passionate about R&D collaboration cutting across cross-functional domains, both within and outside the company, including external subject matter experts. He also has a number of patent publications to his credit and has delivered talks on emerging frontiers in Pharma Science and Technology.

Rakeshwar Bandichhor studied chemistry and obtained Ph.D. degree from the University of Lucknow and worked at the University of Regensburg, Germany, for a year during his Ph.D. tenure. He did postdocs at the University of Regensburg, University of Pennsylvania, and Texas A&M University. He published more than 170 papers including patents and book chapters. He has won various awards and honors in his career and serves as BoS (Board of Studies) and BoG (Board of Governors) members of Institute of Science and Engineering, Jawharlal Nehru Technological University, Hyderabad (JNTU-H), starting from December 2017. He has also been appointed as a honorary visiting professor at the University of Delhi. He has recently been appointed as the international advisory board member of European Journal of Organic Chemistry. He has also edited a book entitled Hazardous Reagent Substitution: A Pharmaceutical Industry Perspective published by Royal Society of Chemistry.

References

1 (a) Johansson Seechurn, C.C.C., Kitching, M.O., Colacot, T.J., and Snieckus, V. (2012). *Angew. Chem. Int. Ed.* 51: 5062–5085. (b) Colacot, T. (2014). *New Trends in Cross-Coupling: Theory and Applications*. RSC Book. (c) Dunetz, J.M.J.R. (2003). *Transition Metal-Catalyzed Couplings in Process Chemistry*. Wiley-VCH. (d) Stephan, E., Kathrin, J., and Matthias, B. (2008). *Angew. Chem. Int. Ed.* 47: 3317. (e) Miyaura, N. (ed.) (2002). Cross-coupling reactions. A practical guide. In: *Topics in Current Chemistry*, vol. 219. Berlin: Springer.
2 de Meijere, A. and Diederich, F. (eds.) (2004). *Metal-Catalyzed Cross-Coupling Reactions*, 2e, vol. 1. Weinheim, Germany: Wiley-VCH.
3 (a) Kharasch, M.S. and Fields, E. (1941). *J. Am. Chem. Soc.* 63 (9): 2316–2320. (b) Kharasch, M.S. and Fuchs, C.F. (1943). *J. Am. Chem. Soc.* 65: 504. (c) Kharasch, M.S. and Reinmuth, O. (1954). *Grignard Reagents of Nonmetallic substances*. New York: Prentice Hall.
4 (a) Tamura, M. and Kochi, J.K. (1971). *J. Am. Chem. Soc.* 93: 1487. (b) Tamura, M. and Kochi, J. (1971). *Synthesis* 6: 303. (c) Neumann, S.M. and Kochi, J.K. (1975). *J. Org. Chem.* 40: 599. (d) Smith, R.S. and Kochi, J.K. (1976). *J. Org. Chem.* 41: 502. (e) Kochi, J.K. (1974). *Acc. Chem. Res.* 7: 351.
5 (a) Sherry, B.D. and Furstner, A. (2008). *Acc. Chem. Res.* 41: 1500–1511. (b) Alois, F. and Ruben, M. (2005). *Chem. Lett.* 34: 624–629. (c) Mako, T.L. and Byers, J.A. (2016). *Inorg. Chem. Front.* 3: 766–790. (d) Martin, R. and Furstner, A. (2004). *Angew. Chem. Int. Ed.* 43: 3955–3957. (e) Carpenter, S.H. and Neidig, M.L. (2017). *Isr. J. Chem.* 57: 1106–1116. (f) Bauer, I. and Knolker, H.-J. (2015). *Chem. Rev.* 115: 3170–3387. (g) Furstner, A. (2016). *ACS Cent. Sci.* 2: 778–789.
6 (a) Majid, M.H. and Parvin, H. (2012). *Monatsh. Chem.* 143: 1575. (b) Majid, M.H., Vahideh, Z., Parvin, H., and Hoda, H. (2019). *Monatsh. Chem.* 150: 535.
7 (a) Tamao, K., Sumitani, K., and Kumada, M. (1972). *J. Am. Chem. Soc.* 94: 4374. (b) Corriu, R.J.P. and Masse, J.P. (1972). *J. Chem. Soc. Chem. Commun.* (3): 144. (c) Tamao, K., Sumitani, K., Kiso, Y. et al. (1976). *Bull. Chem. Soc. Jpn.* 49: 1958.
8 (a) Yamamura, M., Moritani, I., and Murahashi, S.-I. (1975). *J. Organomet. Chem.* 91: C39. (b) Negishi, E. and Baba, S. (1976). *J. Chem. Soc. Chem. Commun.* (15): 596. (c) Baba, S. and Negishi, E. (1976). *J. Am. Chem. Soc.* 98: 6729. (d) Negishi, E., King, A.O., and Okukado, N. (1977). *J. Org. Chem.* 42: 1821. (e) Negishi, E. and van Horn, D.E. (1977). *J. Am. Chem. Soc.* 99: 3168.
9 (b) Tamao, K. (2002). *J. Organomet. Chem.* 23: 653.
10 Selected reviews include: (a) Chincilla, R. and Najera, C. (2007). *Chem. Rev.* 107: 874. (b) Buchwald, S.L., Mauger, C., Mignani, G., and Scholz, U. (2006). *Adv. Synth. Catal.* 348: 23. (c) Oestreich, M. (ed.) (2009). *The Mizoroki Heck*

Reaction. Chichester, UK: Wiley. (d) Suzuki, A. (1991). *Pure Appl. Chem.* 63: 419. (e) Beletskava, I.P. and Cheprakov, A.V. (2000). *Chem. Rev.* 100: 3009.

11 Czaplik, W.M., Mayer, M., Cvengroš, J., and von Wangelin, A.J. (2009). *ChemSusChem* 2: 396.

12 Piontek, A., Bisz, E., and Szostak, M. *Angew. Chem. Int. Ed.* 57: 11116.

13 (a) Srinivas, G., Uday, K.N., Sudharkar, R.B. et al. (2015). *Org. Process Res. Dev.* 19: 470. (b) Johnson, D.S. and Lee, J.J.L. (2010). *Modern Drug Synthesis*, 141–158. Hoboken, NJ: Wiley.

14 (a) Negishi, E. (ed.) (2002). *Handbook of Organopalladium Chemistry for Organic Synthesis*. New York: Wiley. (b) Tsuji, J. (1996). *Palladium Reagents and Catalysts: Innovations in Organic Synthesis*. New York: Wiley. (c) Trost, B.M. and Verhoeven, T.R. (1982). Organopalladium compounds in organic synthesis and in catalysis. In: *Comprehensive Organometallic Chemistry*, vol. 8 (eds. G. Wilkinson, F.G.A. Stone and E.W. Abel), 799. Oxford, UK: Pergamon.

15 European Medicines (2008) Guideline on the specification limits for residues of metals catalysts or metal reagents, EMEA/CHMP/SWP/4446/2000.

16 Hidetoshi, M., Chihiro, S., Eriko, T. et al. (2015). *Org. Process Res. Dev.* 19 (8): 1054.

17 ICH Harmonized Guideline, Guideline For Elemental Impurities Q3D, 2014.

18 Yue, S.L. (1994). *J. Loss Prev. Process Ind.* 7: 413.

19 Lai, Y.-H. (1981). *Synthesis* 8: 585.

20 Ulf, T. and Hilmar, W. (2002). *Org. Process Res. Dev.* 6: 906.

21 Tilstam, U. and Weinmann, H. (2002). Activation of mg metal for safe formation of grignard reagents on plant scale. *Org. Process Res. Dev.* 6: 906–910.

22 Cahiez, G. and Avedissian, H. (1998). *Synthesis* 8: 1199.

23 (a) Ottesen, L.K., Ek, F., and Olsson, R. (2006). *Org. Lett.* 8: 1771–1773. (b) Cahiez, G., Gager, O., Buendia, J., and Patinote, C. (2012). *Chem. Eur. J.* 18: 5860–5863. (c) Gotta, M., Lehnemann, B.W., Jaboi von Wangelin, A. et al. (2015). Process for preparing styrene derivatives, US Patent 9, 024,045 B2. (d) Gulak, S., Gieshoff, T.N., and Jacobi von Wangelin, A. (2013). *Adv. Synth. Catal.* 355: 2197–2202. (e) Malhotra, S., Seng, P.S., Koenig, S.G. et al. (2013). *Org. Lett.* 15: 3698–3701. (f) Gartner, D., Stein, A.L., Grupe, S. et al. (2015). *Angew. Chem. Int. Ed.* 54: 10545–10549.

24 Furstner, A., Martin, R., Krause, H. et al. (2008). *J. Am. Chem. Soc.* 130: 8773.

25 (a) Freeze, H.L. and Glover, W.B. (1979). *Chem. Eng.Prog.* 75: 53. (b) Mutzenberg, A.B. (1965). *Chem. Eng.* 175: 165.

26 Herold, P., Stutz, S., and Spindler, F. (2002). Process for the preparation of substituted octanoylamides, WO 02002508.

6

Development and Scale-Up of a Palladium-Catalyzed Intramolecular Direct Arylation in the Commercial Synthesis of Beclabuvir

Collin Chan[1], *Albert J. DelMonte*[1], *Chao Hang*[1], *Yi Hsiao*[2], *and Eric M. Simmons*[1]

[1] *Bristol-Myers Squibb, Chemical and Synthetic Development, 1 Squibb Drive, New Brunswick, NJ 08903, USA*
[2] *TEDA Tianjin, No. 71, 7th Avenue, Tianjin 300457, China*

6.1 Introduction

The commercial route to Bristol-Myers Squibb's HCV NS5B inhibitor BMS-791325 (beclabuvir) [1] assembles the central seven-membered ring of the active pharmaceutical ingredient (API) via a palladium-catalyzed intramolecular direct arylation [2] of indole-containing aryl bromide **1** (Scheme 6.1) (Braem et al., *manuscript in preparation*). Following workup and crystallization, this process affords the final isolated intermediate (or penultimate) **2** as the monopotassium salt, hemi-DMAc solvate, which is converted to the API through an EDAC-mediated amidation with bicyclic diamine **3**. In this chapter, we detail the evolution of the direct arylation chemistry used to prepare **2**, which produced over one metric ton of the penultimate **2** over the course of this project.

The enabling kilogram-scale synthesis of beclabuvir formed **2** via a direct arylation of aryl bromide ethyl ester **4**, followed by hydrolysis of the ester moiety and isolation as the monopotassium salt (Scheme 6.2) [3]. In the development of the reaction conditions for the direct arylation of **4**, initial ligand screening identified tricyclohexylphosphine (PCy_3) as an effective ligand for this cyclization, with the air-stable tetrafluoroborate salt ($PCy_3{\bullet}HBF_4$) being employed on scale, while the use of $KHCO_3$ led to short reaction times among the bases examined. Under the optimized conditions, the cyclization was catalyzed by 5 mol% palladium(II) acetate ($Pd(OAc)_2$), and 10 mol% $PCy_3{\bullet}HBF_4$, with 4 equiv of $KHCO_3$ as base in a 1 : 1 mixture of *N,N*-dimethylacetamide (DMAc) and toluene at 120 °C. This process was successfully executed on 40 kg scale and used to prepare 139 kg of **2** with >99.6 HPLC area percent purity (AP) [3].

In developing the commercial synthesis of beclabuvir, a more robust pre-penultimate intermediate was sought as the crystallization purity and yield of ethyl ester **4** was highly sensitive to low levels of impurities. This was

Organometallic Chemistry in Industry: A Practical Approach, First Edition.
Edited by Thomas J. Colacot and Carin C.C. Johansson Seechurn.

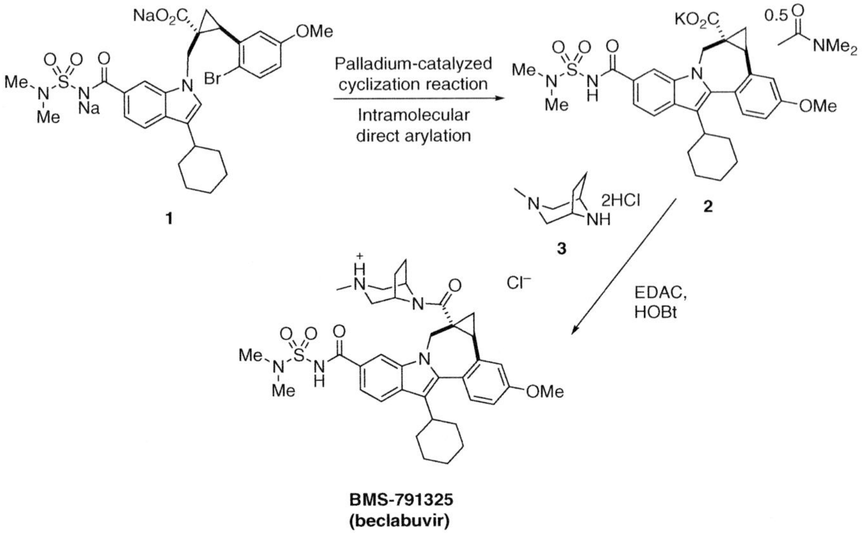

Scheme 6.1 Endgame of the commercial synthesis of HCV NS5B inhibitor beclabuvir.

especially challenging given its preparation via a long telescope sequence that unites alcohol **5** and indole **6** (Scheme 6.3). It was ultimately determined that by moving the hydrolysis step earlier in the sequence, aryl bromide di-sodium salt **1** could be isolated as a highly crystalline compound that effectively purged excess starting materials as well as most other impurities (Braem et al., *manuscript in preparation*). However, this new disodium salt **1** necessitated further process development for the final intermediate step (or penultimate step).

With the identification of a new substrate for the intramolecular direct arylation process, we reevaluated each of the reaction parameters in detail using high-throughput experimentation (HTE) techniques [4]. Over 60 unique ligands were examined, revealing a trend in which optimal results were observed with phosphine ligands bearing a dicyclohexylphosphino group, including DCEphos, XPhos, $PCy_2(o\text{-}SO_3H\text{-}Ph)$, $PPhCy_2$, and PCy_3 (Figure 6.1). Although the fastest reaction rate was observed with $PPhCy_2$, we elected to use $PCy_3{\bullet}HBF_4$ for further development given the low cost, wide availability, and ease of handling of this air-stable phosphonium salt, as well as the fact that it had also been used previously on scale for the cyclization of ethyl ester **4**. It is worth noting that reactions conducted with free PCy_3 or with $PCy_3{\bullet}HBF_4$, under otherwise identical conditions, were consistently found to perform identically. In addition to $Pd(OAc)_2$, other palladium precatalysts such as $PdCl_2(MeCN)_2$, $Pd(TFA)_2$, and $[(Allyl)PdCl]_2$ were also effective but did not provide any significant advantage over $Pd(OAc)_2$.

The direct arylation reaction gave optimal results in dipolar aprotic solvents, such as DMSO, DMF, DMAc, and NMP, due in large part to the poor solubility of di-sodium salt **1** in solvents with low polarity. The use of a DMAc:toluene-mixed solvent system, which was successfully employed for the cyclization of ethyl ester **4**, gave slower reaction rates for **1** compared to pure DMAc, prompting us to

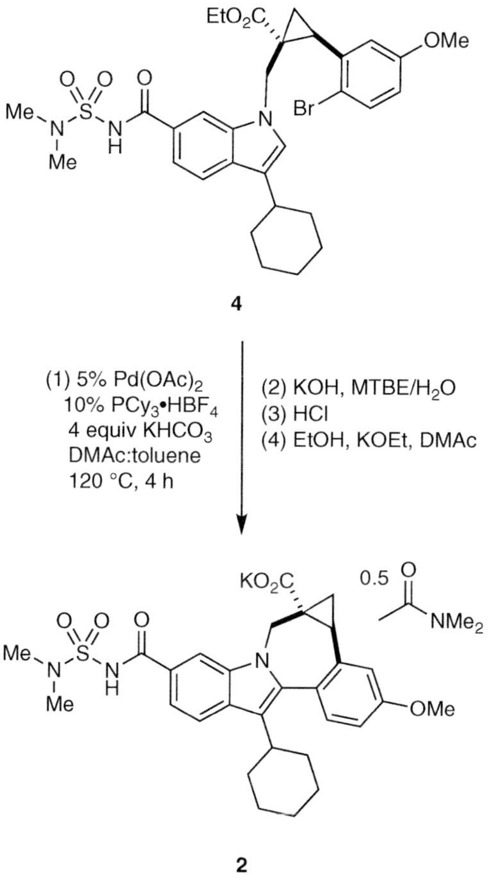

4

(1) 5% $Pd(OAc)_2$
10% $PCy_3 \cdot HBF_4$
4 equiv $KHCO_3$
DMAc:toluene
120 °C, 4 h

(2) KOH, MTBE/H_2O
(3) HCl
(4) EtOH, KOEt, DMAc

2

Scheme 6.2 Palladium-catalyzed intramolecular direct arylation of aryl bromide ethyl ester **4**.

focus our efforts on developing a process utilizing a single solvent. Although DMSO consistently gave the fastest rate and cleanest impurity profile among all solvents investigated, it was not pursued because of safety concerns around exposure, and thus, we focused on DMF, DMAc, and NMP. It should be noted that a DMAc:toluene mixture was chosen for the cyclization of **4** in order to facilitate the workup; before this, the reaction was run in pure DMAc and a subsequent high vacuum distillation was conducted to remove it, whereas the mixed system enabled a much faster extraction to be conducted. Thus, the use of a pure dipolar aprotic solvent for the cyclization of **1** was expected to complicate the workup, which would need to be addressed.

The final critical class variable for this cyclization reaction was the identity of the base. Numerous inorganic bases were investigated, with various carbonates (Cs_2CO_3, K_2CO_3, and $KHCO_3$) and carboxylates proving effective. Stronger bases, such as K_3PO_4, led to increased impurity levels. During the course of these studies, it was found that catalytic amounts of tetrabutylammonium additives, such as TBACl and TBAOH, in combination with stoichiometric inorganic

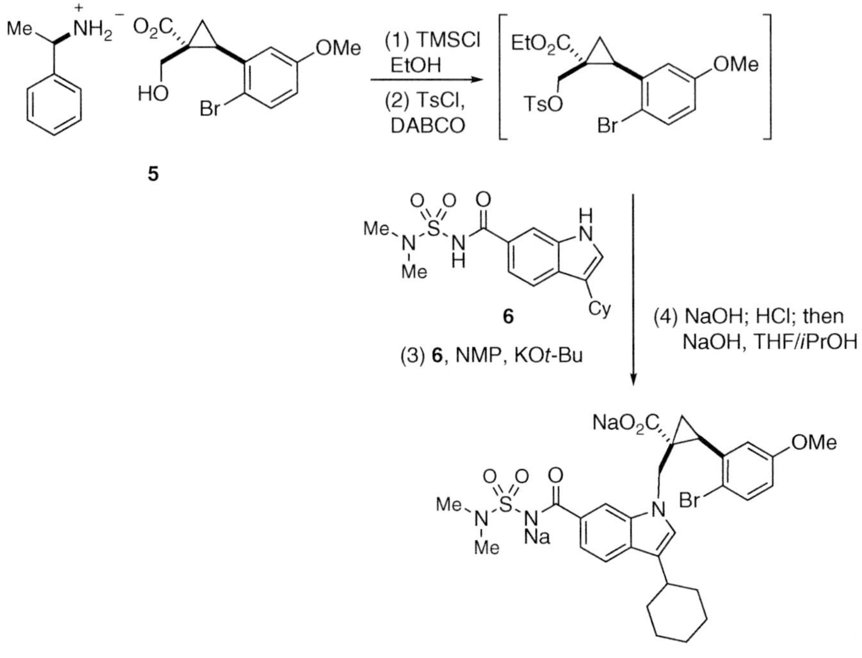

Scheme 6.3 Four-step telescope sequence to prepare aryl bromide disodium salt **1**.

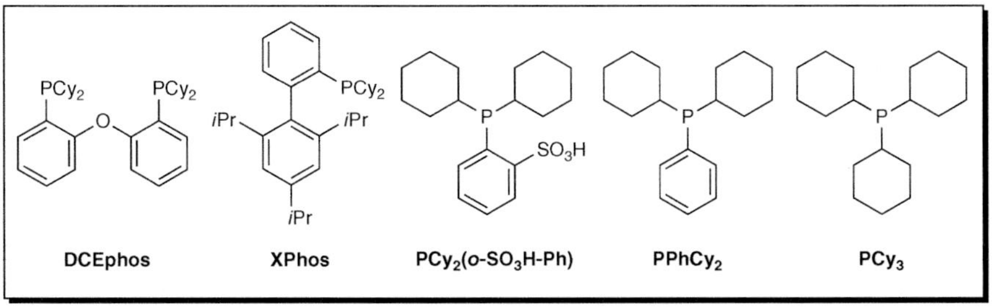

Figure 6.1 Top performing ligands in the palladium-catalyzed cyclization of **1**.

bases, gave significant increases in conversion compared to reactions with inorganic bases alone. This finding led us to investigate tetraalkylammonium carboxylates as stoichiometric bases, and we discovered that both tetrabutylammonium acetate (TBAOAc) and tetramethylammonium acetate (TMAOAc) gave significantly faster reactions compared to more commonly utilized inorganic carboxylates such as NaOAc, KOAc, and KOPiv, presumably because of the increased solubility of tetraalkylammonium carboxylates in the reaction medium.

As detailed herein, three iterations of the intramolecular direct arylation of **1** were developed and successfully implemented on multi-kilogram scale. The first-generation process utilized KOAc in DMAc at 125 °C and rapidly delivered 180 kg of penultimate **2**, while process development work continued in parallel. Capitalizing on the faster reaction rate afforded by tetraalkylammonium carboxylates, a second-generation process was developed that utilized TMAOAc in DMF at 110 °C. The lower reaction temperature led to reduced levels of two key reaction impurities, and this process delivered 280 kg of **2**. However, during the course of the development of the TMAOAc/DMF process, we identified a process robustness risk as a result of the hydrolytic instability of DMF. This finding prompted the development of a third-generation process that employed TMAOAc in DMAc, which mitigated the robustness risk associated with DMF and was successfully validated in the final commercial beclabuvir process.

6.2 KOAc/DMAc Process

Because of rapid project timelines, the team was faced with translating a process with which we had limited experience from gram scale to pilot plant scale in just a few months. The conditions employed in the first-generation KOAc/DMAc process are shown in Scheme 6.4. The cyclization was catalyzed by 5 mol% $Pd(OAc)_2$ and 10 mol% $PCy_3{\cdot}HBF_4$, with 3.5 equiv of KOAc as base in 10 l/kg (volumes or vol, relative to **1**) of DMAc at 125 °C. Following reaction completion, the reaction mixture was cooled to 20 °C and then diluted with toluene and quenched with aqueous KOH. After filtration to remove Pd black, the PCy_3 ligand and other organic-soluble impurities were removed with the toluene layer, and then the aqueous layer was acidified with HCl and the product extracted into methyl *tert*-butyl ether (MTBE). In order to minimize loss of the product to the aqueous layer, as a result of the large amount of DMAc present, an additional back-extraction was performed with MTBE. After two water washes, the organic stream was concentrated and the product was crystallized from EtOH as the monopotassium salt hemi-DMAc solvate (**2**).

Although this process performed well across six batches (93.7–95.5AP in process) and delivered 180 kg of **2** with >99.7AP and 78% average yield, a number of areas were identified for additional development. The high reaction temperature of 125 °C negatively impacted the impurity profile, which in turn impacted the yield. Additionally, it was desirable to streamline the workup operations to avoid the need for multiple solvents as well as to eliminate the back-extraction. We also wanted to re-examine the reaction solvent in light of the European Union (EU) Registration, Evaluation, Authorisation and Restriction of Chemicals (REACH) regulation (*vide infra*). Each of these aspects were addressed in the subsequent development of the second-generation TMAOAc/DMF process.

6.3 TMAOAc/DMF Process

The initial work that led to the identification of TMAOAc as a superior stoichiometric base was conducted in mid-2012. At this time, both DMAc and NMP, two of the most common polar aprotic solvents, were included on the Candidate List

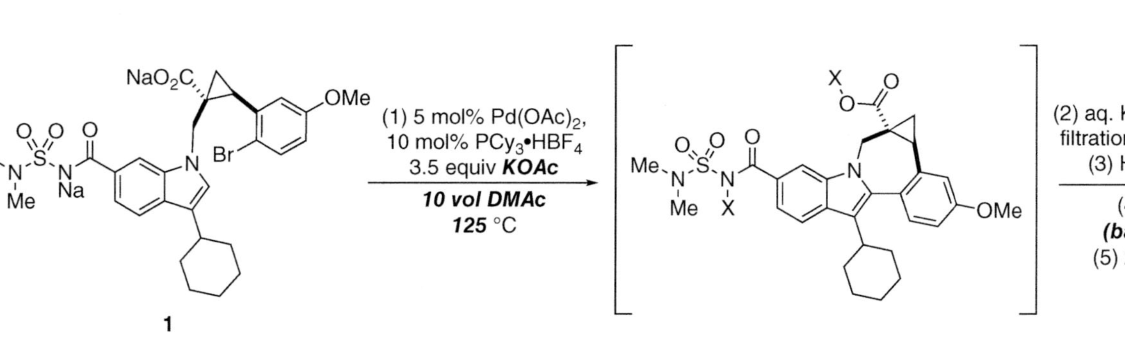

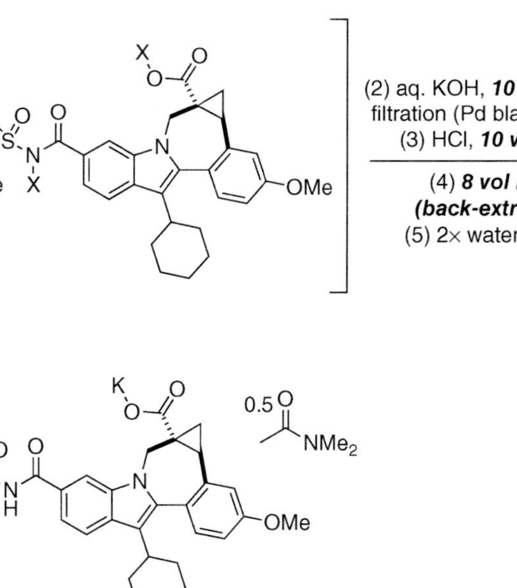

Scheme 6.4 KOAc/DMAc process of final intermediate **2**.

of substances of very high concern for Authorisation under the REACH regulation effective in the EU.[1] As a result of the uncertain implications of the REACH regulation on pharmaceutical manufacturing in the EU, we elected to focus our commercial process development efforts on DMF, which had not been included on the Candidate List at this time.[2]

The conditions employed in the second-generation TMAOAc/DMF process are shown in Scheme 6.5. By taking advantage of the higher solubility of TMAOAc relative to KOAc, we were able to lower the reaction temperature to 110 °C, which resulted in an improved impurity profile compared to reactions conducted at 125 °C. The increased solubility of TMAOAc also enabled the reaction solvent to be reduced to 5 l/kg and the base charge to be lowered to 2.2 equiv, as well as allowing for a slight reduction in the catalyst loading to 4 mol% $Pd(OAc)_2$. As in the KOAc/DMAc process, the reaction mixture was cooled to 20 °C following reaction completion and quenched with aq. KOH. To avoid the use of multiple workup solvents, the organic-soluble impurities were removed by extraction with MTBE instead of toluene as was used previously. The aqueous layer was acidified with HCl and the product extracted into MTBE, but as a result of the decreased charge of the cyclization reaction solvent (5 l/kg of DMF vs 10 l/kg of DMAc in the KOAc/DMAc process), the loss of the product to the aqueous layer was minimal and a back-extraction was no longer needed. Following two water washes, the organic stream was concentrated and the product was crystallized from EtOH as the monopotassium salt hemi-DMAc solvate.

The four main impurities formed in the cyclization reaction are shown in Figure 6.2. In addition to des-bromo **7**, a common impurity type observed in numerous palladium-catalyzed transformations, the reaction also generates an alternative cyclization product resulting from C—C bond formation between the indole moiety and the cyclopropyl ring that was termed the "C–H" by-product (**8**), as well as phosphonium species **9**. Additionally, methyl ester **10**, resulting from alkylation of **2** with a tetramethylammonium cation, is also formed under the conditions that utilize TMAOAc as base. As will be detailed later, under the optimized conditions for the direct arylation reaction, each of these impurities was well controlled and could be effectively purged in the subsequent crystallization. It should also be noted that methyl ester **10** was readily hydrolyzed back to the desired cyclization product (**2**) during workup, and thus, its transitory formation had no impact on yield or quality. The hydrolysis of methyl ester **10** was ensured by an in-process control (IPC).

Initial process development work using TMAOAc as base was conducted with lab-generated batches of aryl bromide **1**, and little difference in the reaction rate or impurity profile was observed for reactions conducted in DMF or DMAc. However, as we started to work with batches of **1** that had been produced in the pilot plant, we began to observe unexpectedly high levels of des-bromo **7** (up to 13AP), for cyclization reactions that were conducted with the TMOAc/DMF conditions. GC analysis of these batches of **1** showed a large, early-eluting peak with *m*/*z* 45 that was identified as dimethylamine (Me_2NH), but its origin

1 https://echa.europa.eu/candidate-list-table

2 Dates of inclusion: NMP – 20 June 2011; DMAc – 19 December 2011; DMF – 19 December 2012.

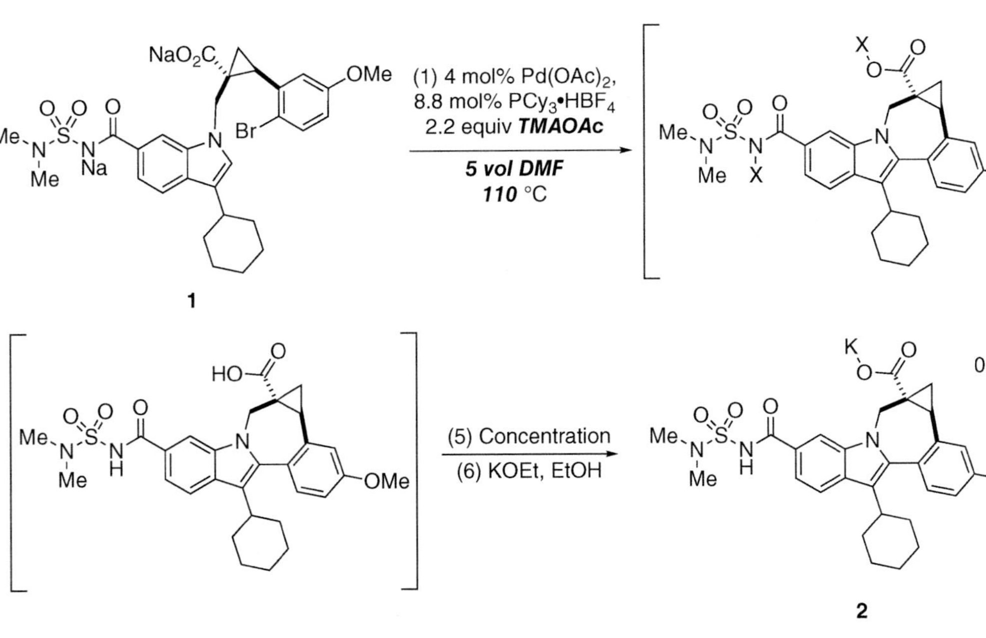

Scheme 6.5 TMAOAc/DMF process of final intermediate **2**.

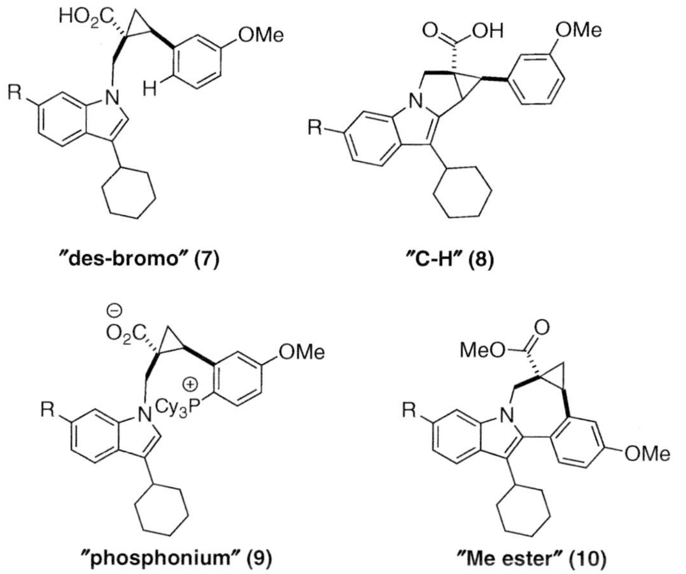

"des-bromo" (7)

"C-H" (8)

"phosphonium" (9)

"Me ester" (10)

Figure 6.2 Main impurities formed in the cyclization of aryl bromide **1** with TMAOAc as base.

was unclear. An early hypothesis for the observation of Me_2NH was that partial hydrolysis of the sulfamoylamide moiety of **1** could occur during the crystallization. Depending on the nature of the sulfamoylamide cleavage, this hydrolysis could lead to the formation of either sodium dimethylsulfamate and the corresponding amide (Path A, Scheme 6.6), or to sodium dimethylsulfamide and the corresponding di-acid (Path B). The sulfamide, but not the sulfamate, was found to decompose during GC analysis to generate Me_2NH; however, the addition of either the sulfamide or the sulfamate to the cyclization reaction mixture did not afford higher levels of 7. Similar spiking experiments demonstrated that Me_2NH had no impact on the cyclization reaction, thus casting doubt on

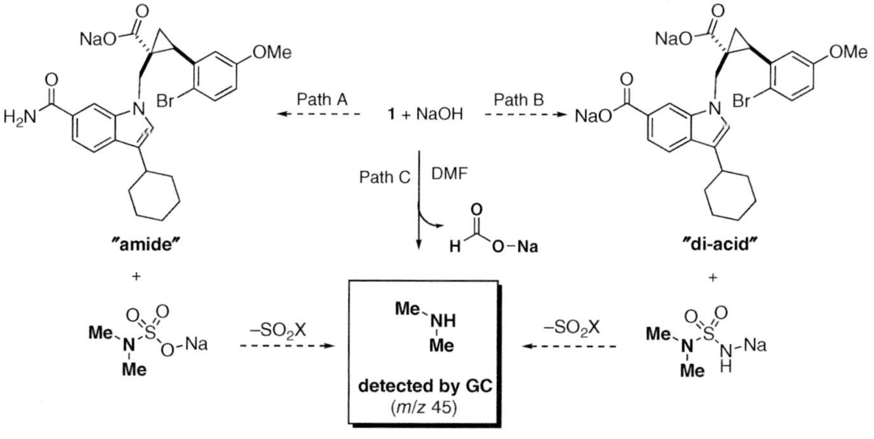

"amide"

"di-acid"

1 + NaOH

Path A

Path B

Path C

DMF

$-SO_2X$

detected by GC (m/z 45)

Scheme 6.6 Potential hydrolyses of aryl bromide **1** during crystallization.

this initial hypothesis for the anomalous des-bromo levels. Although spiking of sodium chloride (0.20 equiv) led to a marked decrease in the reaction rate, it did not lead to elevated levels of des-bromo impurity; furthermore, sodium chloride was considered to be an unlikely contaminant in the isolated batches of **1**.

After extensive investigation, we were ultimately able to trace the root cause of the high des-bromo levels to the presence of residual NaOH, which was used for the final crystallization of the aryl bromide di-sodium salt (**1**), in the isolated material. Rather than a reaction involving the sulfamoylamide moiety, dissolution of the NaOH-containing **1** in DMF led to a rapid hydrolysis reaction of DMF that generated sodium formate and Me_2NH (Path C, Scheme 6.6). The latter species could be detected by GC as noted previously, while both Me_2NH and sodium formate could be observed by ^1H NMR analysis of DMF solutions of **1**. Notably, sodium formate was observed by ^1H NMR in all DMF-dissolved plant batches of **1** in varying amounts (0.02–0.10 molar equivalents) but was not observed in any lab-generated batches of **1**, likely because of higher efficiency of cake washes in the lab in removing residual NaOH, as compared to cake washes in the plant. A titration method was ultimately developed, which indicated that later plant batches of **1** contained as much as 0.28 equiv of NaOH. It is worth noting that when batches of **1** containing high levels of NaOH were re-crystallized or re-slurried in the lab, the reworked batches contained only trace levels of NaOH and performed identically to lab-generated batches in the cyclization reaction.

With an understanding that the key contaminant in **1** was NaOH and that it rapidly reacted with DMF to generate sodium formate, we shifted our focus to understand the impact of formate on the cyclization reaction. Given that sodium formate is a well-known stoichiometric reductant, its presence in the reaction stream was expected to lead to a higher in-process level of the des-bromo impurity **7**. Indeed, when the reaction was run with stoichiometric substitution of TMAOAc with sodium formate, almost exclusive formation of **7** was observed (Scheme 6.7). Similarly, deliberately spiking NaOH into reactions run under TMAOAc/DMF conditions also leads to increased levels of **7** (*vide infra*). Taken together, these findings should serve as a cautionary note for developing transition-metal-catalyzed reactions of aryl halides in DMF in the presence of strong bases, either in the form of intentional additives or in the form of potential contaminants.

Several approaches were explored to mitigate the impact of residual NaOH on the cyclization reaction. Although preliminary laboratory experiments suggested that the addition of 0.5 equiv of acetic acid to the reaction mixture was effective at buffering the residual NaOH contained in plant batches of **1**, this would have required a process change that would necessitate additional development work before it could be implemented on scale. Fortunately, a straightforward solution to partially mitigate the detrimental effect of residual NaOH was quickly realized through a modification of the reaction charge order (Figure 6.3). In the initial charge order, the aryl bromide and DMF solvent were charged first to the reactor, followed by $Pd(OAc)_2$ and $PCy_3 \cdot HBF_4$. Because the NaOH-mediated hydrolysis of DMF is rapid, the majority of the residual NaOH that is introduced by **1** has already been converted to sodium formate by the time the ligand is charged to the reactor. In the extreme case of a long delay between the $Pd(OAc)_2$ charge and

Scheme 6.7 Palladium-catalyzed debromination of **1** with stoichiometric sodium formate.

Figure 6.3 Revised charge order to mitigate the effect of residual NaOH on des-bromo **7**.

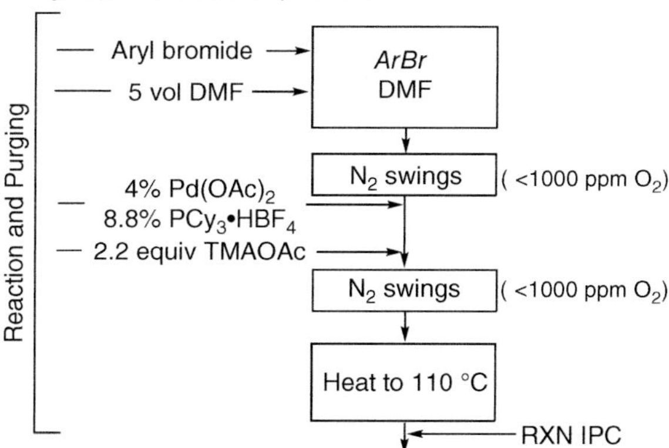

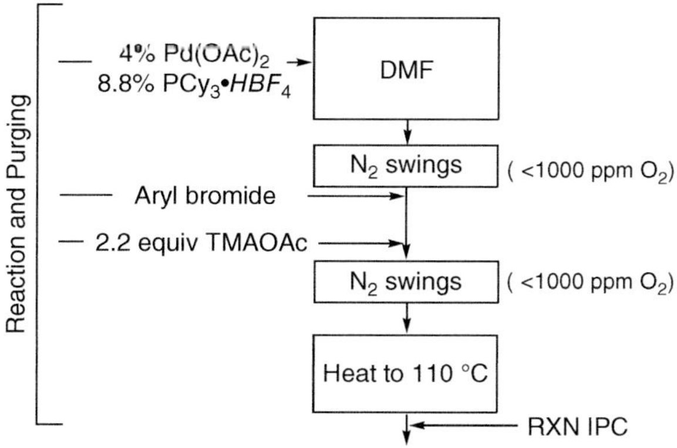

the subsequent $PCy_3{\bullet}HBF_4$ charge, this can result in a sodium formate-mediated reduction of $Pd(OAc)_2$ that generates variable levels Pd black, leading to a darker colored reaction stream and unpredictable reaction profiles. However, modifying the charge order so that the palladium and ligand are added first enabled us to take advantage of tetrafluoroboric acid that is fortuitously introduced by the $PCy_3{\bullet}HBF_4$ ligand salt as a buffer against any NaOH present in **1**. Although this protocol only allows for neutralization of 0.088 equiv of NaOH, charging the palladium and ligand before **1** prevents the formation of Pd black regardless of the amount of NaOH that is present.

With this charge-order modification, the TMAOAc/DMF process was successfully implemented across six batches and delivered 280 kg of penultimate **2** with $>$99.7AP and 83% average yield. Nonetheless, it was clear that the possibility of high levels of residual NaOH in the aryl bromide input, coupled with the facile NaOH-mediated hydrolysis of DMF and detrimental effect of sodium formate on the cyclization reaction, presented a potential risk to process robustness, prompting us to re-evaluate solvent selection for development of the final commercial process. As noted previously, by the end of 2012, three of the most common polar aprotic solvents, DMF, DMAc, and NMP, had all been included on the Candidate List for REACH regulation,[2] thus eliminating the main driver that led to the selection of DMF for the initial TMAOAc process. As depicted in Figure 6.4, the hydrolysis of both DMF and DMAc can be facilitated by hydroxide, but the by-products that are afforded differ; although sodium formate was demonstrated to lead to increased levels of the des-bromo impurity, sodium acetate had no impact on the cyclization reaction.

To further evaluate the impact of residual sodium hydroxide on the cyclization reaction, a NaOH spiking study was performed with both DMF and DMAc. These experiments were conducted using the modified charge order (Figure 6.3) and thus represent a "best-case scenario" for the TMAOAc/DMF process. As depicted in Figure 6.5, spiking NaOH in DMF as the cyclization reaction solvent led to a gradual increase in des-bromo impurity **7**, up to 8.5AP with 0.35 equiv of NaOH. In contrast, at the same NaOH spiking level, the DMAc reaction afforded approximately 2.7AP of **7**, providing a wide safety margin as impurity **7** purges efficiently during crystallization (8.98AP purged at pilot scale to an acceptable level (spec $\leq$0.40AP) in isolated **2**).

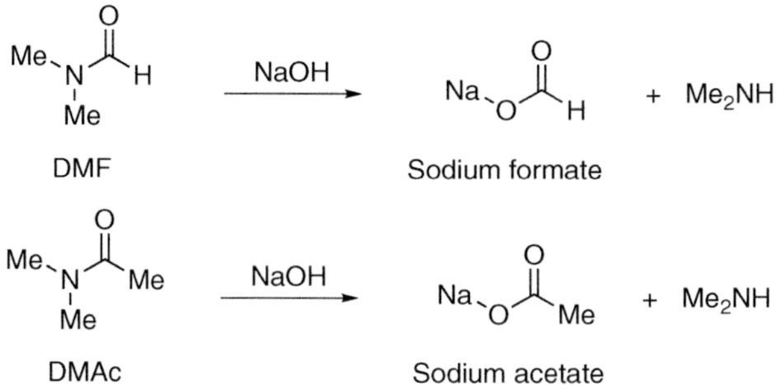

DMF —NaOH→ Sodium formate + Me_2NH

DMAc —NaOH→ Sodium acetate + Me_2NH

Figure 6.4 Hydrolysis of DMF and DMAc with NaOH.

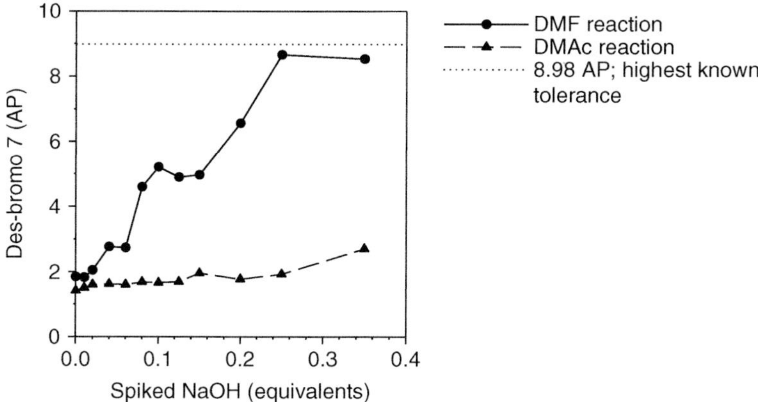

Figure 6.5 Impact of spiked NaOH on the level of **7** in DMF vs DMAc cyclization reactions.

In addition to the potential impact to the in-process impurity profile and isolated yield resulting from NaOH-mediated hydrolysis of DMF, two additional robustness risks are introduced by the use of DMF as the solvent for the cyclization reaction. The first risk results from the fact that it carries through to the reaction quench and presents a competing pathway for consumption of the KOH used for the hydrolysis of methyl ester **10**. Accordingly, with DMF as the reaction solvent, longer age times were needed to reach the hydrolysis IPC completion as compared to reactions conducted in DMAc. The second risk results from the isolation of **2** as the hemi-DMAc solvate, which is facilitated by the back-addition of DMAc to the EtOH crystallization stream. In an effort to streamline the process for cyclization reactions run with TMAOAc/DMF conditions, laboratory development work toward the isolation of a **2**•DMF solvate was evaluated. Based on a limited data set, the target crystallization protocol was modified to include DMF rather than DMAc, which appeared to maintain impurity purging comparable to the DMAc/EtOH system. However, because of processability concerns from product crusting on the reactor walls, further efforts to isolate a **2**•DMF solvate were suspended. Thus, integrating the DMF cyclization reaction process streams into the target DMAc/EtOH crystallization relies on the workup conditions to remove DMF from the process stream, which is desirable to avoid the isolation of **2** as a mixture of DMF and DMAc solvates.

6.4 TMAOAc/DMAc Process

Throughout the scale-up of the KOAc/DMAc and TMAOAc/DMF processes, process development continued with the ultimate goal to increase both the efficiency and ruggedness of the original synthetic route first implemented on scale and to define the final commercial process. Based on the considerations outlined in the previous sections, DMAc was selected as the reaction solvent for the commercial manufacturing direct arylation process to convert aryl bromide **1** to the final intermediate **2**. An overview of this process is shown in Scheme 6.8 and a

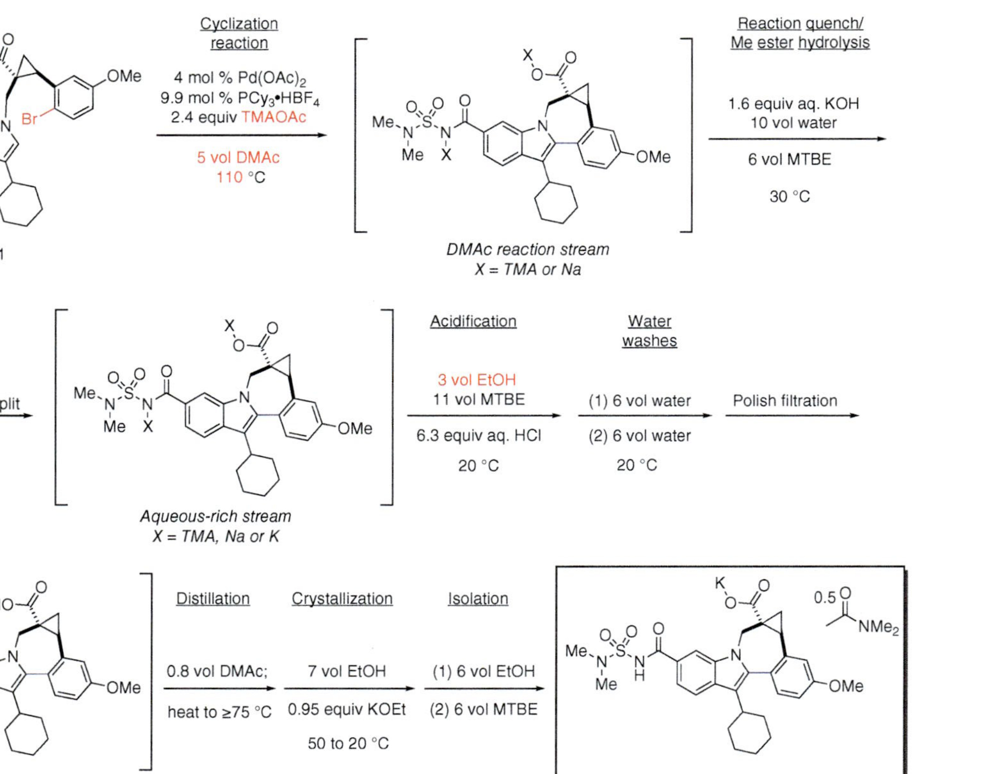

Scheme 6.8 Commercial process for the synthesis of final intermediate **2** from aryl bromide **1**.

detailed description of each unit operation are outlined in the following sections. As will be discussed, the final TMAOAc/DMAc process, which also incorporated modified quench, workup, and crystallization conditions, worked flawlessly in all 18 batches of the internal pilot plant campaign as well as 4 batches at the commercial manufacturing site.

As shown in Scheme 6.8, $Pd(OAc)_2$, PCy_3•HBF_4, **1**, and TMAOAc are mixed in DMAc and the mixture is heated and aged to form the tetramethylammonium or sodium salt of **2** via intramolecular direct arylation. The reaction is quenched with water followed by the addition of an aq. KOH solution, which mediates the hydrolysis of methyl ester **10** to **2**. MTBE is added, the resulting mixture is filtered, and the phases are separated to remove the spent palladium catalyst. To the aqueous solution, EtOH, MTBE, and aq. hydrochloric acid are added and the phases are separated. The organic solution containing the free acid of **2** is washed with water twice and filtered to remove inorganic salts and insolubles. DMAc is added and a facile solvent exchange from MTBE to DMAc is performed. EtOH and a solution of potassium ethoxide in EtOH are charged to crystallize the final intermediate **2** as the monopotassium salt, hemi-DMAc solvate. The resulting slurry is cooled and aged, then **2** is isolated, washed with EtOH followed by MTBE, and dried. The typical overall yield for this step ranged from 75 to 91M%.

6.4.1 Cyclization Reaction

The operational order for the cyclization reaction is shown in Figure 6.6. As noted previously, this charging order should be maintained for maximum process robustness. To an inert reactor, the initial charges of DMAc, $Pd(OAc)_2$, PCy_3•HBF_4, and aryl bromide **1** were added in this order to allow for facile precatalyst formation and to mitigate risk due to residual NaOH that may be present in **1**. TMAOAc was then charged followed by an inertion to purge headspace oxygen before heating of the reaction mixture. Here, it should be emphasized that, like a majority of palladium-catalyzed transformations, the cyclization reaction is sensitive to O_2 and precautions should be taken to ensure that the reaction occurs under an inert environment. O_2 facilitates the oxidation

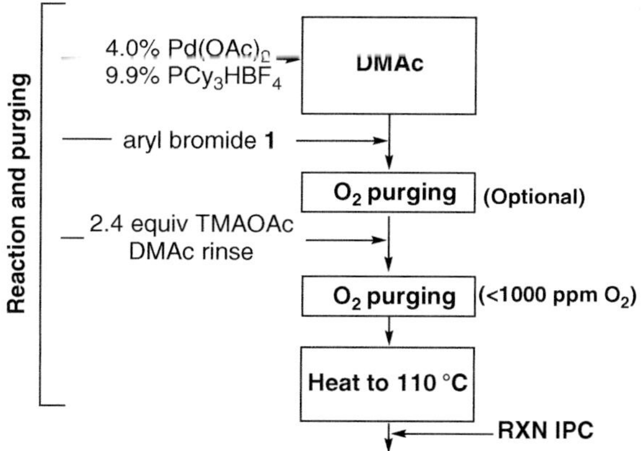

Figure 6.6 Reagent charge order for the cyclization reaction of the commercial process.

of the PCy_3 ligand to the corresponding phosphine oxide, which impacts the stabilization of the palladium catalyst and can lead to reaction stalling. Although it was demonstrated that a stalled reaction could be driven to completion by the addition of catalyst kicker charges, reaction stalling was not observed when appropriate controls were taken to ensure that the headspace O_2 content was ≤1000 ppm, and in all of the pilot plant campaigns, kicker charges were never required. In fact, in order to study the impact of adding kicker charges, cyclization reactions had to be designed to stall by purposely using nitrogen doped with high levels of O_2. It should also be noted that in order to effectively limit the amount of O_2 in the reactor headspace during the reaction, the level of residual O_2 in the nitrogen supply was controlled by specifications to ≤10 ppm (≥99.999% N_2), which was easily achievable using the typical nitrogen boil-off from a liquid N_2 supply.

Some additional commentary regarding reactor inertion is warranted. Because the reaction vessels used in most pilot plant and commercial manufacturing settings are well sealed, a variety of inertion protocols are generally effective for removing oxygen from the headspace of these reactors, including sub-surface N_2 sparging, N_2 pressurization/depressurization, vacuum/N_2 backfill, or any combination of these methods. Indeed, throughout the various pilot plant campaigns for this project, we ultimately employed all of these degassing methods (either alone or in various combinations) successfully in different batches conducted on scale and were able to routinely achieve ≤1000 ppm headspace O_2 levels in every case. On the other hand, many of the reaction vessels used in a typical laboratory setting, including process chemistry and engineering labs, medicinal chemistry labs, as well as academic labs, are not nearly as well sealed. It is thus much more likely that air may leak into a laboratory vessel that is at or below ambient pressure, and consequently, vacuum/N_2 backfill inertion protocols are often highly inconsistent in the laboratory because they depend on both the setup of the equipment and the skill of the scientist. For this direct arylation reaction, as well as for numerous other catalytic transformations that have been developed at Bristol-Myers Squibb, we have consistently found that sub-surface N_2 sparging is a superior method of degassing at laboratory scale. It is also important to note that the *concentration* of O_2 and the *total amount* (i.e. moles) of O_2 in the headspace of a reaction vessel are related, yet distinct, parameters; at equivalent *concentrations* of headspace O_2, a reaction vessel that is 20% full will contain four times the amount of *total* headspace O_2 as a vessel that is 80% full. In general, oxygen-sensitive catalytic transformations tend to perform better in a laboratory when they are run in vessels with smaller headspace because of both the total amount of O_2 and the molar ratio of O_2/catalyst being lower compared to vessels with larger headspace [5].

Figure 6.7 describes the relationship between oxygen content and in-process product purity, as well as residual palladium in the isolated product. At higher O_2 concentrations, the in-process purity of **2** starts to decrease because of increased formation of impurities **7**, **8**, and **9**, which in turn impacts the isolated yield. Another detrimental effect of increased levels of O_2 in the reactor headspace is rising levels of palladium found in isolated product. To minimize the undesired effects, an upper limit of O_2 was set to ≤1000 ppm. At pilot scale, the typical

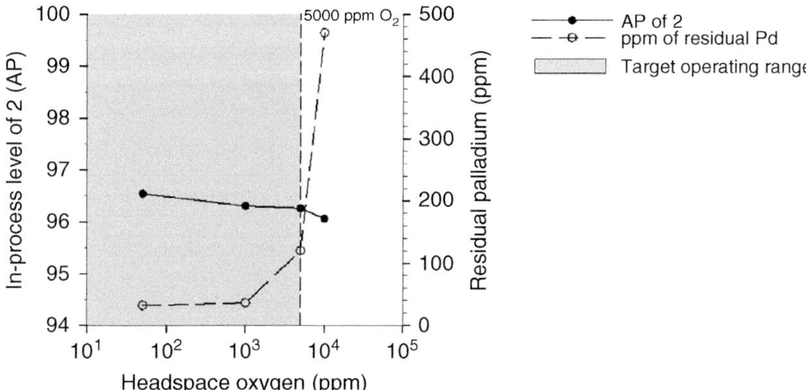

Figure 6.7 Impact of oxygen on impurity profile and residual palladium content in isolated **2**.

observed level of the in-process impurities were <4AP (total) for an O_2 level of approximately 300 ppm in the headspace. All impurities at this level purged efficiently in the process crystallization affording **2** with acceptable quality.

Setting the appropriate IPC for reaction conversion is always an important consideration. Conversion must be high enough to ensure good yield and minimize potential contamination from unreacted starting material. However, it must also be readily attainable to avoid the needs for unnecessary kicker charges. The IPC for the cyclization reaction completion was established based on tolerance data with a criterion of ≤2.0 HPLC relative area percent (RAP) of **1** to **2**. At target conditions, the typical level of **1** at reaction completion was <0.5 RAP. Spiking studies demonstrated that up to 5.4 RAP **1** can be tolerated and purged to the crystallization, with no impact on the quality of **2**. As high conversion was always achieved, limited laboratory development went into the optimization of the kicker reagent charge amounts; however, an important observation was that **2** isolated from these kicker-charged cyclization reactions consistently had higher levels of residual Pd (450–650 ppm). This posed a potential risk to API quality, as it was demonstrated on laboratory scale that **2** with 650 ppm of residual Pd taken through the API process afforded 22 ppm residual Pd in the isolated API (above the 20 ppm Pd target). To reduce the amount of residual Pd in **2** to an acceptable level for reactions that might require a kicker charge, an optional reprocess of **2** was developed and demonstrated on laboratory scale. It was shown to effectively reduce the residual Pd level by approximately 70%. As mentioned previously, the cyclization process proved to be very robust and no kicker charges were required during large-scale processing, so the demonstration of the reprocess was merely to support a contingency strategy.

In order to understand the factors that affect the formation of the four key cyclization reaction impurities (**7**, **8**, **9**, and **10**, see Figure 6.2) and determine the optimal values for the center-point cyclization reaction conditions, a statistically designed experiment was conducted and multivariate analysis was performed on the results. The goals of this design of experiments (DoEs) study were to (i) identify the important factors impacting the formation of each impurity and the

Table 6.1 Factors and ranges used in the cyclization reaction DoE.

Factor	Low	Center	High
$Pd(OAc)_2$ charge (mol%)	2.0	3.0	6.0
Ligand-to-metal molar ratio (L/M)	2.05	2.30	3.00
TMAOAc charge (equiv)	1.70	2.10	3.00
Water content (equiv)	0.55	1.50	3.30
DMAc volume (l/kg)	4.0	5.0	6.0
Reaction temperature (°C)	95	105	120

overall reaction kinetics, (ii) determine the relative risks for formation of each impurity, and (iii) identify process parameter ranges and verify the ranges at pilot scale. The DoE was run to examine six continuous factors that affect the formation of the four key impurities. The experimental design was a fractional factorial with four replicate center-point runs, using the ranges and levels as shown in Table 6.1.

The DoE reactions were conducted on 0.45 mmol scale and assembled in a N_2-filled glove box and sampled using a Freeslate CM3 automation system as described by Selekman et al. [4b]. Complete details of the individual experimental conditions, along with the results for each impurity (all in-process), are shown in Table 6.2. The factor ranges selected allowed for the impurity levels to remain below the highest levels shown to purge in the crystallization, thus ensuring that all batches of product would be of acceptable quality after crystallization.

The analysis of the experiments was performed using JMP®, a statistical software program, to calculate the p-values for the main effect of each factor. The p-value was used to determine if a reaction parameter is statistically significant ($p < 0.05$), marginally significant ($p < 0.10$), or not statistically significant ($p > 0.10$). Table 6.3 summarizes the calculated p-values and directionality of each factor with respect to the measured responses (**7**, **8**, **9**, and **10**). The directionality of response is denoted as positive or negative to indicate the impact of the response as the factor is increased. For instance, the response of L/M molar ratio is positive for phosphonium **9**, indicating that increasing the L/M molar ratio will increase the level of **9**. Conversely, the response of L/M molar ratio is negative for the C–H by-product (**8**), indicating that increasing the L/M molar ratio will decrease the level of **8**.

Within temperature range of 95–125 °C explored in this DoE, all reactions reached >98% conversion of **1** in <15 hours (see Table 6.2). Additionally, the levels of the four in-process impurities (**7**, **8**, **9**, and **10**) were within purgeable limits for all reactions, and thus none of the six factors were quality impacting, only yield impacting. The three factors that had the largest impact on the levels of **7**, **8**, and **9** were L/M molar ratio, TMAOAc equivalents, and reaction temperature. In addition to these three factors, $Pd(OAc)_2$ loading also had a significant (although indirect) impact on the phosphonium impurity **9**. Table 6.4 summarizes the impact of the statistically significant factors on impurity formation during the cyclization reaction.

Table 6.2 Experiments and results for the DoE of the cyclization of aryl bromide **1** to **2**.

	Factors							In-process cyclization impurities			
Reaction ID	$Pd(OAc)_2$ (mol%)	L/M	TMAOAc (equiv)	DMAc (l/kg)	Water (equiv)	Temperature (°C)	Time[a] (h)	7 (AP)	8 (AP)	9 (AP)	10[b] (AP)
Target conditions	4.0	2.48	2.40	5.0	—	110	2.50	1.36	0.98	0.36	0.46
Highest amount tolerated in the crystallization[c]								8.98	3.80	3.84	(IPC)
A1	6.0	2.05	2.99	4.0	3.40	95	1.59	1.51	1.91	0.04	0.05
A2	6.0	2.05	1.70	6.0	3.30	95	4.60	2.19	0.62	0.04	0.16
A3	2.0	3.00	1.70	4.0	3.28	95	12.50	2.23	0.47	0.76	0.38
A4	2.0	2.05	1.71	4.0	0.52	95	8.88	1.70	0.75	0.06	0.42
A5	6.0	3.00	3.02	4.0	0.64	95	2.99	0.99	0.57	0.88	0.14
A6	2.0	2.05	3.01	6.0	0.65	95	3.03	0.77	1.94	0.05	0.25
B1	2.0	3.00	3.00	6.0	3.41	95	4.68	1.38	1.07	0.45	0.24
B2	6.0	3.00	1.70	6.0	0.54	95	9.85	1.63	0.50	2.04	0.55
B3	3.0	2.30	2.11	5.0	1.52	105	2.33	1.30	0.95	0.14	0.38
B4	3.0	2.30	2.11	5.0	1.52	105	2.22	1.27	0.97	0.15	0.45
B5	3.0	2.30	2.10	5.0	1.52	105	2.11	1.27	1.00	0.16	0.45
B6	3.0	2.30	2.12	5.0	1.52	105	2.13	1.23	0.99	0.17	0.44
C5	2.0	2.05	1.72	6.0	3.30	120	1.44	1.70	1.51	0.04	0.79
C6	2.0	3.00	1.71	6.0	0.54	120	1.99	1.44	1.08	0.77	1.77
D1	2.0	2.05	3.02	4.0	3.39	120	0.50	1.31	3.01	0.06	0.96
D2	6.0	3.00	1.72	4.0	3.28	120	1.26	3.05	0.71	3.07	0.52
D3	6.0	3.00	3.02	6.0	3.41	120	0.40	1.07	1.18	1.09	1.22
D4	6.0	2.05	3.07	6.0	0.65	120	0.36	1.22	3.37	0.06	0.98
D5	6.0	2.05	1.70	4.0	0.52	120	0.68	2.49	1.09	0.08	0.58
D6	2.0	3.00	3.00	4.0	0.63	120	0.67	0.88	1.86	0.57	1.32

a) Time to reach 98% conversion.
b) IPC controlled during the hydrolysis.
c) Highest amount of each impurity in the precrystallization stream (observed in-process or spiked) that purged to acceptable levels in the isolated **2**.

A number of important learnings were obtained from the results of the DoE. Temperature is the most straightforward as all four impurities are increased at high reaction temperature, particularly **8** and **10**. At low temperature, impurity levels are improved; however, the rate of the reaction is decreased. Although impurities **7**, **8**, and **9** can no longer form once aryl bromide **1** is fully consumed, methyl ester **10** is unique in this respect as it will continue to grow if the reaction mixture is held at the reaction temperature beyond the target time of 2.5 hours (Figure 6.8). However, because this impurity is hydrolyzed back to product (**2**) during the workup, this does not present a risk to either quality or yield.

Table 6.3 Fractional factorial DoE analysis for the cyclization reaction DoE.

Response	Des-bromo (7)		C–H (8)		Phosphonium (9)		Me ester (10)	
Factor	Response	*p*-value	Response	*p*-value	Response	*p*-value	Response	*p*-value
$Pd(OAc)_2$ charge	+	0.0151[a)]	–	0.2367	+	0.0053[a)]	–	0.0228[a)]
L/M molar ratio	+	0.8644	–	0.0010[a)]	+	0.0002[a)]	+	0.0077[a)]
TMAOAc charge	–	0.0008[a)]	+	0.0003[a)]	–	0.0508[b)]	–	0.5463
Water content[c)]	+	0.0106[a)]	–	0.7398	+	0.3517	–	0.0335[b)]
DMAc volume	–	0.0320[b)]	+	0.4876	–	0.4249	+	0.0325[b)]
Reaction temperature	+	0.3243	+	0.0014[a)]	+	0.2269	+	<0.0001[a)]

a) Response considered statistically significant based on a p-value criterion <0.05.
b) Response considered marginally significant based on a p-value criterion <0.10.
c) Controlled by input specifications.

Table 6.4 Summary of the impact of cyclization process parameters on impurity formation.

Parameter variation from target		Reaction temperature		Ligand-to-metal molar ratio		TMAOAc charge		
		Low	High	Low	High	Low	High	Other
Impurity level vs target reaction	7 (des-Br)	Low	High	(Minimal impact)		High	Low	High water content will lead to higher des-Br **7**
	8 (C–H)	Low	High	High	Low	Low	High	—
	9 (phosphonium)	Low	High	Low	High	High	Low	Higher $Pd(OAc)_2$ loading with the same L/M molar ratio will lead to higher phosphonium **9**
	10 (Me ester)	Low	High	Low	High	Low	High	**10** is hydrolyzed to **2** during the hydrolysis and is IPC controlled

Note: A high L/M molar ratio will result from either a high $PCy_3 \bullet HBF_4$ charge or a low $Pd(OAc)_2$ charge; a low L/M ratio will result from either a low $PCy_3 \bullet HBF_4$ charge or a high $Pd(OAc)_2$ charge.

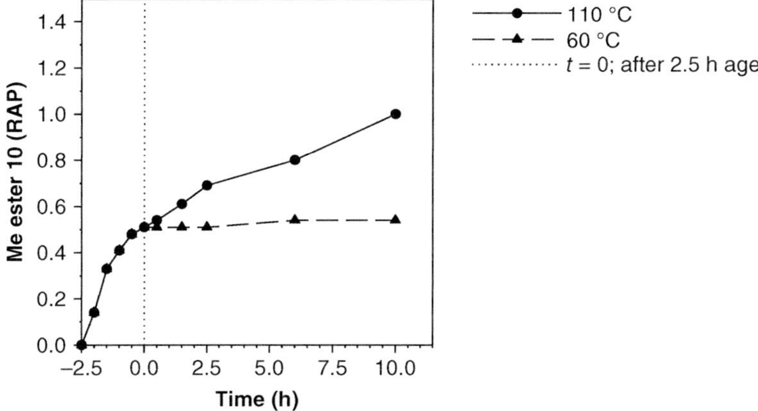

Figure 6.8 Impact of temperature and hold time post-reaction age on the formation of **10**.

The impact of ligand-to-metal molar ratio as well as quantity of TMAOAc offered more mixed results. However, the data could be used to select center-point conditions and set the parameter ranges. The values for the center-point cyclization reaction parameters were selected as 4 mol% $Pd(OAc)_2$, 9.9 mol% $PCy_3{\bullet}HBF_4$ (2.48 L/M), and 2.4 equiv of TMAOAc in 5 l/kg of DMAc at 110 °C. It is important to note that these values were not necessarily chosen based on what would give the absolute fastest reaction rate or lowest overall impurity levels but rather to afford the maximum process robustness. That is, the center-point reaction conditions were selected such that a modest deviation in either direction from any of these values would not have a significant impact on the reaction rate or impurity profile, thereby establishing a wide operating space for the process.

The DoE results were also used to guide parameter selection during an internal pilot plant campaign. Because none of the six cyclization parameters were quality impacting within the design space, emphasis was placed on identifying conditions that would generate the highest acceptable levels of each in-process impurity in order to demonstrate the robustness of the workup and crystallization process. Additionally, with guidance from our manufacturing colleagues, the ranges were tightened for $Pd(OAc)_2$ (3.6–4.4 mol%) and $PCy_3{\bullet}HBF_4$ (9.0–10.8 mol%) relative to the ranges of the multivariate experiments. This was done as a matter of practical convenience in order to ensure that the L/M molar ratio in the plant would be within the proven acceptable range (PAR) of 2.05–3.00 at both potential extremes of either charge. Maintaining the L/M ratio range obviates the need to perform any additional calculations during manufacturing. However, it is important to note that the cyclization reaction can be run at lower than 4.0 mol% catalyst loading provided that the charges of both $Pd(OAc)_2$ and $PCy_3{\bullet}HBF_4$ are adjusted accordingly to maintain a target L/M molar ratio of 2.48 (*vide infra*).

Based on the results of the DoE and subsequent modeling using JMP, the conditions shown in Table 6.5 were selected to generate acceptably high in-process levels of the indicated impurities on scale. The in-process levels of each impurity that were observed during the pilot plant campaign are shown for each set of

Table 6.5 Verification data for cyclization process parameters.

Batch number	$Pd(OAc)_2$ (mol%)	PCy_3HBF_4 (mol%)	L/M ratio	TMAOAc (equiv)	Temperature (°C)	2	7 (≤0.40)	8 (≤0.15)	10 (≤0.20)	9 (≤0.40)
PAR	3.6–4.4	9.0–10.8	2.04–3.00	2.0–2.8	95–125	In-process impurity level (AP)				
						Isolated product quality (wt%)				
1										
(Target)	4.0	9.9	2.48	2.4	110	95.55	1.36	0.98	0.46	0.36
						99.93[a]	**<0.05**	**<0.05**	**<0.05**	**<0.05**
2	3.6	10.8	3.00	2.8	125	92.15	1.38	1.79	2.39	1.12
						100.00	**<0.05**	**<0.05**	**<0.05**	**<0.05**
3	3.6	10.8	3.00	2.0	125	94.05	2.10	0.78	0.79	1.50
						99.71	**0.10**	—	—	**0.19**
4	4.4	9.0	2.05	2.8	125	92.89	1.28	3.05	1.31	0.09
						99.93[a]	**<0.05**	**<0.05**	**<0.05**	**<0.05**

The bold values represent isolated product quality (wt%).

a) Daughter impurity detected at 0.07 wt%, which is a result of an input impurity.

conditions. As expected from the DoE experiments, none of the conditions gave rise to levels of **7**, **8**, **9**, or **10** that were outside of the purgeable limits, and none of these parameter variations had an impact on the quality of the isolated final intermediate **2**. In addition, the pilot-scale batches of **2** were further converted to beclabuvir drug substance that met all specifications.

In an effort to reduce the amount of catalyst used in the cyclization reaction, a Pd loading study was performed at laboratory scale (**1**: 0.5 mmol). All other cyclization reaction parameters were kept at target conditions: 2.48 L/M ratio, 2.4 equiv of TMAOAc in 5.0 l/kg DMAc at 110 °C. Table 6.6 summarizes the study results; the IPC specification for the cyclization reaction was ≤2.0 RAP **1** to **2** and in-process HPLC sampling was performed for a maximum duration of 18 hours. In this study, reactions with 0.4–4.0 mol% Pd met the IPC criterion within 10 hours, and the in-process impurity profiles of these reactions were consistent with typical impurity levels seen on pilot scale. As expected, the Me ester impurity **10** level increases with the reaction time at 110 °C.

At pilot scale, the $Pd(OAc)_2$ charge was successfully reduced to 3.0 mol% while maintaining target conditions of all other cyclization reaction parameters (2.48 L/M ratio, 2.4 equiv of TMAOAc in 5.0 l/kg DMAc at 110 °C), including reaction age time (2.0 hours). The cyclization reaction afforded a consistent impurity profile leading to isolated **2** of high quality (Table 6.7).

6.4.2 Mechanistic Understanding of the Cyclization Reaction and Impurity Formation

Based on all of the data collected, the intramolecular cyclization of aryl bromide **1** to form final intermediate **2** is consistent with the general mechanistic

Table 6.6 Experiments and results for the palladium loading study.

ID	$Pd(OAc)_2$ (mol%)	Time to meet IPC spec (h)	In-process impurity profile (area percent)				
			2	**1**	**7**/**9**[a]	**8**	**10**
Control	4.0	2.0	96.49	0.59	1.49	0.82	0.46
A2	3.0	2.0	96.86	0.29	1.34	0.91	0.56
A3	2.5	2.1	96.86	0.20	1.31	0.95	0.61
A4	2.0	2.2	96.66	0.29	1.27	1.01	0.77
A5	1.5	3.0	96.31	0.35	1.31	1.01	1.03
A6	1.25	3.2	96.42	0.20	1.29	1.00	1.09
B1	1.0	5.3	95.97	0.14	1.48	1.13	1.29
B2	0.8	5.3	94.57	0.09	1.89	1.23	2.22
B3	0.6	5.3	95.35	0.08	1.52	0.95	2.09
B4	0.4	10.4	93.25	0.11	1.90	1.02	3.72
B5	0.2	—	72.49	17.95	2.42	1.08	6.02
B6	0.1	—	37.12	54.23	2.06	0.73	5.86

a) The HPLC column used for this analysis did not separate **7** and **9**.

Table 6.7 Results for the reduced Pd loading at pilot scale.

Conditions	$Pd(OAc)_2$ (mol%)	2	7 (spec ≤ 0.40)	8 (spec ≤ 0.15)	10 (spec ≤ 0.20)	9 (spec ≤ 0.40)
		In-process impurity level (AP) / Isolated product quality (wt%)				
Target	4.0	95.55	1.36	0.98	0.46	0.36
		99.93[a]	**<0.05**	**<0.05**	**<0.05**	**<0.05**
Reduced Pd charge	3.0	95.99	1.26	1.18	0.67	0.32
		100.00	**<0.05**	**<0.05**	**<0.05**	**<0.05**

The bold values represent isolated product quality (wt%).
a) A daughter impurity is present in **2** that results from an input impurity in **1**.

pathway previously described for the Pd-catalyzed functionalization of aromatic, heteroaromatic, and sp^3 C—H bonds with aryl halide coupling partners [6]. Additionally, detailed kinetic and mechanistic studies of the related cyclization of aryl bromide ethyl ester **4** have been conducted (Simmons, E.M., Hang, C., Ramirez, A., Chan, C., Hsiao, Y., and DelMonte, A.J., *manuscript in preparation*), which support the proposed mechanism for the cyclization **1**. This catalytic cycle is depicted in Figure 6.9 and is described in the following steps:

(1) The catalytic cycle is initiated by the reduction of the *in situ* generated, bis-ligated PCy_3 adduct of the Pd^{II} precatalyst, $Pd(OAc)_2$ to a mono-ligated Pd^0 species. This reduction process is mediated by tricyclohexylphosphine (PCy_3) and water in the presence of TMAOAc and generates 1 equiv of tricyclohexylphosphine oxide relative to Pd [7].
(2) Oxidative addition to the Ar—Br bond of **1** to the mono-ligated (Cy_3P)-Pd^0 complex (L_1-Pd^0) leads to the mono-ligated oxidative addition complex (L_1-Pd(Ar)Br), which can undergo exchange of Br for OAc, mediated by TMAOAc, to generate the mono-ligated complex (L_1-Pd(Ar)(OAc).
(3) In the presence of excess PCy_3, the mono-ligated complexes L_1-Pd(Ar)Br and L_1-Pd(Ar)(OAc) are in equilibrium with bis-ligated complexes L_2-Pd(Ar)Br and L_2-Pd(Ar)(OAc) that lie off the main catalytic cycle but can re-enter the cycle via reversible dissociation of PCy_3.
(4) The mono-ligated complex L_1-Pd(Ar)OAc undergoes intramolecular C–H cleavage of the indole C2 proton, which after subsequent reductive elimination leads to formation of the final intermediate and regeneration of L_1-Pd^0.

As noted previously, four in-process impurities are formed during the cyclization step (see Figure 6.2): **7** (des-Br), **8** (C–H), **9** (phosphonium or Ar-PCy_3), and **10** (Me ester). With the exception of methyl ester **10**, these impurities are formed via shunt pathways off of the main catalytic cycle. The phosphonium (**9**) is believed to be formed from the bis-ligated catalyst resting state (L_2-Pd(Ar)X) via C—P bond-forming reductive elimination. The des-Br (**7**) and C–H (**8**) impurities are formed following an alternative intramolecular C–H cleavage of the cyclopropane C—H bond that is *syn* to the aryl group. The resulting palladacycle

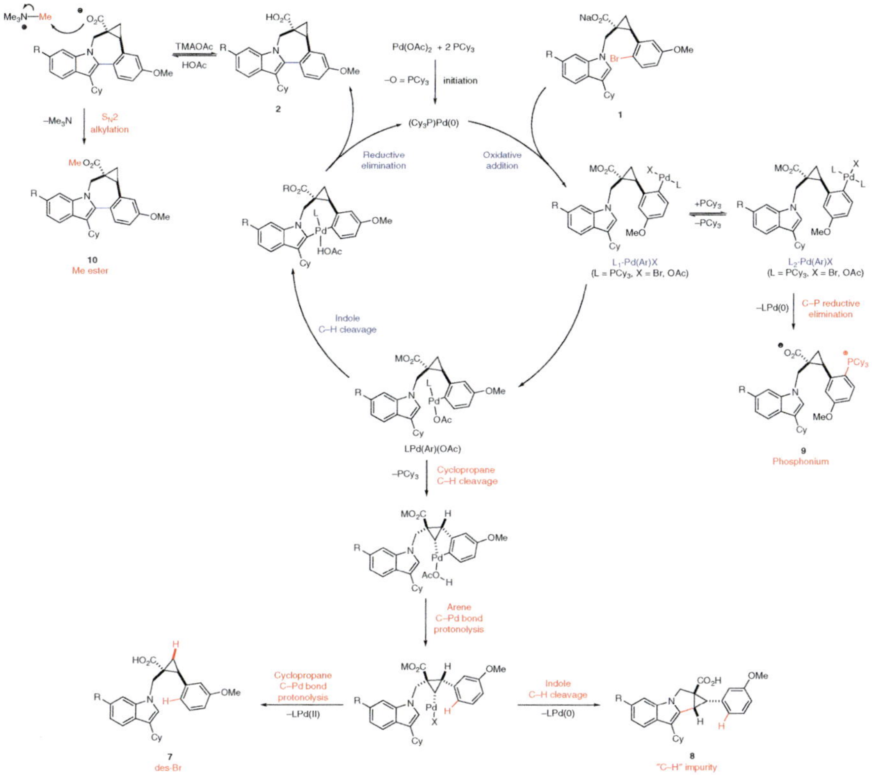

Figure 6.9 Proposed catalytic cycle for the cyclization of **1** to **2** and formation of impurities **7–10**.

then undergoes C–Pd protonolysis of the aryl-Pd bond to generate a cyclopropylpalladium species that can either undergo a second C—Pd bond protonolysis to generate des-Br **7** or indole C–H cleavage and reductive elimination to generate **8**. Me ester **10** is formed by an uncatalyzed S_N2 alkylation of the final intermediate **2** with the tetramethylammonium cation of the TMAOAc base.

To probe the relationship between the des-Br (**7**) and C–H (**8**) impurities and the potential role of water in their formation, a series of water spiking reactions were conducted. As shown in Table 6.8, increasing amounts of water led to a gradual increase in the amount of des-Br impurity (**7**), as well as a corresponding gradual decrease in the level of the C–H impurity (**8**); however, the *total* amount of **7** + **8** was unchanged. These results are consistent with the des-Br (**7**) and C–H (**8**) arising from one or more common intermediates (presumably the metallacycle and cyclopropylpalladium species at the bottom of Figure 6.9), with water impacting the eventual partition to either **7** or **8**. The consistent level of **7** + **8** observed in these experiments also implies that the amount of water present in the reaction mixture does not impact the rate of the initial irreversible step that leads to **7** and **8**, presumably either the cyclopropane C–H cleavage or the protonolysis of the aryl-Pd bond. Consistent with the proposed common pathway for the formation of **7** and **8**, solvent proton incorporation into the cyclopropane

Table 6.8 Effect of water on relative amounts of Des-Br **7** and C–H **8**.

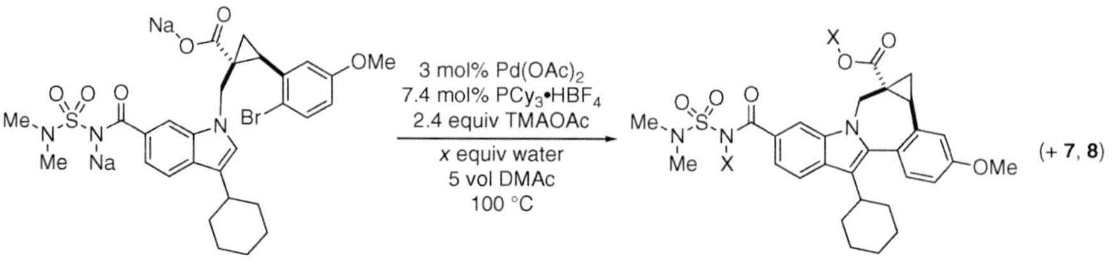

Reaction	Water (equiv)		In-process impurity profile (area percent)			
ID	Added	Total[a)]	2	7	8	7 + 8
1	—	0.6	97.04	1.21	1.07	2.29
2	1.0	1.6	97.07	1.27	1.03	2.30
3	2.0	2.6	97.10	1.32	0.98	2.30
4	3.0	3.6	97.02	1.41	0.92	2.33
5	4.0	4.6	97.22	1.41	0.85	2.26
6	5.0	5.6	97.24	1.49	0.80	2.30

a) Includes water introduced by **1** and TMAOAc.

moiety of the des-Br by-product was observed when the cyclization reaction was conducted with an aryl bromide substrate deuterated at both of the cyclopropane methylene positions. Similarly, deuterium incorporation into the aryl group of the C–H impurity was observed when the reaction was conducted in the presence of D_2O, ruling out alternative processes in which the cyclopropane C–H is directly transferred to the arene moiety.

In addition to studying the relationship between water and the des-Br (**7**) and C–H (**8**) impurities, we also investigated the impact of L/M ratio on the level of phosphonium impurity **9** (Table 6.9). At a L/M ratio of 2.05, only a trace amount of phosphonium **9** was observed, but there was a steady increase in **9** as the L/M ratio increased. Both the overall trend with respect to L/M ratio as well as the near absence of phosphonium **9** at L/M 2.05 suggest that this impurity is formed from a bis-ligated catalyst species (either L_2-Pd(Ar)Br or L_2-Pd(Ar)(OAc)). As expected based on the results of the DoE, increasing the L/M ratio led to a decrease in the level of the C–H impurity (**8**) but had a negligible impact on the level of des-Br **7**.

6.4.3 Hydrolysis and Workup

Upon completion of the cyclization reaction, the reaction mixture was cooled to 15 °C and quenched by the addition of water (10.0 l/kg), followed by the addition of 45 wt% aq. KOH (1.6 equiv). The addition of the water for the reaction quench is exothermic, with $T_{ad} = +22$ °C. In order to ensure that the methyl ester **10** was hydrolyzed to the final intermediate **2**, the mixture was aged for at least one hour at 30 °C, before a sample was taken and analyzed for hydrolysis completion (IPC: **10** $\leq$0.5 RAP relative to **2**). At target conditions, the typical level of **10** at

Table 6.9 Effect of L/M ratio on amount of phosphonium **9**.

3 mol % $Pd(OAc)_2$
x mol % PCy_3•HBF_4
2.4 equiv TMAOAc
5 vol DMAc
100 °C

1 → **2**•TMA or **2**•Na (+ **7**, **8**, **9**)

Reaction ID	PCy_3•HBF_4 mol%	PCy_3•HBF_4 L/M	In-process impurity profile (area percent) 2	7	8	9
1	6.1	2.05	95.76	1.04	1.78	0.05
2	6.7	2.25	96.50	1.15	1.36	0.15
3	7.4	2.48	96.95	1.21	1.07	0.37
4	8.4	2.80	96.99	1.17	0.81	0.68
5	10.2	3.40	81.82[a)]	1.01	0.61	1.30

a) Reaction incomplete after five hours with 13.68AP ArBr **1** remaining.

Table 6.10 In-process level of Me ester **10** at pilot scale.

Cyclization reaction temperature (°C)	Cyclization reaction (RAP)	Hydrolysis completion (RAP)
95 °C	0.21	<0.1
110 °C (target)	0.29–0.66	<0.1–0.2
125 °C	0.80–1.54	<0.1–0.2

hydrolysis completion was <0.2 RAP. At pilot scale, using target hydrolysis conditions, up to 1.54 RAP of **10** was successfully hydrolyzed to <0.1 RAP (Table 6.10). Spiking studies demonstrated that up to 70% of **10** could be purged during the crystallization, with no impact on the quality of isolated **2**. If the hydrolysis completion IPC was to fail, a kicker charge of aqueous KOH (target 0.2 equiv) could be added; however, in the pilot plant campaigns, the kicker charge of KOH was not required.

The pathway by which the active palladium [Cy_3P-Pd^0] catalyst degrades at the end of the cyclization reaction when all the limiting reactant (**1**) has been consumed follows a mechanism previously described by Wei and coworkers [7]. At the cyclization reaction temperature, the mono-ligated catalyst undergoes disproportionation to [$(Cy_3P)_2$-Pd^0] and Pd black. To control palladium levels in the isolated final intermediate, we had to remove sources of both soluble and insoluble palladium. The inherent solubility differences of these two Pd species were leveraged to remove the palladium in the final intermediate process (Table 6.11). The post-hydrolysis reaction mixture was cooled to 20 °C and MTBE (7.0 l/kg) was added, which facilitated the dissolution of soluble palladium species and had

Table 6.11 Removal of palladium throughout the process.

Palladium species	$[(Cy_3P)_2\text{-}Pd^0)]$	Pd black
Solubility characteristics	Miscible in organic solvents	Insoluble in aqueous or organic phases
Control point	1. MTBE waste stream – discarded 2. Crystallization – EtOH mother liquor	1. Filtration of the biphasic stream post-hydrolysis completion 2. Polish filtration of the MTBE-rich stream before crystallization

the added benefit of improving the filtration of the insoluble palladium species (Pd black).

The resulting biphasic mixture was agitated and filtered to remove insoluble palladium by-products. Post-filtration, the biphasic mixture was allowed to separate and the upper organic phase was discarded, retaining the product-rich aqueous phase. Maintaining a temperature of 20 °C, EtOH (3.0 l/kg), MTBE (11.0 l/kg), and 37 wt% aqueous HCl (6.3 equiv) were charged to generate the free acid of final intermediate **2**. The HCl charge is exothermic, with $T_{ad} = +7\,°C$.

During the acidification step in an early pilot plant campaign, the unexpected precipitation of a **2**•DMAc free acid solvate during the phase split was observed in several initial batches. Although this had no impact to quality and the **2**•DMAc-free acid solvate partitioned to the upper organic phase, minimizing losses to the phase split, the lack of robustness was rapidly addressed during inter-campaign laboratory development with the evaluation of the impact of phase split temperature and the addition of cosolvents. Increasing the temperature of the phase split up to 50 °C was evaluated but had minimal impact. The selection of cosolvents was constrained to those solvents currently used in the final intermediate or API process solvents, and EtOH was identified as being an attractive option. The solubility of the **2**•DMAc-free acid solvate was determined in an MTBE process stream with increasing amounts of added EtOH. As indicated in Figure 6.10, the addition of EtOH increases the solubility above the 60 mg/ml concentration at which precipitation may occur in the acidification phase split. Within the EtOH PAR of 2.5–3.5 l/kg, minimal yield loss was incurred.

Following acidification and phase split, the product-rich MTBE stream was washed twice with water (6.0 l/kg). The conductivity of each water wash was measured throughout the pilot plant campaign; the first water wash was 1200–2000 μs/cm (microsiemens per centimeter), indicating that residual salts may be present. A second water wash was <500 μs/cm and implementation afforded batches with consistent potency.

6.4.4 Crystallization and Drying

Following the water washes, the product-rich organic phase was polish filtered to remove any extraneous insoluble particulates. In preparation for the

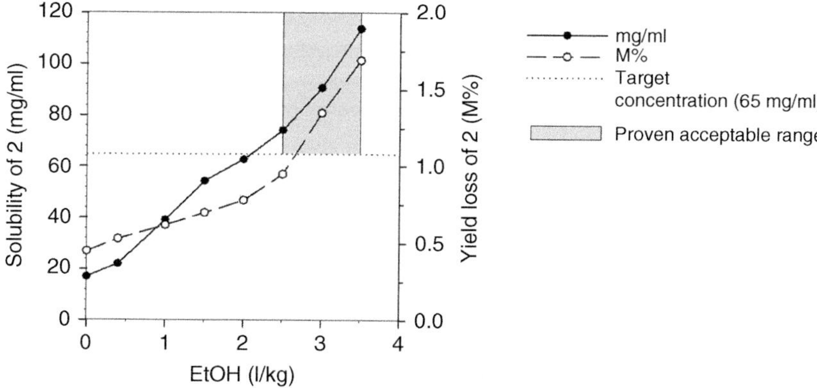

Figure 6.10 Solubility impact of added EtOH to the Acidification phase split.

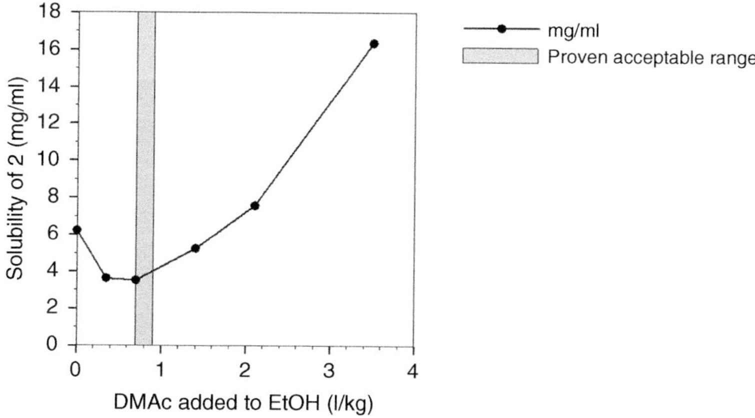

Figure 6.11 Solubility of **2** with respect to DMAc (l/kg) in EtOH (target 7.0 l/kg).

crystallization of **2** as a hemi-DMAc solvate, DMAc (0.8 l/kg) was charged to the polish-filtered product-rich organic phase to facilitate the removal of MTBE by distillation. The optimal value for the DMAc charge was informed by the measured solubility of **2** as a function of the relative amount of DMAc in EtOH (Figure 6.11), which reflects the composition of the subsequent crystallization stream (*vide infra*). The polish-filtered stream was concentrated by atmospheric distillation, removing MTBE under an initial reflux temperature of approximately 55 °C. The distillation was continued until a batch temperature of ≥75 °C was reached with the jacket temperature ≤100 °C. At pilot scale, under target conditions, the typical level of residual MTBE was <4 wt%. Spiking studies have demonstrated that up to 14 wt% residual MTBE could be tolerated with no impact to quality.

After the distillation was complete, the batch was cooled and then EtOH (7.0 l/kg) was charged. EtOH was selected as the primary crystallization solvent because it achieved sufficient purging of the impurities and afforded **2** in high quality and yield. At 50 °C, 24 wt% potassium ethoxide (KOEt) in EtOH

(0.95 equiv) was charged to deprotonate the free acid of **2**, generating the monopotassium salt. The KOEt in EtOH was typically added over a period of two hours to facilitate crystal agglomerate growth. Following the KOEt addition, the batch was subsequently cooled to 20 °C over a period of one hour to crystallize the product in a controlled manner, and then the slurry was aged at 20 °C for at least three hours to fully desaturate the solution. Aging the batch over this time period and temperature range was found to reduce product loss to the mother liquor, as shown in Figures 6.12 and 6.13.

Upon completion of the desaturation age, the batch was filtered to remove the mother liquor. The wet cake was initially washed with EtOH (6.0 l/kg) to displace the mother liquor, followed by a MTBE wash (6.0 l/kg) to displace the EtOH from the cake. The operating parameters of the centrifuge (e.g. spin speed and feed rate) or filter dryer (e.g. pressure) are equipment specific and do not impact the quality; the isolation of **2** has been demonstrated in both a filter dryer and

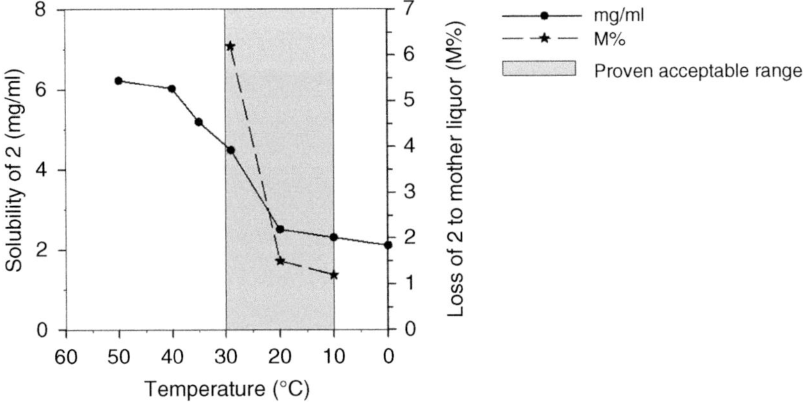

Figure 6.12 Solubility of **2** as a function of temperature.

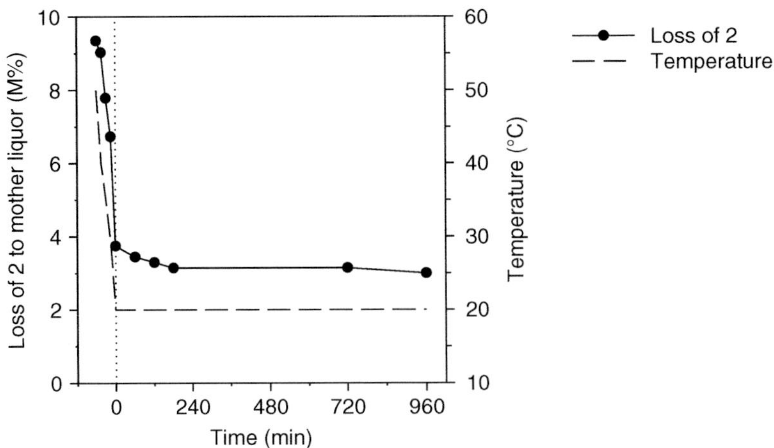

Figure 6.13 Desaturation of **2** over 16 hours during the crystallization.

Table 6.12 Average cake properties of dried **2**.

Material	D50 (μm)	Bulk density untapped (g/ml)	Bulk density tapped (g/ml)	Bulk density wet cake (g/ml)
Tray dryer without agitation	<50	0.44	0.55	0.41
Conical dryer with agitation	<12	0.37	0.39	0.41
Filter dryer with agitation	<25	0.29	0.33	0.41

centrifuge without issue. The final intermediate **2** is considered a fast filtering solid with filtration fluxes as high as 900 l/m^2/h at pilot scale. Filtration data measured for **2** during laboratory development determined a specific cake resistance (α_0) of 6.6×10^9 m/kg and a compressibility (n) of 0.49.

The wet cake was dried under vacuum with agitation and a jacket temperature set to 65 °C to ensure the removal of residual solvents. A sample was taken for drying completion (IPC: residual EtOH <4.0 wt%) and the batch was dried until the IPC was met. The time required to achieve dried product is equipment specific and may vary significantly with scale, equipment configuration, and operating parameters (e.g. agitation speed); however, at pilot scale, the typical drying time observed was 8–12 hours using a variety of agitation protocols. Stability studies have demonstrated that **2** was stable at 75 °C for at least four weeks, with no impact on quality and with the hemi-DMAc solvate maintained. Lower jacket temperatures did not impact quality; however, they may impact cycle time. Attrition was observed after agitated drying but had no downstream impact. Table 6.12 summarizes the cake properties of dried **2**.

6.5 Conclusion

Three generations of a palladium-catalyzed intramolecular direct arylation process were developed and successfully implemented on multi-kilogram scale. Key developments and learnings include the faster reaction rate afforded by the use of TMAOAc as base, compared to inorganic bases such as KOAc and K_2CO_3 that are more commonly employed for these transformations, as well as the identification of a process robustness risk resulting from the hydrolytic instability of DMF, both of which may have broader implications in palladium-catalyzed transformations. The third-generation, commercial TMAOAc/DMAc process worked flawlessly in all 18 batches of the internal pilot plant campaign, including batches that were run with parameter variations, which served to verify the PARs for the process parameters on scale and helped in the subsequent risk assessment analysis during the technology transition to manufacturing. The cumulative manufacturing experience for **2** using TMAOAc/DMAc conditions comprises 22 batches at 37–50 kg scale at two different manufacturing sites, demonstrating that the process is robust and is scale independent, and producing final intermediate **2** with acceptable quality and yield on large scale. Including the earlier KOAc/DMAc and TMAOAc/DMF processes, this direct arylation chemistry produced >1100 kg of **2** in >80% yield with high quality (>99.79AP). To the best of our knowledge, this is the largest scale implementation of a palladium-catalyzed direct arylation process in a pharmaceutical manufacturing setting [8]. We believe that the work described herein demonstrates that this chemistry can be highly robust, provided that there is a thorough understanding and control strategy for both the reaction parameters and any potential catalyst poisons.

Biography

Collin Chan received his B.S. degree in Chemistry and Molecular Biology from the University of Wyoming in 2001 and his Ph.D. degree in Organic Chemistry with Professor Samuel J. Danishefsky Columbia University in 2007. In 2007, he joined the process group at Bristol-Myers Squibb in New Brunswick, NJ. He is currently in the Pharma Manufacturing Science and Technology department focused on actives and intermediates.

Albert J. DelMonte received his Ph.D. degree from Professor Daniel Singleton at Texas A&M in 1998 in Physical Organic Chemistry. He then performed his postdoctoral studies in the area of asymmetric catalysis with Professor Eric N. Jacobsen at Harvard. In 2001, Dr. DelMonte joined the process group at Bristol-Myers Squibb in New Brunswick, NJ. He is currently a chemistry head in the Chemical and Synthetic Development department.

Chao Hang received his Ph.D. degree from Professor Daniel Singleton at Texas A&M in 2000 in Physical Organic Chemistry. He then performed his postdoctoral studies in the area of modifying antibiotics with Professor Daniel Kahne at Princeton. In 2003, Dr. Hang joined the process group at Bristol-Myers Squibb in New Brunswick, NJ. He is currently a senior investigator in the Chemical and Synthetic Development.

Dr Yi Hsiao graduated from Sun Yet-Sen University, then moved to Japan and joined Professor Noyori's research group at Nagoya University, and studied asymmetric catalytic hydrogenation of enamides and asymmetric synthesis of complex target molecules. After receiving his Ph.D. degree in 1994, Yi performed postdoctoral studies at ERATO, Japan (Exploratory Research for Advanced Technology) on hydrogenation of supercritical carbon dioxide and at Colorado State University (US) with Professor Louis Hegedus. In 1997, he joined the Process Research department of Merck, where he has defined and developed processes for numerous drug candidates. In 2006, Yi moved to Process R&D in Bristol-Myers Squibb to create an innovative and productive catalysis research group. In 2018, Yi returned to China and joined Asymchem as the senior vice president, in charge of process development, introduction, and implementation of new technologies and integration of research, analytical, and technology functions. Dr. Yi Hsiao has published over 50 research papers and book chapters, over 40 presentations, and received several awards. He is an organizing committee member of the 2018 International Forum of Reactions and Processes for Pharmaceutical Development in China and the co-chair of 2019 GRC-organic reactions and processes.

Eric M. Simmons received his B.S. degree in Chemistry and Spanish from Tufts University in 2004 and his Ph.D. degree in Organic Chemistry with Professor Richmond Sarpong at the University of California at Berkeley in 2009. He then performed postdoctoral studies in the areas of organometallic chemistry and catalysis with Professor John F. Hartwig at the University of Illinois at Urbana-Champaign. In 2011, he joined the process group at Bristol-Myers Squibb in New Brunswick, NJ, working in the Catalyst R&D group with Dr. Yi Hsiao. He is currently the Catalyst R&D group lead in the Chemical and Synthetic Development department.

References

1 Meanwell, N.A. (2016). *J. Med. Chem.* 59: 7311.

2 For reviews, see: (a) Alberico, D., Scott, M.E., and Lautens, M. (2007). *Chem. Rev.* 107: 174. (b) Ackermann, L., Vicente, R., and Kapdi, A.R. (2009). *Angew. Chem. Int. Ed.* 48: 9792.

3 Bien, J., Davulcu, A., DelMonte, A.J. et al. (2018). *Org. Process Res. Dev.* 22: 1393–1408.

4 For reviews, see: (a) Berritt, S. (2013). *Aldrichimica Acta* 46: 71–80. (b) Selekman, J.A., Qiu, J., Tran, K. et al. (2017). *Ann. Rev. Chem. Biomol. Eng.* 8: 525–547.
5 Simmons, E.M., Mudryk, B., Lee, A.G. et al. (2017). *Org. Process Res. Dev.* 21: 1659.
6 (a) Liégault, B., Petrov, I., Gorelsky, S.I., and Fagnou, K. (2010). *J. Org. Chem.* 75: 1047. (b) Rousseaux, S., Gorelsky, S.I., Chung, B.K.W., and Fagnou, K. (2010). *J. Am. Chem. Soc.* 132: 10692. (c) Wakioka, M., Nakamura, Y., Hihara, Y. et al. (2013). *Organometallics* 32: 4423.
7 Wei, C.S., Davies, G.H.M., Soltani, O. et al. (2013). *Angew. Chem. Int. Ed.* 52: 5822.
8 For a palladium-catalyzed intermolecular direct arylation on 2.7 kg scale, see: (a) Gauthier, D.R., Limanto, J., Devine, P.N. et al. (2005). *J. Org. Chem.* 70: 5938. For a ruthenium-catalyzed intermolecular direct arylation on 3 kg scale, see: (b) Ouellet, S.G., Roy, A., Molinaro, C. et al. (2011). *J. Org. Chem.* 76: 1436.

7

Ruthenium-Catalyzed C—H Activated C—C/N/O Bond Formation Reactions for the Practical Synthesis of Heterocycles and Pharmaceutical Agents

Anita Mehta[1], *Naresh Kumar*[2], *and Biswajit Saha*[1]

[1] *Chicago Discovery Solutions LLC, 23561 West Main Street, Plainfield, IL 60544, USA*
[2] *GVK BIO, Plot No. 28 A, Industrial Development Area, Nacharam, Hyderabad, Telangana 500076, India*

7.1 Introduction

Some of the most significant reactions in organic syntheses are C—C (carbon–carbon), C—N (carbon–nitrogen), and C—O (carbon–oxygen) bond formations to generate biologically active compounds such as pharmaceuticals and natural products. The diversity of molecular structures in pharmaceuticals and natural products with various pendant groups as well as new frameworks is constantly increasing, thus requiring new and efficient methods to build them. Transition metal catalysts are commonly used as homogeneous catalysts for these cross-coupling reactions because of their ability to switch their oxidation states. Typically, C—C/N/O bond forming reactions are done by palladium (Pd), although rhodium (Rh), nickel (Ni), and ruthenium (Ru) complexes are being increasingly used [1a–d].

Recently, there has been an interest in transition-metal-catalyzed direct C—C, C—N, and C—O bond formation involving activation of normally unreactive C—H bonds. Typically, activation of an unreactive bond is accomplished by coordination of a substrate with a catalytic complex of a transition metal such as palladium (Pd), rhodium (Rh), and ruthenium (Ru) [2a–e]. Because of the inherent high stability of C—H bonds and often the presence of multiple C—H bonds in the reactants, these processes are nonspecific and most often require harsh reaction conditions or a specific directing group. With the discovery of new catalysts and protocols that proceed efficiently under mild conditions, improvement on functional group tolerance and selectivity has been accomplished [2d–f]. Compared to palladium (Pd) and rhodium (Rh), ruthenium (Ru) stands out in terms of low cost [3]. Direct C–H functionalization via ruthenium complex catalysts has been explored in the past two decades [2d, f, g]. Although ruthenium(0) complexes are air- and water-sensitive, several ruthenium(II) and ruthenium(III) complexes are air-stable and are shown to have better tolerance to a variety of oxidants and work in various media including greener solvents such as water [2f, g, 4a–c].

Organometallic Chemistry in Industry: A Practical Approach, First Edition.
Edited by Thomas J. Colacot and Carin C.C. Johansson Seechurn.

In this chapter, we have summarized the selected examples on the use of ruthenium complexes for C—C/N/O bond formation through C—H bond activation with an emphasis on practicability and scalability of the reactions for the synthesis of heterocycles and pharmaceuticals. This chapter highlights the fact that ruthenium-catalyzed C–H activation reactions are very practical and can be performed in the absence of inorganic additives and surfactants and can be conducted in sustainable solvents such as water.

7.2 C–H Activation Followed by C—C Bond Formation

The development in C—C bond formation in the past few decades and its wide use in several pharmaceuticals have undoubtedly established its significance. The 2010 Nobel Prize in Chemistry for such C—C bond formation reactions further substantiates its relevance. This work has been extensively reviewed by several research groups and covered by several books as well as reviews on this topic [1c, d]. However, these reactions must be carried out with prefunctionalized nucleophiles, requiring alkali or alkaline earth metal reagents, which generate lots of wastes, hence making this process less friendly for the environment. On the other hand, C—H activation with transition metal catalysts directly conducts C—C bond formation without requiring any further derivatization. These reactions, with the use of catalytic amounts of transition metal catalysts, activate specific C—H bonds and lead to different types of C—C bond formation. Although several transition metals such as palladium (Pd), rhodium (Rh), and nickel (Ni) are known for this purpose [1], the section below deals with the practical examples of such reactions with the use of low-cost ruthenium (Ru) complexes as the catalyst.

7.2.1 C–H Activation Followed by C—C Bond Formation: Biaryl/Heterobiaryl Synthesis in Organic Solvents

Many biaryl- and heterobiaryl-containing motifs are among the most valuable synthetic scaffolds in organic chemistry because of their prevalence in natural products, advanced materials, and pharmaceuticals [5a–e].

Biaryl compounds are commonly prepared using well-known aryl–aryl coupling reactions such as the Suzuki and Stille couplings [1c, d]. Typically, conventional coupling reactions involve C—C bond formation between an aryl halide and an organometallic reagent (Scheme 7.1). Aryl boronic acids, organotin, and organozinc derivatives are very commonly used for the cross-coupling reactions, which can require multiple synthetic steps from the arene starting material.

Thus, C—H bond activation for direct arylation is a better strategy compared with conventional cross-coupling reactions as it does not require stoichiometric amount of prefunctionalized organometallic reagents. This approach is highly desirable because of readily available starting materials and clean reaction conditions as it does not generate large amount of waste containing toxic metals (i.e. Sn, B, and Zn). Besides, HX is the only side product that can be conveniently trapped by a base [2e,f].

As early as 2001, Oi, Yoshio, and coworkers [6a] reported successful direct arylation and alkenylation at the *ortho*-position of the aromatic ring of

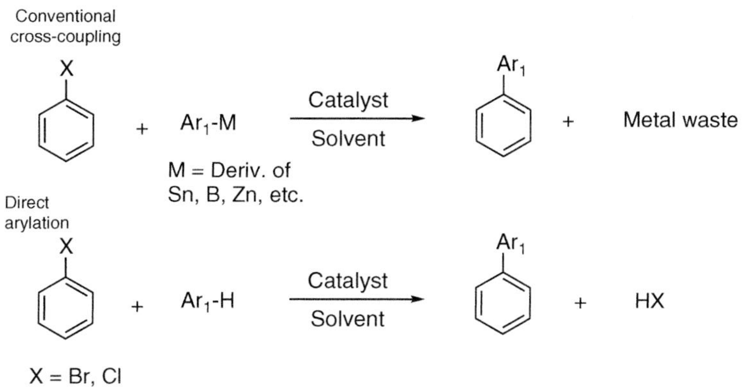

Conventional cross-coupling

X + Ar_1-M —Catalyst / Solvent→ Ar_1 + Metal waste

M = Deriv. of Sn, B, Zn, etc.

Direct arylation

X + Ar_1-H —Catalyst / Solvent→ Ar_1 + HX

X = Br, Cl

Scheme 7.1 Conventional cross-coupling vs. direct arylation.

pyridyl-substituted aromatic compounds with organic halides in the presence of a catalytic amount of a ruthenium(II)–phosphine complex. Significant progress has since been reported in transition-metal-catalyzed *ortho*-directed C—H bond activation and functionalization reactions [2]. Nitrogen-containing substrates such as 2-phenylpyridine, 2-phenyl-2-oxazoline, benzo[*h*]quinoline, and *N*-phenylpyrazole have been exploited to act as *ortho*-DG groups for aromatic C–H activation in the presence of transition metal complexes for the synthesis of hetero-biaryl compounds [2c, d, 6b]. Even simple ruthenium compounds such as $RuCl_3{\cdot}H_2O$ and $RuCl_2$(*p*-cymene) dimer have been used for direct arylation reactions [6c, d]. Bogdan Štefane et al. [6e] reported C—H bond functionalization of arylpyrimidines catalyzed by an *in situ*-generated ruthenium(II) carboxylate system for the synthesis of tris(heteroaryl)-substituted benzenes using $RuCl_2$(*p*-cymene) dimer.

A practical approach for aryl substitution on heterocycles such as furan has been reported with aryl bromide/iodide using $[RuCl_2(p\text{-cymene})]_2$ (5.0 mol%) in the presence of potassium carbonate (K_2CO_3) and *N*-methylpyrrolidine (NMP) as a solvent. Aryl groups such as phenyl or substituted phenyl or heteroaryls have been used by Cao et al. [7] to achieve up to 89% yields (Scheme 7.2). The reaction also demonstrated the tolerability of active groups such as esters and aldehydes under the reaction conditions.

Several angiotensin (II) receptor antagonists (ARBs) are clinically used, which include azilsartan, candesartan cilexetil, eprosartan, irbesartan, losartan,

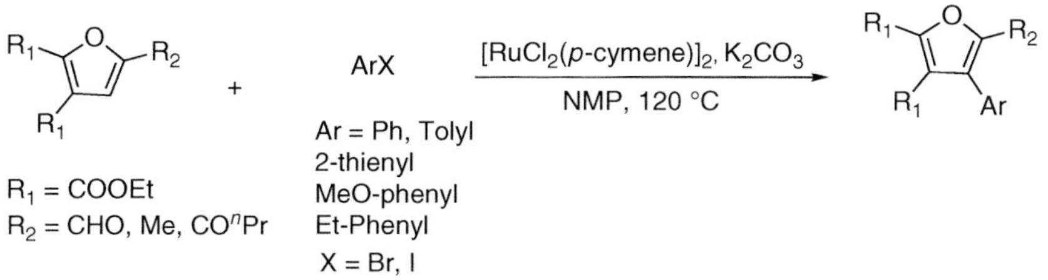

R_1, R_2, R_1 (furan) + ArX —$[RuCl_2(p\text{-cymene})]_2$, K_2CO_3 / NMP, 120 °C→ R_1, R_2, R_1, Ar (furan)

R_1 = COOEt
R_2 = CHO, Me, CO^{n}Pr

Ar = Ph, Tolyl
2-thienyl
MeO-phenyl
Et-Phenyl
X = Br, I

Scheme 7.2 C—H arylation of heterocycles.

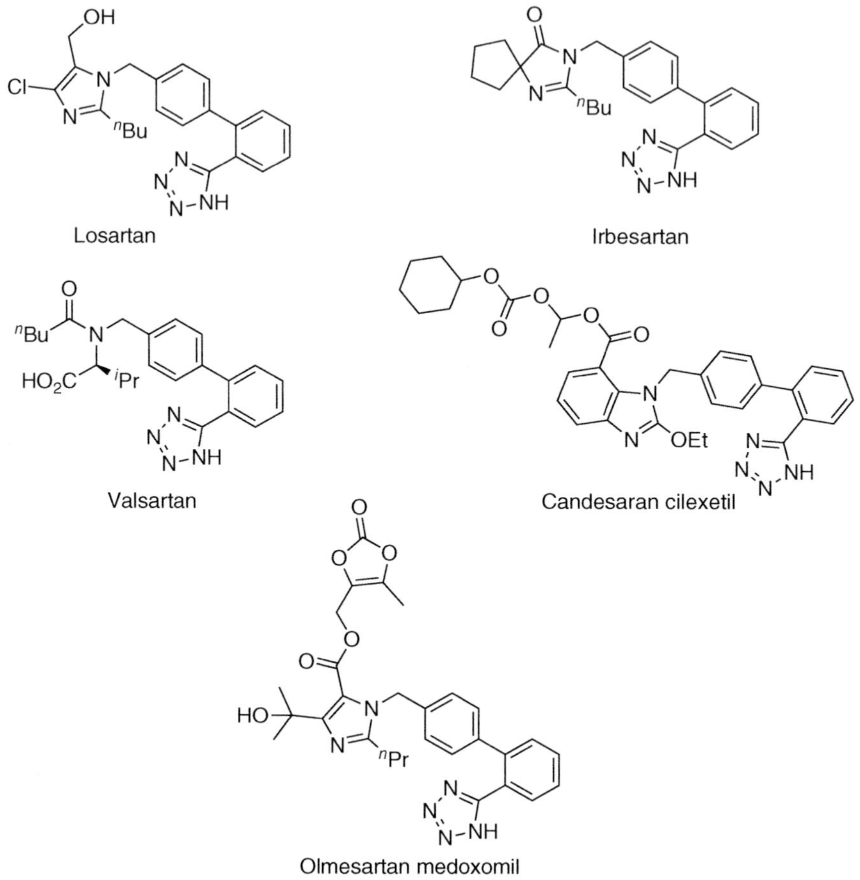

Losartan

Irbesartan

Valsartan

Candesaran cilexetil

Olmesartan medoxomil

Figure 7.1 Angiotensin(II) receptor antagonists (ARBs).

olmesartan medoxomil, telmisartan, and valsartan. Representative structures are given below (Figure 7.1).

The common feature of several important ARBs is the presence of biphenyl methyl group with heterocycle in the 4-position and a polar acidic functionality, i.e. —COOH, —$NHSO_2CF_3$, —SO_3H, or isosteres, i.e. tetrazole [8] in the 2′ position (Figure 7.2).

Several commercially important ARBs such as candesartan cilexetil, eprosartan, irbesartan, losartan, olmesartan medoxomil, telmisartan, and valsartan contain biphenyltetrazole fragment (Figure 7.1). Syntheses of a common starting

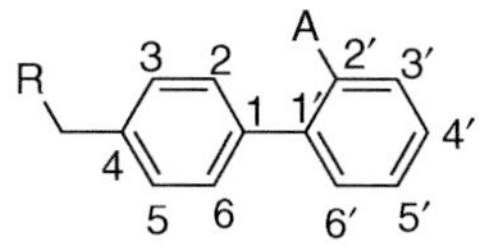

A = –COOH, –$NHSO_2CF_3$, –SO_3H, –CN_4H, etc.
R = Heterocycle

Figure 7.2 Biphenyl core of angiotensin(II) receptor antagonists (ARBs).

material biphenyltetrazole fragment of ARBs have been accomplished in the past by employing Ullmann coupling between 4-iodotoluene and 2-iodobenzoate where nucleophilic aromatic substitutions at the *ortho*-position of a suitable benzoic acid derivative was performed [9]. Furthermore, chemistry of Meyers et al. was also applied for the synthesis of several biphenyl intermediates by using phenyl oxazoline and 4-methyl phenyl magnesium bromide [10]. Ni(0)-catalyzed biaryl coupling was also used for the synthesis of nitrile-substituted biphenyls [11]. The problem with these approaches was that benzonitrile conversion to tetrazole required the need of hazardous chemicals such as trialkyltin azide.

In order to avoid the use of trialkyltin azide, Suzuki coupling was performed to obtain biphenyltetrazole fragment starting from phenyltetrazole by Larsen et al. [12]. The fundamental drawback of this approach for the synthesis of substituted biphenyls was that the process generated waste besides utilizing costly organometallic reagents (e.g. boronic acid/esters).

In designing an alternate synthesis of ARBs, Seki [13] utilized C–H activation technology using cheaper ruthenium catalysts such as ruthenium trichloride or $[RuCl_2(benzene)]_2$ with triphenylphosphine (PPh_3) and other phosphine ligands such as BINAP and Xphos to achieve the arylation of 1-*N*-protected phenyltetrazole.

After the key step of aryl–aryl coupling reactions to form the biphenyl intermediate, the protected halide intermediate was alkylated with appropriate heteroaryls and finally deprotected to give common intermediate of several angiotensin receptor blockers such as valsartan, candesartan cilexetil, losartan, irbesartan, and olmesartan medoxomil as shown below (Scheme 7.3).

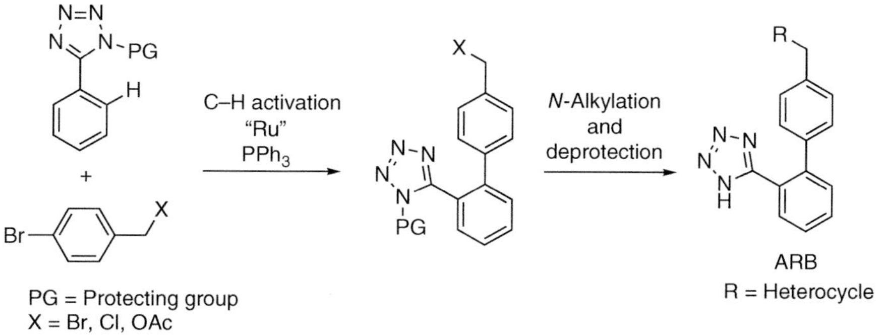

C–H activation "Ru" PPh_3; *N*-Alkylation and deprotection; ARB, R = Heterocycle

PG = Protecting group
X = Br, Cl, OAc

Scheme 7.3 Model for Ru-catalyzed angiotensin(II) receptor antagonists (ARBs) synthesis.

Synthesis of commercially important ARBs such as losartan, valsartan, and olmesartan was accomplished by using a common intermediate as shown below (Scheme 7.4).

Losartan was synthesized by starting with methoxybenzoyl-protected biphenyltetrazole intermediate with aryl halide [13a, b]. This intermediate was converted to the corresponding alcohol by hydrolyzing the acetate group using sodium methoxide. Next, bromination reaction was easily performed to convert the alcohol to the bromide. This intermediate bromide was converted to

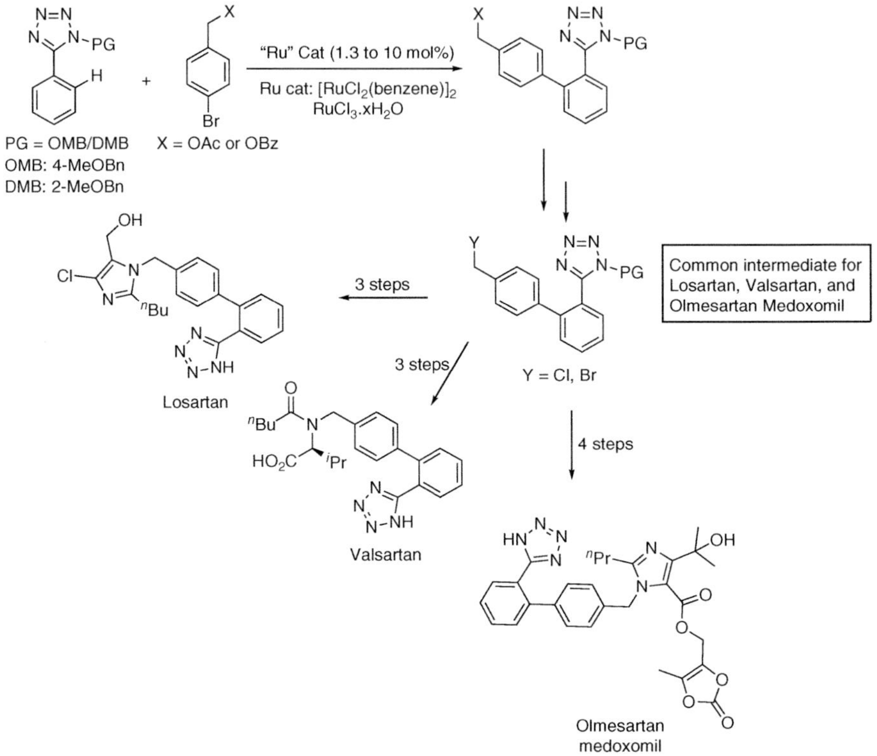

Scheme 7.4 Seki's method for Ru-catalyzed synthesis of selected angiotensin(II) receptor antagonists (ARBs).

losartan in three steps as shown in Scheme 7.4. In order to avoid the formation of undesired side products, removal of the *p*-methoxybenzyl group was done before the final reduction of the aldehyde group using sodium borohydride.

Another important ARB, valsartan, was also synthesized using the C–H activation methodology [13a, b]. Simple ruthenium complex, $RuCl_3 \cdot xH_2O$, was used in the presence of triphenyl phosphine and potassium carbonate to couple 2-methoxybenzyl-protected phenyltetrazole with aryl bromide (Scheme 7.4). Next, the acetate group of biphenyltetrazole intermediate was converted to the corresponding alcohol by performing the hydrolysis reaction using sodium methoxide in 90% yield. The *in situ* bromination of the alcohol was performed by using trimethylsilyl bromide (TMSBr) followed by the esterification with *L*-valine benzyl ester in the presence of *N,N*-diisopropylethylamine in acetonitrile. Next, the direct acylation was performed by using *n*-butylacetyl chloride and pyridine to prepare the corresponding ester moiety. The final deprotection step was successfully done by Pd/C-catalyzed hydrogenation in high yield (76%) to afford valsartan. Valsartan is known to exist in several crystalline forms, but control of the polymorphs was not examined in this study.

In 2014, Seki et al. improved the synthesis of losartan and valsartan from the common intermediate (biphenyltetrazole core) by slight modification of the protecting group [13c]. The performance of three types of Brønsted acid potassium salts, potassium carboxylates, potassium phosphates, and potassium sulfonates, were tested as the cocatalyst in ruthenium-catalyzed C–H arylation. Overall, the basic synthetic strategies were very similar as mentioned before for the synthesis of common ARBs. Thus, for the synthesis of losartan, the protected group of the biphenyltetrazole moiety was modified from PMB (*p*-methoxybenzoyl) to benzyl unit. Next, the regioselective *N*-alkylation was performed between heterocyclic aldehyde and benzyl-protected biphenyltetrazole. The final product losartan was synthesized in high yield (71%).

Subsequently, the synthesis of valsartan was also improved by a coupling reaction between the biphenyl intermediate and the *L*-valine benzyl ester to generate the corresponding *N*-alkylated product. In the next step, amidation was successfully performed by using *n*-butylacetyl chloride and *N*, *N*-diispropylethylamine to form the corresponding amide in quantitative yield. Debenzylation of the amide was achieved smoothly with $Pd/BaSO_4$ to form valsartan in 71% yield.

In 2015–2016, the Seki group reported the synthesis of another important ARB, olmesartan medoxomil (Scheme 7.4), from the common intermediate, which was used for the synthesis of losartan and valsartan [13d, e]. Unlike the synthesis of losartan and valsartan, the medoxomil ester group was not stable under the Pd-catalyzed debenzylation reaction step. Therefore, the deprotection of *N*-benzyl group was done, followed by *N*-tritylation, hydrolysis, and medoxomil ester formation. In the final step, detritylation was performed to form olmesartan medoxomil in quantitative yield. Thus, 2,4-dimethoxybenzyl (DMB) group was found to be an ideal protecting group to reduce the number of steps. The chloro compound was coupled with the ethyl ester to get the *N*-alkylated product. In the next step, medoxomil ester was synthesized in two steps (64% yield). Finally, trifluroacetic acid was used to deprotect the DMB group, which leads to the formation of olmesartan medoxomil in quantitative yield in the final step (98%).

For the synthesis of candesartan cilexetil, the scope of the synthesis of biphenyltetrazole intermediate via C–H activation was improved by using the bulky coupling partner 4-bromobenzyl benzoate via ruthenium-catalyzed C–H activation reaction [13f]. The use of a bulky group promoted faster crystallization in comparison to the smaller acetyl derivative in their previously reported synthesis. The benzoyl group was easily converted to chloro group using debenzoylation and subsequent chlorination in the presence of sodium hydroxide and thionyl chloride. In order to avoid impracticality with respect to the previous synthesis, the coupling reaction was conducted between benzimidazole and corresponding chloro compound to prepare the N-alkylated product, which leads to candesartan cilexetil through sequential steps [13d]. A series of potassium salts of sulfonic acids were tested as a cocatalyst for Ru-catalyzed C–H arylation. Among them, K-2,4,6-trimethylbenzenesulfonate (TMBSK) was found to be most active, and the TMBSK was applied to a practical synthesis of candesartan cilexetil [13g].

In the case of irbesartan, benzylated phenyl tetrazole was used as a starting material (Scheme 7.5). Debenzylation was performed in the last step using Rosenmund catalyst to afford irbesartan in excellent yield [13d, h].

Ruthenium-catalyzed direct arylation of 1-*N*-protected phenyltetrazole with aryl bromides in NMP with ruthenium trichloride in combination with

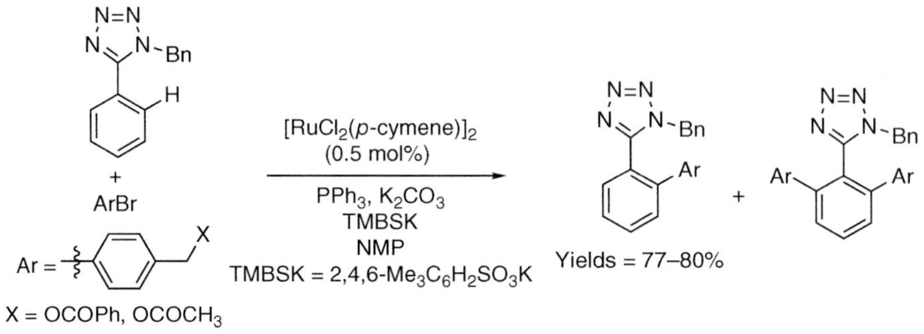

ArBr

$[RuCl_2(p\text{-cymene})]_2$ (0.5 mol%)

PPh_3, K_2CO_3
TMBSK
NMP
TMBSK = 2,4,6-$Me_3C_6H_2SO_3K$

Yields = 77–80%

X = OCOPh, $OCOCH_3$

Scheme 7.5 Ruthenium-catalyzed synthesis of irbesartan.

triphenylphosphine (PPh_3) was achieved successfully in small scale to provide 81% yield. However, at a scale of more than 100 g, it proved difficult to scale up as it lacked reproducibility because of the effect of different amounts of minute impurities present in NMP and formation of di-arylation [13d].

To address the challenges of reproducibility of ruthenium-catalyzed C–H activation chemistry for the formation of pharmaceutically useful ARBs, Seki successfully used cocatalysts such as potassium carboxylates and potassium phosphates [13d]. Thus, synthesis of the common biphenyl-tetrazole intermediate using C–H arylation coupling reactions between 1-benzyl-2-phenyl-1*H*-tetrazole with 4-bromobenzyl acetate in the presence of a cocatalyst. The Ru-catalyzed C–H coupling reaction was efficiently performed using $[RuCl_2(p\text{-cymene})]_2$ (0.5 mol%), with PPh_3 (1.0–2.0 equiv to Ru), and potassium carboxylates and potassium phosphates as cocatalysts (2.0 equiv to Ru), in the presence of K_2CO_3 as a base (1.0 equiv). After screening various cocatalysts (Brønsted acid potassium salts, potassium carboxylates, potassium phosphates, and potassium sulfonates), it was found that potassium carboxylates such as 2,4,6-$Me_3C_6H_2SO_3K$ (TMBSK), $(PhO)_2P(O)$ (OK), and *k*-bis(2 ethylhexyl)phosphate (BEHPK) were the best additives [13b, d, e].

Ackermann et al. [14] successfully utilized potassium carboxylates for ruthenium-catalyzed C–H arylation of *N*-1-protected aryltetrazoles. It was disclosed that ruthenium catalyst $[RuCl_2(p\text{-cymene})_2]$ in combination with potassium carboxylates (KOAc or a combination of pivalic acid and potassium carbonate) gave 57–75% yields of the key intermediate for ARB synthesis as shown in the following scheme for a combination of pivalic acid and potassium carbonate (Scheme 7.6). The reaction could be conducted in toluene as well as in 1,4-dioxane, with NMP or DMA as a solvent, while water was not a suitable reaction media.

Both Seki and Ackermann groups used a similar ruthenium complex in the presence of organic solvents, but the choice of ligand and cocatalyst made both processes unique for C–H arylation reactions. Seki systematically improved the Ru-catalyzed C–H arylation reactions for the formation of biphenyltetrazole core by selecting the particular ruthenium complex and tuning cocatalysts

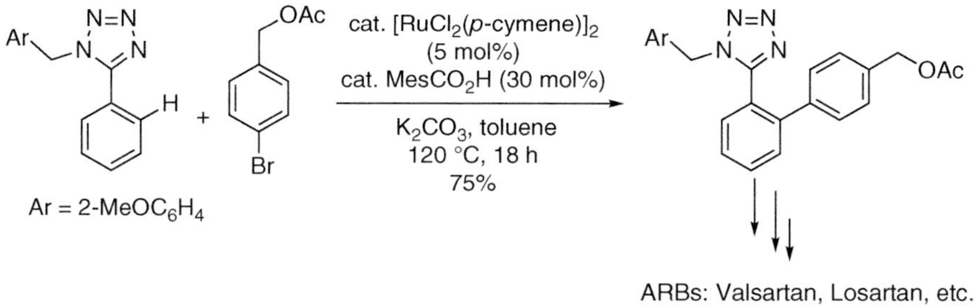

Scheme 7.6 Ackermann's method for Ru-catalyzed synthesis of selected angiotensin(II) receptor antagonists (ARBs).

(i.e. Brønsted acid potassium salts, potassium carboxylates, potassium phosphates, and potassium sulfonates) [13b, d, e]. In Ackermann's case, the presence of carboxylate was considered enough to provide assistance by bidentate coordination with ruthenium complex, which facilitated the C–H transfer process followed by C–X (X = halogen) bond cleavage and reductive elimination to promote the final aryl transfer [14]. The presence of PPh_3 or other phosphine ligands did not lead to significant rate acceleration.

Certainly, the current approach for the Ru-catalyzed C–H arylation reaction and its direct application to efficient synthesis of ARBs has tremendous influence in the process development of active pharmaceutical ingredient (API) from the pharmaceutical standpoint. The practicality of this convergent method is of commercial interest for the synthesis of API under mild condition without any hazardous reagents. Furthermore, it was demonstrated that the synthesis would be readily applicable to a multi-ton scale. However, the replacement of hazardous organic solvents (i.e. NMP and toluene) with a greener alternative (i.e. water) still remains the biggest challenge for the next-generation reactions (*vide infra*). Both Seki [13] and Ackermann and coworkers [14] reported that water was not a suitable media as a reaction solvent.

Ruthenium-catalyzed direct arylation was also applied as one of the key reactions for the large-scale preparation of another API, i.e. anacetrapib (MK-0859) [15], which was considered to be useful for the treatment of hypercholesteromia [16, 17]. Anacetrapib was identified as a promising inhibitor of cholesteryl ester transfer protein (CETP) and was tested in clinical phase III.

In 2011, Ouellet et al. [18] from Merck Co. developed the synthesis of anacetrapib for the preparation of kilogram quantities for studying its pharmacological properties. The synthetic analysis approach was supported by a Ru-catalyzed direct arylation for the formation of biaryl alcohol, which was used as the intermediate of anacetrapib synthesis (Scheme 7.7) Ru-catalyzed C–H activation was successfully achieved with high efficiency by combining $[RuCl_2(benzene)]_2$ (1.0 mol%), triphenylphosphine (PPh_3) (2.0 mol%), and potassium phosphate (K_3PO_4) (2.0 equiv). According to the reaction protocol, NMP in the presence of low-level impurity (i.e. γ-butyrolactone) was desired for getting quantitative conversion (>98%). On the other hand, the reaction failed to produce the desired

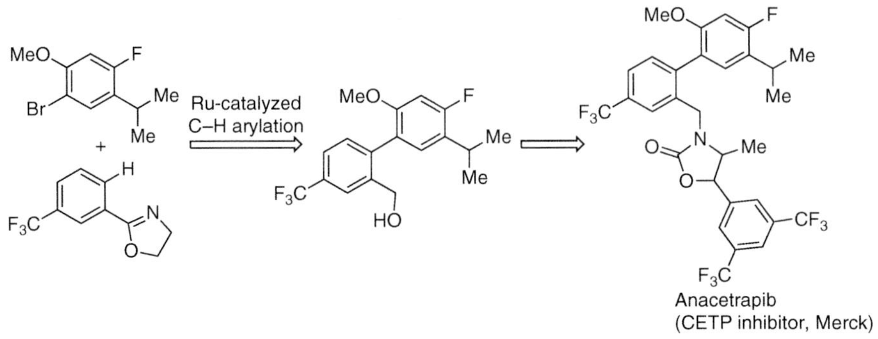

MeO F Br Me Me + H F3C N O Ru-catalyzed C–H arylation MeO F Me Me F3C HO MeO F Me Me F3C O N Me O CF3 F3C Anacetrapib (CETP inhibitor, Merck)

Scheme 7.7 Retrosynthetic analysis of synthesis of anacetrapib.

coupling product in comparable yield when γ-butyrolactone-free NMP was used. The amount of this impurity in NMP was found to be responsible for batch-to-batch variations.

Seki [13d] also faced this problem of lack of reproducibility in the synthesis of arylation of phenyltetrazole in NMP and observed that the presence of a small amount of impurity in the NMP had major influence on the performance of catalysts used in the synthesis of phenyl tetrazole intermediate at preparative scale.

It is noteworthy that the success of the Ru-catalyzed coupling reaction is enhanced by the cocatalyst (i.e. carboxylate additives) as reported by the other groups [19]. Next, the key intermediate (i.e. biaryl alcohol) of anacetrapib was synthesized using one pot reaction as shown in the above scheme (Scheme 7.8). Thus, biaryl alcohol was prepared from the corresponding chloride, which was obtained by the ring-opening reaction of the corresponding oxazolidinone [18, 20].

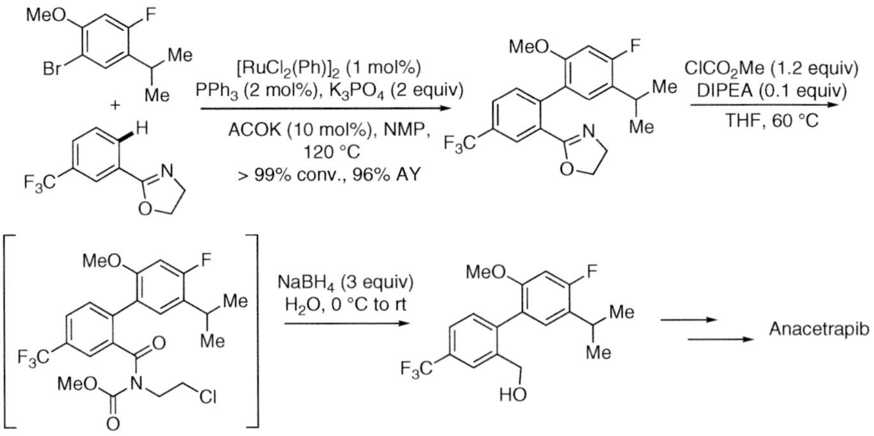

MeO F Br Me Me + H F3C N O $[RuCl_2(Ph)]_2$ (1 mol%) PPh_3 (2 mol%), K_3PO_4 (2 equiv) ACOK (10 mol%), NMP, 120 °C > 99% conv., 96% AY MeO F Me Me F3C N O $ClCO_2Me$ (1.2 equiv) DIPEA (0.1 equiv) THF, 60 °C

MeO F Me Me F3C O MeO N Cl O $NaBH_4$ (3 equiv) H_2O, 0 °C to rt MeO F Me Me F3C HO Anacetrapib

Scheme 7.8 C—H activated carbon-carbon coupling for the synthesis of anacetrapib.

7.2.2 C–H Activation Followed by C—C Bond Formation: Biaryl/Heterobiaryl Synthesis in Green Solvents

Decreasing the amount of toxic material is one of the major green chemistry challenges for the pharmaceutical industry [21]. Therefore, it is highly desirable to have water-friendly synthetic methods that can be used in the synthesis of the important class of heterocycles and pharmaceuticals. C–C coupling reaction via C–H activation is an atom-economical route of synthesis [2] that does not require extra steps of derivatization. Most of these C–H activation reactions are done in expensive, flammable, as well as toxic organic solvents such as NMP, DMF, and toluene (see Section 7.2.1). These solvents have high boiling points, and washing/quenching with water is the most popular method for the reaction workup. Because of the miscibility of NMP in water, workup requires a large amount of water to isolate the products. The affluent contains a mixture of solvents, reactants, and products. The problem gets bigger at the large-scale synthesis. It is highly desirable to carryout C–H-activated couplings in water as water is natural, cheap, and a nontoxic alternative to organic solvents.

Synthetic chemists have traditionally shied away from the water and carried out the reactions in organic medium for most of the molecules and often resort to extreme length to keep all traces of water out of their reaction flasks or reactors [22]. Reagents such as *n*-butyllithium useful in *ortho*-metalation are extremely water sensitive [12]. Sharpless et al. [23] noted that many reactions (e.g. Claisen rearrangement) are dramatically accelerated when performed in aqueous suspension ("on water") relative to organic solvents or even neat conditions. Low miscibility of organic compounds with water is not detrimental; in fact, it facilitates isolation of the product.

A few examples of palladium-catalyzed C—H bond functionalization in water are already known. In 2009, palladium-catalyzed C—H-activated coupling of aryl iodides with anilides and ureas in the presence of oxidants such as AgOAc in 2 wt% surfactant water solutions had been described by Lipshutz and coworkers [24]. Surfactants such as Brij 35 or TPGS-750 M in water were considered to be necessary for the success of the reaction. An oxidant was also required. Furthermore, in 2009, Ackermann et al. [25a] also demonstrated palladium-catalyzed direct arylations in polyethylene glycol (PEG) in the presence of air with a recyclable phosphine ligand-free palladium complex.

Several ruthenium(II) complexes are tolerant to water and air, making them more practical for industrial applications toward the synthesis of valuable heterocycles and pharmaceuticals in a cost-effective manner. As early as 2005, Ackermann et al. [25b] performed the reaction of chlorobenzene with 2-phenylpyridine by using 2.5 mol% of $[RuCl_2(p\text{-cymene})]_2$ with 10 mol% of $Ad_2P(O)H$ in NMP containing water (NMP/H_2O in a 2 : 1 ratio) at 120 °C for 20 hours and obtained 61% yield of di-arylated product. This experiment was important as it revealed that the ruthenium(II) C—H bond activation catalyst system thus formed was tolerant to water.

In 2009, Arockiam et al. [26a] showed that diethyl carbonate (DEC) can be used as a solvent for C—H bond arylation of heteroamines instead of NMP. For example, arylation reactions of 2-phenylpyridine and other heteroarenes were performed with $[RuCl_2(p\text{-cymene})]_2$ and potassium pivalate (KOPiv) as a cocatalyst in DEC. The workup procedure involved the distillation of DEC under vacuum. The use of DEC facilitates the workup procedure, thus reducing the amount of wastewater. The slight loss of activity because of the use of DEC is counterbalanced by the improvement of the catalyst efficiency achieved by a judicious choice of additives.

In 2010, Dixneuf and coworkers [26b] also showed that in a sealed tube, ruthenium(II)-catalyzed C—H bond arylation could be done using water as a solvent. *In situ*-generated $[Ru(O_2CR)_2(\text{arene})]$ catalysts efficiently perform the direct *ortho*-arylation of functional arenes with chloroarenes or chloroheterocycles in water, without the need for a surfactant. The activity of these catalysts was higher in water than in organic solvents. Although di-arylation was the major product, mono-arylation could be achieved best in the presence of phosphines. The catalyst was prepared *in situ* by using $[RuCl_2(p\text{-cymene})]_2$ and KOPiv or potassium acetate (KOAc).

Dixneuf and coworkers [26c, d] also successfully demonstrated that *ortho*-diarylation of arylimines can be performed in water by using $[RuCl_2(p\text{-cymene})]_2$/KOAc/$PPh_3$ as a catalyst and with K_2CO_3 as a base. Although water was shown not to increase di-arylation of arylaldimines as compared to NMP, it was shown to promote di-arylation of arylketimines and phenyl oxazolines.

Although these and other related literature examples [6b] exist to show that C–H-activated C—C bond forming coupling reactions can be carried out in an aqueous medium, still the practice of doing organic reactions in water is not prevalent. Most of the reactions are carried out under sealed tube conditions. Carrying out C–H-activated C–C coupling reactions in water in a selective manner in the presence of multiple functional groups is still a challenge. Ackermann and coworkers [14] reported that arylation of aryltetrazole could be conducted in organic solvents such as toluene, but water was not considered a suitable solvent. Seki [13d] also had the same observation about the detrimental effect of water in Ru-catalyzed C–H activation reactions. Additionally, there are still problems related to di-arylation in C–H arylation reactions. Toxic additives are also required to enforce the reaction [6b]. This gap was phenomenally filled by Koohang and Mehta [27] who disclosed a novel and highly efficient system generated from ruthenium(II) arene precursors with formate sources, i.e. sodium formate and ammonium formate. The C–H-activated arylation reactions were conducted in water as the only solvent and at atmospheric pressure (no need to use sealed tube conditions or to add surfactants, phosphine ligands, or a cosolvent).

Subsequently, Mehta et al. [28] reported the use of first of its class, solid, bench, and air-stable novel ruthenium complex with formato ligands, MCAT-53^{TM}, i.e. $[Ru_2Cl_2(p\text{-cymene})(HCOO)_3]Na$ (sodium η-6-p-cymene dichloro diruthenium triformato) (Figure 7.3), as a catalyst to affect aromatic C—H bond activation and C–C coupling reactions in DI or distilled water.

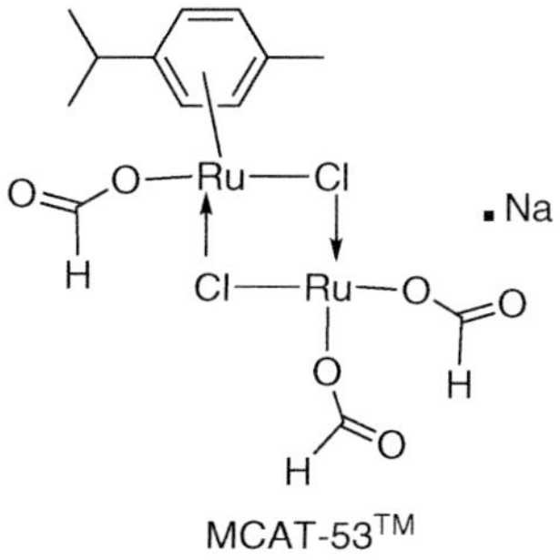

Figure 7.3 [Ru_2Cl_2(*p*-cymene)$(HCOO)_3$]Na (sodium *η*-6-*p*-cymene dichloro diruthenium triformato) MCAT 53™.

C–H-activated cross-coupling reactions were performed in a simple setup using a round-bottom flask equipped with a reflux condenser. DI or distilled water was used as a solvent, and the reaction was done under air and ligand-free condition (no need to add phosphines or other ligands) without further activation of the catalyst. Although the catalyst MCAT-53™ could not be recovered, water used in the reaction could be recycled (reused) for carrying out the same reaction again.

Mehta et al. [28] further demonstrated the usefulness of the air-stable solid ruthenium formato complex, MCAT-53™, by showcasing aryl substitution on a variety of biphenyls and phenyl-heteroaryl systems as shown in Scheme 7.9.

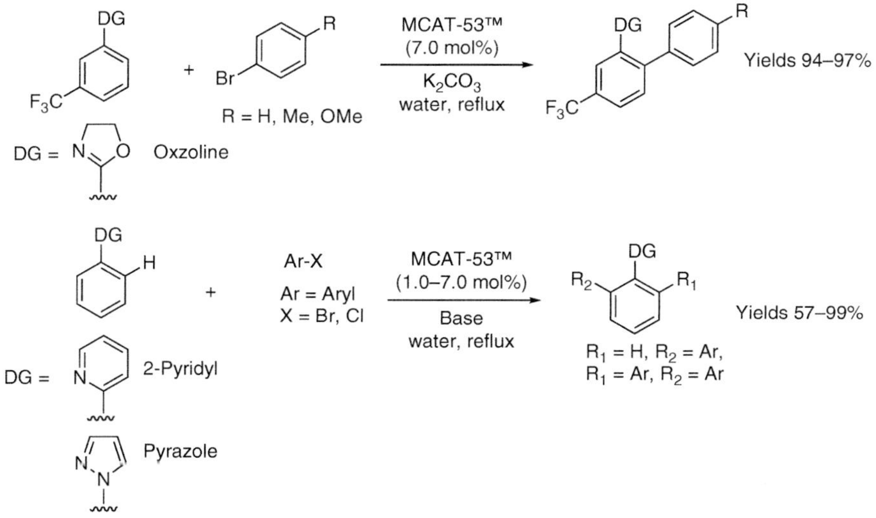

Scheme 7.9 C—H arylation using ruthenium formato complex MCAT-53™.

Although water has been reported to promote di-arylation of aryl-ketimine and phenyl oxazoline by Dixneuf et al. [26d], when ruthenium formato catalyst MCAT-53™ was used for the arylation reaction of 3-(trifluoro)-2-phenyloxazoline in water as a solvent, only mono-arylation products were formed in quantitative yields and no side product was observed.

Several heteroaromatics starting from 3-(trifluoro)-2-phenyloxazoline, 2-phenylpyridine, 2-phenylpyrazole, and benzo[*h*]quinoline were successfully synthesized using MCAT-53™ with various halides. Interestingly, the choice of base used, temperature, and the duration of the reaction was found to determine the ratio of mono- and di-arylated products. In the case of 2-phenylpyridine, mono-arylation was achieved in 9 : 1 ratio with the MCAT-53™ catalyst.

All reactions were done typically in an open atmosphere in water as a solvent under reflux, with or without the presence of a base, i.e. potassium carbonate (K_2CO_3). The reaction also avoids the use of any phosphine or other ligands/reagents or a cosolvent. Water is the sole solvent and the product, if solid, precipitates out of the reaction mixture upon cooling, while in some cases, it needs to be extracted with a small amount of ethyl acetate. The aqueous layer after the reaction could be recycled.

A new ruthenium formato catalyst, MCAT-53™, which is commercially available, promises advantages in simplicity of operation, atom economy (uses C–H activation route), easy processing, regioselectivity, and reuse (recycle) of aqueous layer and water as a benign solvent.

Although Seki and Nagahama [13a] attempted to replace NMP by more water-friendly conditions, it was found that water was detrimental to the reactions. Specifically, water in $RuCl_3{\cdot}xH_2O$ was found to have significant adverse effect on the reaction. Neither anhydrous $RuCl_3$ nor $RuCl_3$ with external addition of water included in $RuCl_3{\cdot}xH_2O$ (2.5 equiv. with respect to Ru) catalyzed the reaction. Under catalytic conditions, it was found that more than 5% water resulted in a lower conversion rate or no conversion.

For the reasons stated above, there is a need to achieve synthetic efficiency through new catalyst development, which will work in an aqueous environment for C–H activation under the ligand-free condition without the use of a cosolvent or surfactant for substrates with multifunctional groups.

The usefulness of water-friendly ruthenium formato catalyst MCAT-53™ was explored for the synthesis of CETP inhibitor anacetrapib [16]. The structure of the catalyst MCAT-53™ and successful examples of C–H arylations are described. The reaction conditions involved the condensation of the side chain of anacetrapib intermediate, i.e. 1-bromo-4-fluoro-2-methoxy-5-(propan-2-yl) benzene in the presence of a catalytic amount of MCAT-53™ and a base in DI water (Scheme 7.10). Using K_2CO_3 as a base in the presence of DI water gave poor yields of the product with heavily substituted 1-bromo-4 fluoro-2-methoxy-5-(propan-2-yl) benzene. When the same reaction was heated for a longer period (24 hours), significant decomposition was seen, and the yield of the desired product did not increase. Even the combination of MCAT-53™ in water in the presence of pivalic acid and K_2CO_3 also resulted in unsatisfactory conversion.

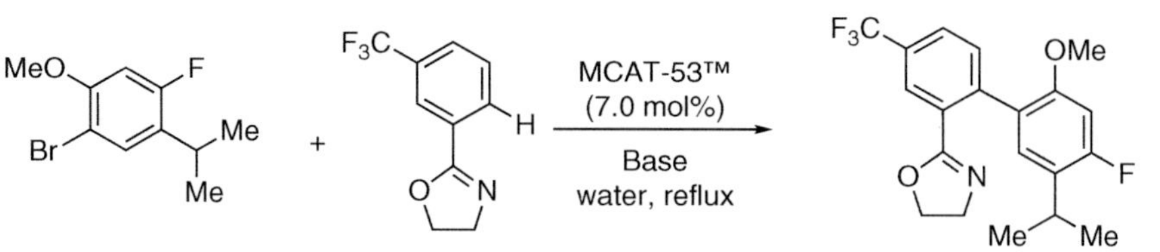

Scheme 7.10 Application of ruthenium formato complex MCAT-53™ toward anacetrapib synthesis.

Interestingly, the presence of simple and very cheap sodium acetate (CH_3COONa) and sodium formate (HCOONa) in combination with ruthenium formato catalyst MCAT-53™ provided the best conditions to obtain the desired product. In a typical reaction, using sodium acetate as a base, the isolated yield of the final product was as high as 73%.

It was reported that when $RuCl_2$(*p*-cymene) dimer was used in the presence of sodium formate under the same condition as mentioned above, the conversion based on NMR was almost quantitative, but the isolated yield was only 40% thus, showing the superiority of MCAT-53™. The use of ruthenium formato catalyst MCAT-53™ for the synthesis of CETP inhibitor has proven that the new catalyst system can promote C–H arylation reactions in DI/distilled water without requiring any cocatalysts, surfactants, or cosolvents. Additionally, the synthesis maintained a minimum number of steps and the product was easily isolated. There were no cryogenic conditions involved, and it also avoided the use of reagents and reactions that are hazardous in nature. Additionally, the reactions are done in water, which solve the problem of reproducibility and batch-to-batch variation reported by Ouellet et al. [18] and Seki and Nagahama [13a] when ruthenium-catalyzed C–H activation reactions were scaled up in NMP.

7.3 Alkyl/Acyl/Alkenyl Substitution on Heterocycles

Efficient cleavage of C—H bonds by an organometallic ruthenium complex for the subsequent addition to alkenes by C—C bond formation was first reported by Murai et al. [29]. The catalyst was shown to operate with a degree of efficiency, selectivity, and generality that made this work extremely valuable in organic synthesis. Although not the first report on C–H activation, the work of Murai et al. was remarkable for demonstrating the ease of ruthenium catalyst to cleave an inert C—H bond. The reaction has been diversified to a variety of substitutions on the arenes and heteroaryls in a simple, practical, and atom-efficient manner. The reaction has also been demonstrated with a variety of common substituents such as alkyl, alkenyl, acyls, and aryls [6b, 30].

Alkylation of aryls is one of the most common reactions in organic synthesis and done in the most practical and efficient way by C–H activation. The first example by Murai et al. describes this in a simple step with just 2 mol% of Ru catalyst in toluene in four hours [29]. Here, as in other ruthenium catalysis, *ortho*-C–H activation is directed and assisted by a directing group (i.e. methyl ketone). In cases where the directing group offers two positions for C–H activation (and hence, alkylation), the regioselectivity can be controlled by the electronic nature of the directing group; for example, in indoles [30a], an electron-donating group on ketone directs insertion at the C-2 carbon of indole while electron-withdrawing substitution promotes a C-4 insertion. The usefulness and practicality of such alkylation has been shown by alkylating nonaromatic moieties, for example, substituting tetrahydropyrans by Murai and coworkers [31a] and Trost et al. [31f]. For example, transformations of acetyl-substituted tetrahydropyrans to 3-alkyl-substituted analogs were

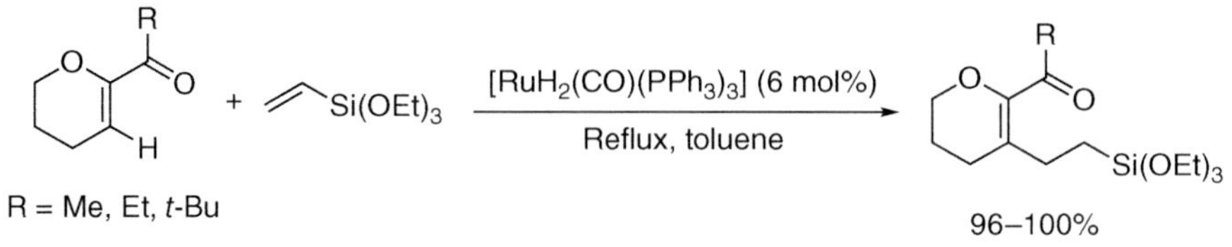

R = Me, Et, *t*-Bu

$[RuH_2(CO)(PPh_3)_3]$ (6 mol%)

Reflux, toluene

96–100%

Scheme 7.11 Examples – Alkylation of nonaromatic moieties.

performed in >96% yields by using 6 mol% of $[RuH_2(CO)(PPh_3)_3]$ in toluene under refluxing conditions for 30 minutes (Scheme 7.11).

Similarly, Moore et al. [31b] have first shown the use of sp^2-hybridized nitrogen in a heterocycle for directing acylation of aromatic rings. They have used CO gas atmosphere as a source of the new carbonyl group, which gets inserted during Ru-catalyzed C–H activation. By using this approach, a variety of heterocycles and bicyclic heterocycles were substituted with functionally useful acyl groups as exemplified by several research groups [31]. The tolerability of a number of functional groups on these heterocycles was also demonstrated.

Acylation of aromatic heterocycles with acyl chlorides is also known with Ru catalysts, i.e. $[RuCl_2(PPh_3)_3]$. Several aryl (and α,β-unsaturated) acid chlorides have been used for acylation at C-10 position of benzoquinolines (Scheme 7.12) [31g]. They make the reaction very practical by using 10 mol% of Ru catalyst in refluxing toluene with K_2CO_3 as the base to get an isolated yield up to 97%.

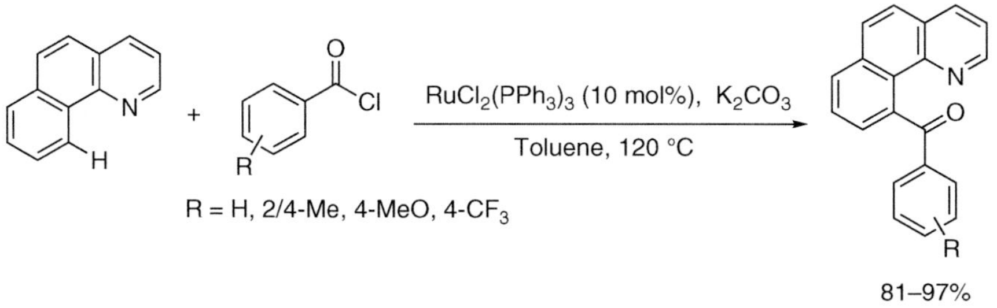

$RuCl_2(PPh_3)_3$ (10 mol%), K_2CO_3

Toluene, 120 °C

R = H, 2/4-Me, 4-MeO, 4-CF_3

81–97%

Scheme 7.12 Example – Acylation of benzoquinoline.

In addition to the direct alkylation and acylation, direct allylation has also been demonstrated by Uemura and coworkers [32] using substituted benzene, furan, thiophene, and pyrrole as substrates. The reaction has yielded predominantly the desired isomer in overall conversion up to 99% by using 5 mol% of diruthenium complex $[CpRuCl(\mu 2\text{-}S^iPr)_2RuCp^*(H_2O)]OTf]$.

Besides the O/S/N-containing directing groups, carboxylic acid has also been demonstrated for directing alkenylation on heterocyclic compounds by Miura and coworkers [33]. Ru(II) catalyst, i.e. $[RuCl_2(p\text{-cymene})]_2$, has been used in 2 mol% loading with conversion up to 95% in vinylation reactions.

In 2018, Korvorapun et al. [34] reported a sequential two fold *meta*-C–H and *ortho*-C–H functionalization by means of ruthenium(II) biscarboxylate catalyst. This showed the future of C–H activation for required alkylation/

acylation/alkenylation, where sequential substitution is also possible. Such substitutions are beneficial for generating complex heterocycles that are part of several molecules of interest in pharmceutical and chemical industries. The examples shown here are good practical examples of conversion and are ready to be exploited for its application in the industry.

7.4 C–H Activation Followed by C—O/N Bond Formation: Heterocycle Synthesis

The classical way of synthesizing saturated heterocycles is by creating an electrophilic center (with the help of coordination of Br^+, I^+, $Hg,^+$ or an organic electrophile such as $PhSe^+$ on π-cloud of either double or triple bond) followed by an intramolecular cycloaddition by nucleophilic $OH/NH_2/COOH$ groups. This has been used traditionally for many heterocycles of interest and natural products [35]. All of these methods involve a stoichiometric amount of electrophile and sequence of reactions to remove any new unwanted substituent generating from this electrophile. This makes the overall synthesis longer and tedious.

In contrast, transition-metal-catalyzed oxidative annulations involving C–H activation followed by heteroatom-H bond cleavages shorten the synthetic route and reduce the undesired waste. Not only saturated heterocycles such as substituted furans, pyrans, and isoquinolines but also many types of unsaturated heterocycles such as dihydrofurans, dihydropyrans, and dihydroisoquinolines can be prepared using C–H functionalization approach.

7.4.1 C–H Activation Followed by C—O/N Bond Formation: Heterocycle Synthesis in Organic Solvents

C–H activation followed by C—O/N bond formation is emerging as a very promising field as this approach shortens the synthetic routes of important heterocycles as demonstrated by selected examples below.

2,3-Dihydrofurans have been made using substituted petene-5-ol with up to 99% conversion by Kondo et al. [36], by using $[RuCl_2(PPh_3)_3]$ or $Ru_3(CO)_{12}$ as a catalyst (Scheme 7.13).

On a similar protocol, Trost and Rhee [37a] have demonstrated annulations of alkynol to give pyrans or lactones using ruthenium catalyst $[RuCp(triarylphosphine)_2Cl]$. For example, bis-homopropargylic alcohols have been shown to convert into dihydropyran(I) or lactone(II) (Table 7.1). The flexibility of the reaction is such that the ratio between two products can be directed toward either of the product via modulating tri-aryl phosphine ligand, ratio of Ru catalysts, and other additives, thus making this a practical route for synthesizing either compound from the same starting material. The additive triaryl phosphine chosen was based on the ligand used in the generation of the ruthenium catalyst for improving the yields. For example, using 10 mol% of $RuCp[(p\text{-}CH_3OC_6H_4)_3P]_2Cl$, 40 mol% of tri(4-methoxyphenyl)phosphine, 30 mol% of tetra-*n*-butylammonium hexafluorophosphate, 6 equiv of *N*-hydroxysuccinimide, and 2 equiv of $NaHCO_3$ in

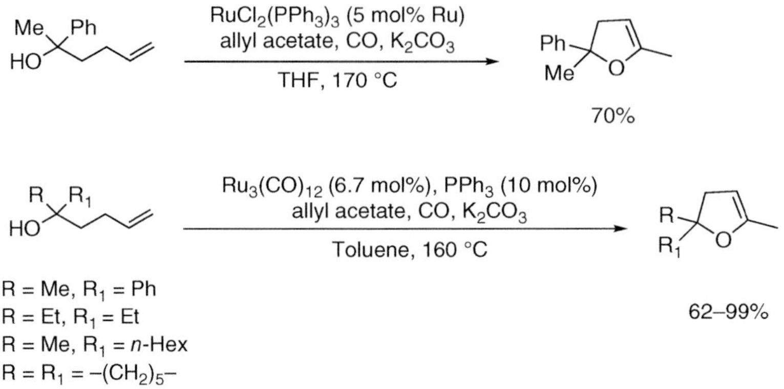

Me Ph HO

$RuCl_2(PPh_3)_3$ (5 mol% Ru)
allyl acetate, CO, K_2CO_3
THF, 170 °C

Ph Me O

70%

R R_1 HO

$Ru_3(CO)_{12}$ (6.7 mol%), PPh_3 (10 mol%)
allyl acetate, CO, K_2CO_3
Toluene, 160 °C

R R_1 O

R = Me, R_1 = Ph
R = Et, R_1 = Et
R = Me, R_1 = *n*-Hex
R = R_1 = $-(CH_2)_5-$

62–99%

Scheme 7.13 Synthesis of dihydrofurans.

Table 7.1 Different condition to switch between pyran and lactone synthesis.

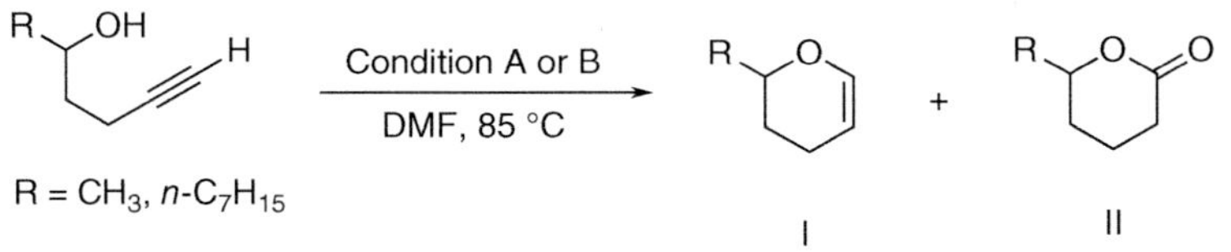

R OH H — Condition A or B, DMF, 85 °C → R O (I) + R O O (II)

R = CH_3, *n*-C_7H_{15}

Condition	Catalyst	Additives	Base	Product	Yield
A	$RuCp[(p\text{-Fluoro-}C_6H_4)_3P]_2Cl$ (5 mol%)	$P(4\text{-F-Ph})_3$ (20 mol%); $nBu_4N^+PF_6^-$ (15 mol%); N-OH-succinimide sodium salt (0.5 equiv)	—	I II	64% 4%
B	$RuCp[(p\text{-}CH_3OC_6H_4)_3P]_2Cl$ (10 mol%)	$P(4\text{-MeO-Ph})_3$ (40 mol%); $nBu_4N^+PF_6^-$ (30 mol%); N-OH-succinimide (6 equiv)	$NaHCO_3$ (2 equiv)	I II	7% 69%

DMF as a solvent, at 85 °C, gave lactone(II) as a major product (10 : 1). On the other hand, a mixture of 5 mol% $RuCp[(p\text{-Fluoro-}C_6H_4)_3P]_2Cl$, 20 mol% tri(4-fluorophenyl)phosphine, 15 mol% tetra-*n*-butylammonium hexafluorophosphate, and 50 mol% *N*-hydroxysuccinimide sodium salt in DMF as a solvent at 85 °C gave dihydropyran(I) as a major product (16 : 1).

Trost and Rhee [37a] further demonstrated the usefulness of this practical conversion by stereoselectively synthesizing narbosine B (Scheme 7.14), an antiviral *sec*-metabolite of *Streptomyces*, with two steps of Ru-catalyzed cyclization with

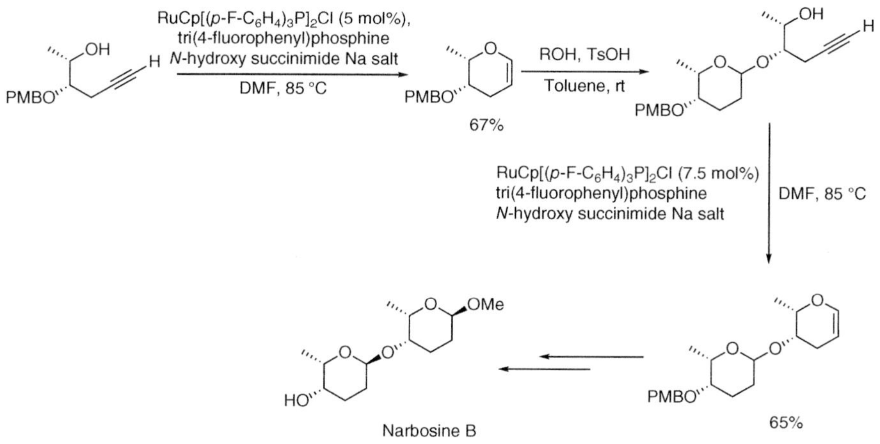

Scheme 7.14 Example – Annulation of alkynol.

similar yields of 65% and 67%. Different groups [37a–f] have used both types of practical transformations extensively.

Taking clue from this practical cycloisomerization, Merck's team [37g–i] demonstrated the use of similar ruthenium catalysts for multikilogram delivery of a key API, omarigliptin, a novel DPP-IV inhibitor for type-2 diabetes treatment. During the development of multikilogram synthesis of omarigliptin, ruthenium (Ru)-catalyzed cycloisomerization was identified as a crucial step. The key substrate propargylic alcohol was made asymmetrically from the precursor ketone using Ru(II) complex (*R,R*)-Ts-DENEB as the transfer hydrogenation catalyst. This reaction gave required propargylic alcohol in >99%ee and 92%de. This alcohol was used (without isolation) in the next step of cycloisomerization in the presence of 2.2 mol% of $RuCp[(C_6H_5)_3P]_2Cl$ and 6.6 mol% of PPh_3 to required dihydropyran in 86% yield over the two steps (Scheme 7.15). The reaction was performed at the 165 kg batch level [37i].

Similarly, alkenyl amines have also been reported [38] to cyclize in a very practical manner to give dihydro-pyrrole, tetrahydro-pyridines in the presence of a ruthenium catalyst. The same group explored several ruthenium catalysts and found that 2 mol% of $[RuCl_2(CO)_3]_2$ with 5 mol% of dppp (1,3-bis(diphenylphosphino)propane) in the presence of K_2CO_3 in NMP as the solvent at 140 °C gave quantitative yields (>99%) of the required cyclic compound (Scheme 7.16). The substitution at C2 facilitates the reaction, as unsubstituted alkenes show almost no cyclization. The geometry of the alkene does not play any role as both *cis*/*trans* gave the same product with a similar conversion.

7.4.2 C-H Activation Followed by C—O and C—N Bond Formation: Heterocycle Synthesis in Green Solvents

Alkynyl amines have been reported [39, 40] to yield cyclic amines such as dihydro-pyrrole, tetrahydro-pyridines, or higher cyclic amines via intramolecular cyclization catalyzed by Ru catalyst. The reaction has been reported in high

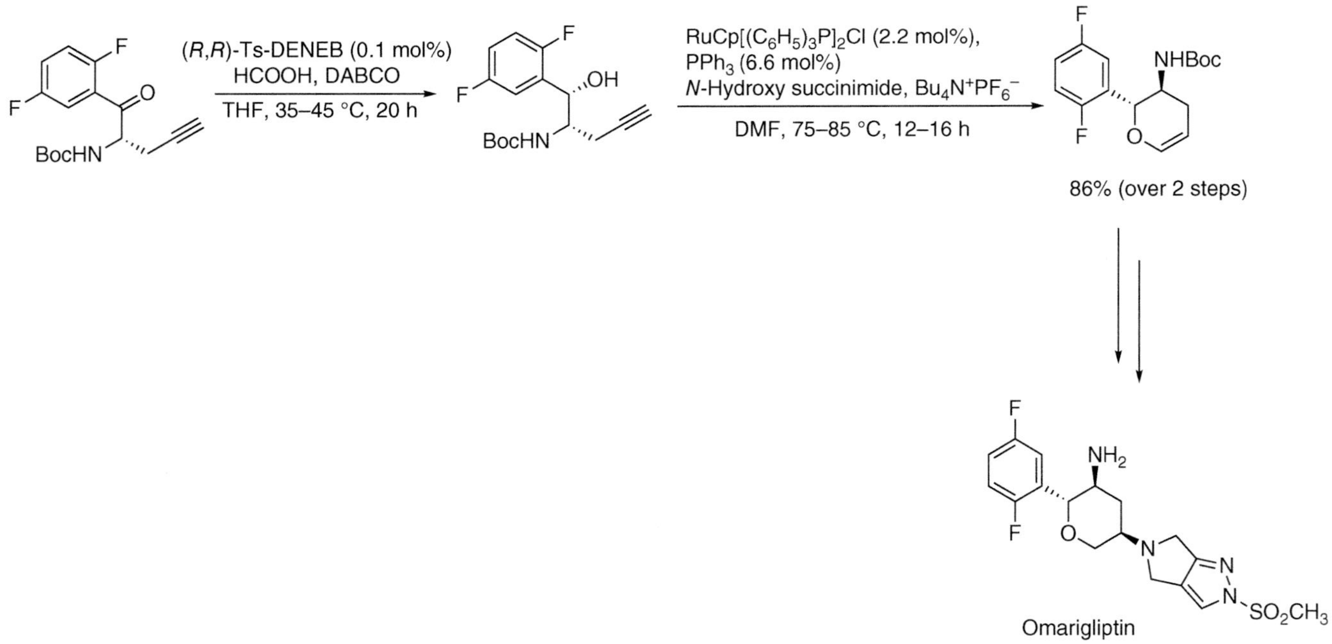

(*R*,*R*)-Ts-DENEB (0.1 mol%)
HCOOH, DABCO
THF, 35–45 °C, 20 h

$RuCp[(C_6H_5)_3P]_2Cl$ (2.2 mol%),
PPh_3 (6.6 mol%)
N-Hydroxy succinimide, $Bu_4N^+PF_6^-$
DMF, 75–85 °C, 12–16 h

BocHN
OH
NHBoc
86% (over 2 steps)

NH_2
SO_2CH_3
Omarigliptin

Scheme 7.15 Synthesis of narbosine B.

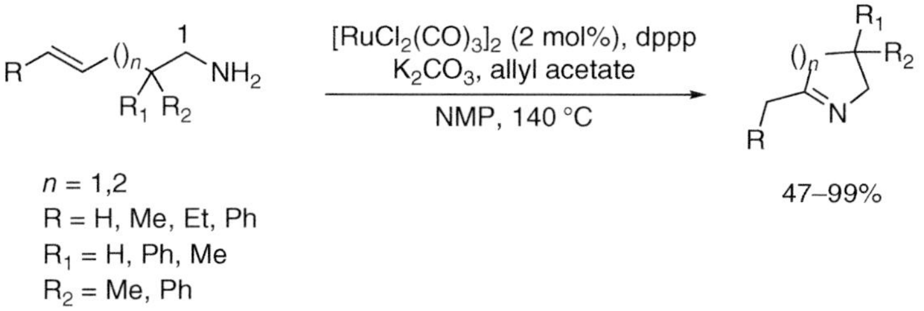

$[RuCl_2(CO)_3]_2$ (2 mol%), dppp, K_2CO_3, allyl acetate, NMP, 140 °C

n = 1,2
R = H, Me, Et, Ph
R_1 = H, Ph, Me
R_2 = Me, Ph

47–99%

Scheme 7.16 Synthesis of omarigliptin.

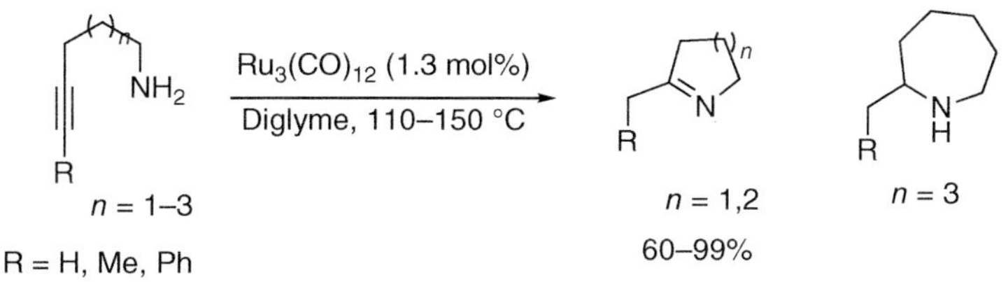

$Ru_3(CO)_{12}$ (1.3 mol%), Diglyme, 110–150 °C

n = 1–3
R = H, Me, Ph

n = 1,2
60–99%

n = 3

Scheme 7.17 Synthesis of N-containing carbocyles.

yields using diglyme as a solvent. The cyclizations have been accomplished by triruthenium dodecacarbonyl ($Ru_3(CO)_{12}$) at 110 °C or above (Scheme 7.17). The reaction is very practical with the simple procedure and yield as high as 100% with an isolated yield of 99%. However, the yield decreased with increasing ring size, and a maximum yield was obtained for dihydropyrrols. In a typical procedure, a mixture of aminoalkyne, ruthenium catalyst (1.3 mol%), and solvent was stirred at 110–150 °C for four hours.

Mechanistically, it has been postulated by the same group that oxidative addition of Ru to N—H bond is followed by alkene/alkyne insertion to Ru—N bond. The sequential reductive elimination of Ru was considered to give the desired cyclic amine.

The research groups of Dixneuf and Polshettiwar have independently reported the formation of substituted furans via a Ru-catalyzed cyclization reaction [41, 42]. Dixneuf et al. employed $[RuCl_2PPh_3(p\text{-cymene})]_2$ as the precatalyst (Table 7.2). Polshettiwar's modification of this reaction involved the use of Ru-arene RAPTA catalyst, containing a water-soluble phosphane (i.e. 1,3,5-triaza-7-phosphatricyclo[3.3.1.1^{3,7}] decane) in water as the reaction medium. They further improved the process by embedding the Ru catalyst on silica gel supported on magnetic (Fe_3O_4) particles. This meant that after completion of the reaction, the catalyst could be removed magnetically from the reaction mixture.

Jeganmohan and coworker [43a] reported a similar cyclization for the synthesis of similar isocoumarins *albeit* by using slightly different conditions. $AgSbF_6$ was chosen instead of KPF_6 and the reaction was done in DCE or *t*-butyl alcohol as a solvent. Recently, Ackermann and coworkers [43b] showed electrocatalysis as a powerful strategy for organometallic catalysis. One of the first examples of electrocatalytic C–H activation by weak O-coordination was presented (Scheme 7.18). The work exploits sustainable electricity as the sole oxidant and

Table 7.2 Synthesis of substituted furans in different condition.

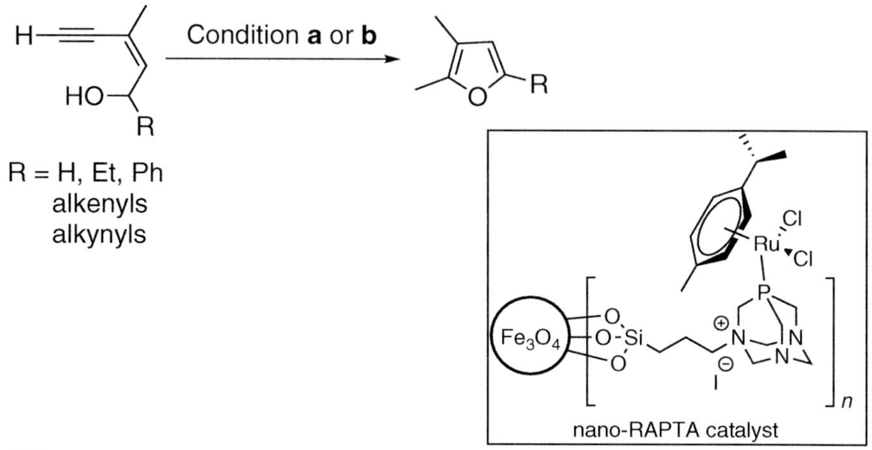

Condition	Catalyst	Solvent	Temperature	Yield
a	$[RuCl_2PPh_3(p\text{-cymene})]_2$ (1 mol%)	None or hexane or toluene	60–110 °C	50–89%
b	Nano-RAPTA catalyst (1.58 mol%)	Water	150 °C	68–99%

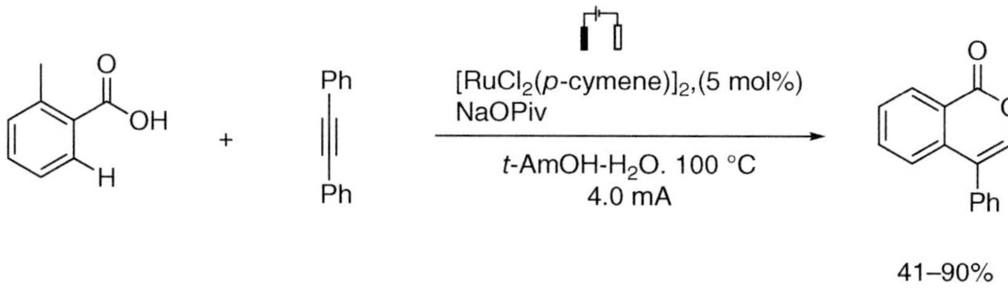

Scheme 7.18 Synthesis of isocoumarins using electrolysis.

thus avoids the use of traditional metal oxidants and yet obtain the practical yield of 90% of the product, in water as a cosolvent.

Reactions of benzoic acid with alkenes in water as a solvent have been shown [44] to go via an intermolecular oxidative alkenylation and a subsequent oxa-Michael reaction to give substituted pthalides. This type of reaction was shown to be higher yielding (up to 97%) and hence more practical with water being a reaction medium than other organic solvents (Scheme 7.19). Such a condition makes it a very practical solution for generating a variety of pthalides.

Indoles is another class of important heterocycles that are widely occurring in natural products or biologically active compounds [45a, b]. These heterocycles have been synthesized by the oxidative annulation of acetylenes with anilines in an atom-efficient manner. Ackermann and Xu groups have demonstrated

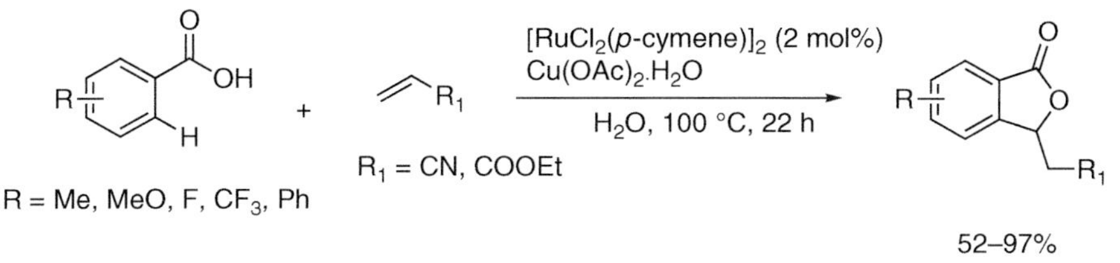

$[RuCl_2(p\text{-cymene})]_2$ (2 mol%)
$Cu(OAc)_2.H_2O$
H_2O, 100 °C, 22 h

R_1 = CN, COOEt

R = Me, MeO, F, CF_3, Ph

52–97%

Scheme 7.19 Synthesis of substituted phthalides.

[45c, d] such a reaction with the help of a cationic ruthenium(II) complex. With the reaction (i) yielding up to 94% of the product (ii) in water as a solvent (iii) using the catalyst in 5 mol% loading, this is a greener option to make indoles. The scope of the reaction has been tested on a variety of anilines as well as acetylenes, and regioselectivity has been obtained (with aryl on carbon attached to N, preferably) for unsymmetrical alkynes. Thus, 2-pyridyl or 2-pyrimidyl group plays the role of directing group on nitrogen of anilines and can be removed easily in 93% yield by simple heating in DMSO with NaOEt.

Liao et al. [45e] further made it more efficient by using 3 mol% of $[Ru_2Cl_3(p\text{-cymene})_2][PF_6]$ in a mixture of PEG-400 and water, which could be recycled up to six times without losing any catalytic activity (Scheme 7.20). The yields of up to 96% make this a good practical option for the synthesis of indoles.

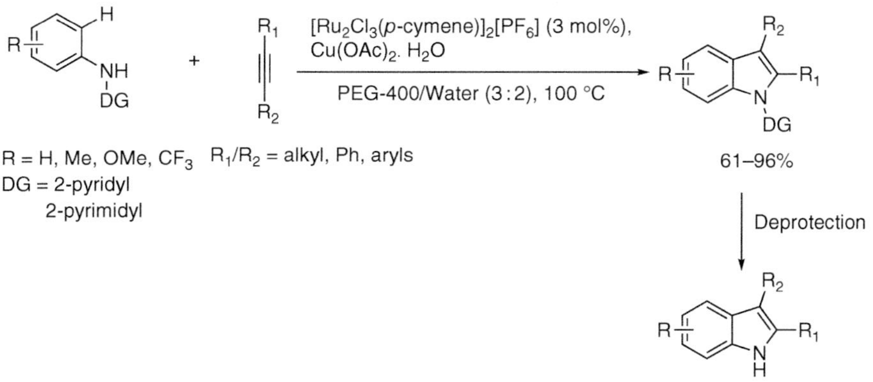

$[Ru_2Cl_3(p\text{-cymene})]_2[PF_6]$ (3 mol%),
$Cu(OAc)_2$. H_2O
PEG-400/Water (3 : 2), 100 °C

R = H, Me, OMe, CF_3 R_1/R_2 = alkyl, Ph, aryls
DG = 2-pyridyl
2-pyrimidyl

61–96%

Deprotection

Scheme 7.20 Synthesis of substituted indoles.

N-heterocycles such as dihydroisoquinolines are ubiquitous building blocks present in alkaloids [46a–c] and drug-like drotaverine [46d]. Their practical synthesis is utmost required for making such compounds on a large scale. Synthesis of 3,4-dihydro-isoquinolines has been achieved by oxidative C–H activation of arenes to get required *ortho*-directed alkenylation via C-Ru(II)-N cyclic intermediate. A subsequent reductive elimination then leads to a required cyclized compound (Scheme 7.21). The presence of internal oxidative directing group (*N*-methoxybenzamide) not only ensured exclusive *ortho*- and mono-substitution of alkenyl group but also avoided the use of external oxidative agents in the reaction. The corresponding alkenylated arenes can be

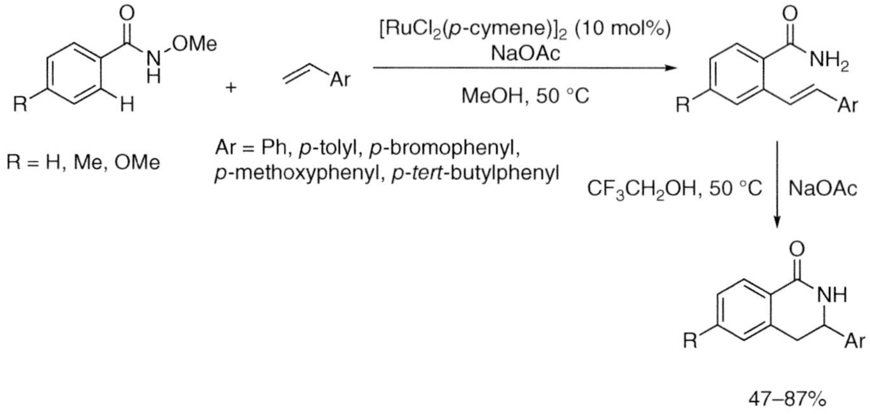

Scheme 7.21 Synthesis of dihydro-isoquinolones.

isolated by using conditions reported by Li et al. [46e]. However, the use of NaOAc and CF_3CH_2OH in the same step yielded the intramolecular cyclized product in up to 80% yield. The scope of R group variation has been studied widely with inclusion of methoxy, fluoro, chloro, bromo, iodo, nitro, ester, and acetyl groups. In a typical procedure, substituted *N*-methoxybenzamide with 10 mol% of $[RuCl_2(p\text{-cymene})]_2$ and NaOAc in methanol at 50 °C gave alkenylated product, which on further stirring with NaOAc and CF_3CH_2OH, at the same temperature in methanol for 24–36 hours, yielded the corresponding 3,4-dihydro-isoquinolines. Using the same reaction, the key core structure of annotinolide-A (a *Lycopodium* alkaloid that shows anti-aggregation activity against Aβ1-42 peptide) has been synthesized [46f].

Similar oxidative annulation of alkynes on aromatic amides has been demonstrated by Ackermann et al. [47a] to make isoquinolones (Scheme 7.22). The regioselectivity and reactivity (in case of substituted benzoyl amide) are decided by acidity of the C–H proton, i.e. higher acidity favors the chance of cyclization. The regioselectivity (in case of unsymmetrical acetylene) is obtained with R_3 on aryl group attaching preferably to N of amide. The reactivity of acetylene changes with the electron-withdrawing nature of the R_2/R_3 substituent, for example, di(*p*-fluorophenyl) acetylene gives better reactivity and conversion than di(*p*-methoxyphenyl) acetylene. Isolated yields of up to 81% were achieved and reactions were performed on a variety of substrates.

The reaction has also been reported [47b] for isoquinolone synthesis by using *N*-methoxybenzamides in water as a solvent and in yields up to 93% (Scheme 7.22), thus making these reactions more practical as well as sustainable. The reaction uses a ruthenium(II) carboxylate complex as the catalyst, which makes the direct use of free hydroxamic acids possible for required annulations. Kumaraswamy's group [47c] also described the practical synthesis of different quinolones from *N*-arylbenzamides.

The same group further modified the above reactions by avoiding the use of an external oxidant such as copper acetate and demonstrated the reaction in water as solvent, thus making the reaction greener and more practical [48]. The

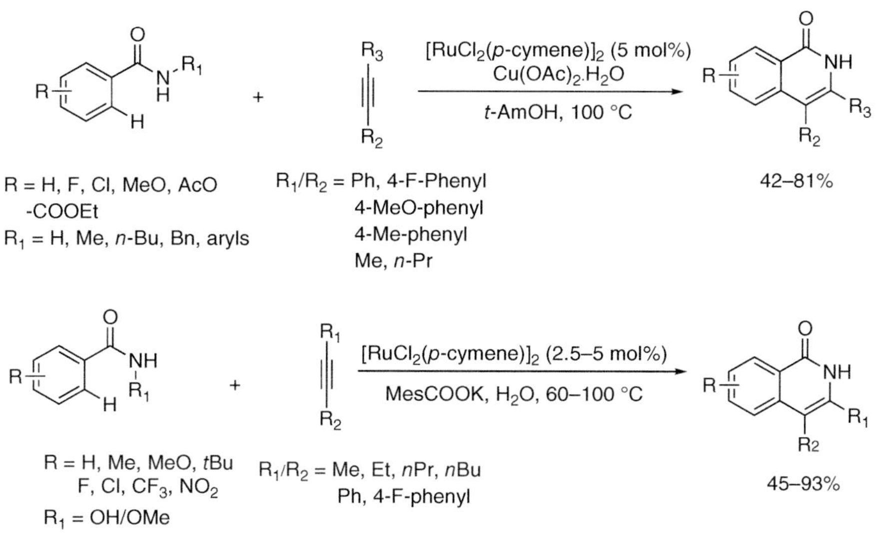

$[RuCl_2(p\text{-cymene})]_2$ (5 mol%)
$Cu(OAc)_2.H_2O$
t-AmOH, 100 °C

R = H, F, Cl, MeO, AcO
-COOEt
R_1 = H, Me, n-Bu, Bn, aryls

R_1/R_2 = Ph, 4-F-Phenyl
4-MeO-phenyl
4-Me-phenyl
Me, n-Pr

42–81%

$[RuCl_2(p\text{-cymene})]_2$ (2.5–5 mol%)
MesCOOK, H_2O, 60–100 °C

R = H, Me, MeO, tBu
F, Cl, CF_3, NO_2
R_1 = OH/OMe

R_1/R_2 = Me, Et, nPr, nBu
Ph, 4-F-phenyl

45–93%

Scheme 7.22 Synthesis of isoquinolones from benzamides.

effective dehydrative reaction between C–H and N–OH generates water as a sole by-product.

Yedage and Bhanage [49a] added value with the use of Ru(II)-PEG400 as a catalyst, which can be recycled. Simple reaction conditions and workup procedure are required, thus making it more sustainable. The same group demonstrated the use of *N*-hydroxamic acids, Weinreb's amides, and benzoic acids to get corresponding cyclized products. Jeganmohan and coworkers [49b] demonstrated the expansion of the substrate scope by making different isoquinolines using keto-oximes as a substrate (Scheme 7.23). In a typical procedure, the oxime was treated with an alkyne in the presence of $[RuCl_2(p\text{-cymene})]_2$ (2.5 mol%) and NaOAc (25 mol%) in MeOH at 100 °C for 16 hours, and the isoquinoline derivatives were obtained in up to 92% isolated yield.

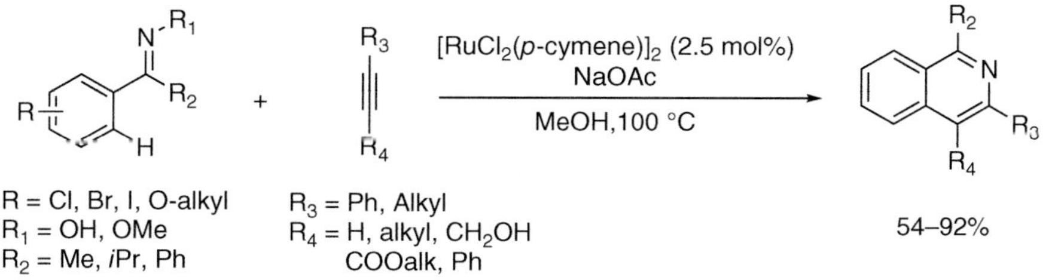

$[RuCl_2(p\text{-cymene})]_2$ (2.5 mol%)
NaOAc
MeOH,100 °C

R = Cl, Br, I, O-alkyl
R_1 = OH, OMe
R_2 = Me, iPr, Ph

R_3 = Ph, Alkyl
R_4 = H, alkyl, CH_2OH
COOalk, Ph

54–92%

Scheme 7.23 Synthesis of isoquinolines from keto oximes.

In line with benzamides/benzoximes, benzoic acids were also reported by Ackermann et al. [50] for oxidative annulation to give *iso*-coumarins. Here, cycloruthenation is assisted by acetate ion, which is then followed by alkyne coordination. Regioselective migratory insertion followed by reductive elimination then gives the desired iso-coumarins. The yields of such coumarins

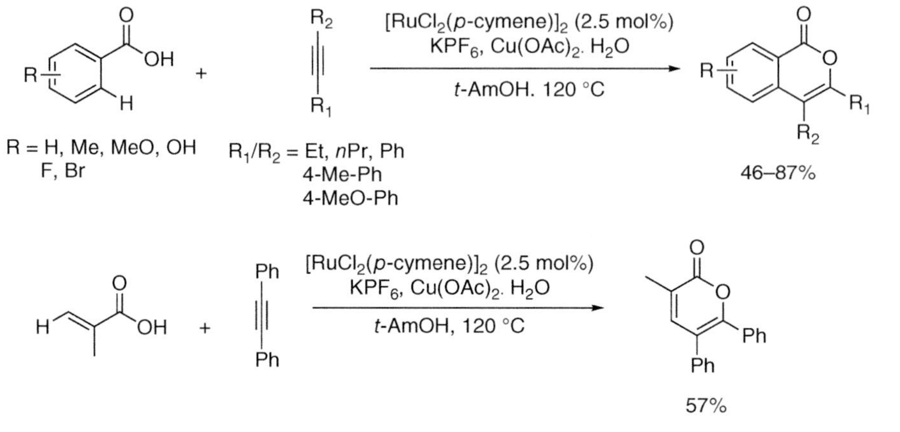

Scheme 7.24 Synthesis of isocoumarin and α-pyrone.

have been very high (up to 87%) (Scheme 7.24). Furthermore, acrylic acid has also been explored to give α-pyrone.

In Section 7.4, we have described several examples of synthesis of heterocycles ranging from saturated heterocyclic compounds to aromatic heterocyclic compounds by direct C–H functionalization followed by C—O/N bond formation using ruthenium catalysts. Some of these methods are sustainable as they involve very low catalyst loading and/or use recyclable version of catalysts. Development of more reliable and efficient ruthenium-based catalysts will make C–H functionalization followed by C—O/N bond formation more practical and useful. As heterocycles listed above are common scaffolds in different pharmaceuticals, these methods can potentially be used in the scale up of various pharmaceutical agents.

7.5 Conclusion

Among all transition metals, ruthenium catalysts can be considered superior with respect to cost, practicability, stability, and tolerance to a variety of functional groups as well as oxidants. Additionally, several ruthenium catalysts are air-stable and can operate in various environmentally friendly (green) reaction media including water.

This chapter has surveyed the significant contributions of ruthenium-catalyzed C–H activation for the synthesis of heterocycles by classifying them on the basis of C–H activation reactions that work in organic solvents (and largely water-free conditions) and others that work in environmentally friendly solvents including water. Examples of practical, multikilo preparation of drug molecules such as valsartan, candesartan cilexetil, losartan, irbesartan, olmesartan medoxomil, anacetrapib, and omarigliptin described in this chapter illustrate that simple and inexpensive ruthenium complexes such as $RuCl_3$, $Ru(arene)Cl_2$ dimers, $Ru_3(CO)_{12}$, and others can match and exceed the activity and efficacy of catalysts derived from more expensive transition metals such as palladium and rhodium.

Widespread success at small-scale reactions with ruthenium complexes such as $[Ru_2Cl_2(p\text{-cymene})(HCOO)_3]Na$ (MCAT-53™) in green solvents such as water under atmospheric pressure also hints at their potential for industrial applications. We hope that this review will be helpful for readers who are interested in developing cheaper, practical, and environmentally friendly synthetic catalytic reactions in this area.

Biography

Anita Mehta is a Ph.D. graduate from the University of Delhi (India) with postdoctoral research experience from SUNY at Stony Brook (USA), Manchester University (UK), and ICSN-CNRS (France). She was employed at Ranbaxy Labs India (now Sun Pharma) for nine years from 1993 till 2002 and moved up the ranks to become the Associate Director, New Drug Discovery Research (NDDR) before working for Saintlife Inc. (USA) as a chief scientific officer. She also worked as VP, R&D at Avocet Polymer Technologies Inc. (USA) till December 2006. Subsequently, she worked as a technical services manager at TFM plant of Freeport McMoRan (USA). She is currently president of Chicago Discovery Solutions (USA) and is actively involved in the design of new active ruthenium metal catalysts for green and sustainable chemistry. She is an inventor/coinventor in 25 US patents (issued) and has 26 publications.

Naresh Kumar carried out his undergraduate to postgraduate studies at the University of Delhi (Delhi, India) in 1993. He did his doctoral studies with Prof. VS Parmar at the University of Delhi in 1997. Then, he went for his postdoctoral work with Prof Eric J Thomas at the University of Manchester (Manchester, UK). During the post doctoral work, he completed the first total synthesis of a complex natural product Pamamycin-607. He returned to India in 2001 and joined a couple of industries to work in the area of medicinal chemistry and synthetic chemistry. He has 28 publications and is a coinventor in 19 patent applications. He is presently working as the Sr. Director with GVK BIO(Hyderabad, India) and managing several chemistry and medicinal chemistry programs.

Biswajit Saha completed his undergraduate studies (B.Sc.) in Chemistry at Presidency College, India. He received M.Sc. in Chemistry from Indian Institute of Technology Kanpur, India, where he worked in the area of small-molecule synthesis for therapeutic applications. Then, he was awarded Monbusho Scholarship to pursue graduate studies in Kyushu University, Fukuoka, Japan, where he completed his Ph.D. in 2003, under Prof. Tsutomu Katsuki. His doctoral research was focused on the development of new environmentally benign chiral catalysts and application toward the synthetic methodology development. In 2003, he moved to The Ohio State University, USA, for postdoctoral research with Prof. T. V. RajanBabu and continued the research work in the area of homogeneous asymmetric catalysis. His research articles are well cited and published in the peer-reviewed international journals. Currently, in the scientist role, he is affiliated with Chicago Discovery Solutions, USA.

References

1 (a) Crabtree, R.H. (2003). *The Organometallic Chemistry of the Transition Metals*, 5e. Wiley-VCH. (b) Fuhrhop, J.H. and Li, G. (2003). *Organic Synthesis*, 3e, 468. Weinheim, Germany: Wiley-VCH. (c) Colacot, T.J. (2014). *New Trends in Cross-Coupling: Theory and Applications*. UK: RSC. (d) Seechurn, C.C.C.J., Kitching, M.O., Colacot, T.J., and Snieckus, V. (2012). *Angew. Chem. Int. Ed.* 51: 5062–5085.

2 (a) Minami, Y. and Hiyama, T. (2018). *Chem. Lett.* 47: 1–8. (b) Bhattacharya, T., Pimparkar, S., and Maiti, D. (2018). *RSC Adv.* 8: 19456–19464. (c) Zweig, J.E., Kim, D.E., and Newhouse, T.R. (2017). *Chem. Rev.* 117: 11680–11752. (d) Davies, H.M.L. and Morton, D. (2016). *J. Org. Chem.* 81: 343–350. (e) Basu, D., Kumar, S., Sudhir, V.S., and Bandichhor, R. (2018). *J. Chem. Sci.* 71: 130–131. (f) Wu, X.-F. (2016). *Transition Metal Catalyzed Heterocyclic Synthesis via C–H Activation*. Wiley-VCH. (g) Duarah, G., Kaishap, P.P., Begum, T., and Gogoi, S. *Adv. Synth. Catal.* https://doi.org/10.1002/adsc.201800755.

3 U.S. Geological Survey (2019). Mineral commodity summaries 2019: U.S. Geological Survey, p. 200, https://doi.org/10.3133/70202434 (accessed 16 September 2019) (Manuscript approved for publication February 28, 2019).

4 (a) Murahashi, S.I. (ed.) (2005). *Ruthenium in Organic Synthesis*. Weinheim, Germany: Wiley. (b) Bruneau, C. (2004). *Ruthenium Catalysts and Fine Chemistry*, vol. 11 (ed. P.H. Dixneuf), 45. Heidelberg, Germany: Springer. (c) Dixneuf, P.H. and Cadierno, V. (2013). *Metal Catalyzed Reactions in Water*. Wiley.

5 (a) Hassan, J., Sevignon, M., Gozzi, C. et al. (2002). *Chem. Rev.* 102: 1359–1496. (b) Horton, D.A., Bourne, G.T., and Smythe, M.L. (2003). *Chem. Rev.* 103: 893–930. (c) Nareddy, P., Jordan, F., and Szostak, M. (2017). *ACS Catal.* 7: 5721–5745. (d) Brown, D.G. and Bostron, J. (2016). *J. Med. Chem.* 59: 4443–4458. (e) Denmark, S.E., Smith, R.C., Chang, W.-T.T., and Muhuhi, J.M. (2009). *J. Am. Chem. Soc.* 131: 3104–3118.

6 (a) Oi, S., Susumu, F., Naoki, H. et al. (2001). *Org. Lett.* 3: 2579–2581. (b) Arockiam, P.B., Bruneau, C., and Dixneuf, P.H. (2012). *Chem. Rev.* 112: 5879–5918. (c) Ackermann, L., Althammer, A., and Born, R. (2007). *Synlett*: 2833–2836. (d) Ackermann, L., Althammer, A., and Born, R. (2008). *Tetrahedron* 64: 6115–6124. (e) Bogdan Štefane, B., Fabris, J., and Požgan, F. (2011). *Eur. J. Org. Chem.* 19: 3474–3481.

7 Cao, H., Zhan, H.Y., Shen, D.S. et al. (2011). *J. Organomet. Chem.* 696: 3086–3090.
8 (a) Kubo, K., Kohara, Y., Yoshimura, Y. et al. (1993). *J. Med. Chem.* 36: 2343–2349. (b) Bernhart, C.A., Perreaut, P.M., Ferrari, B.P. et al. (1993). *J. Med. Chem.* 36: 3371–3380. (c) Wexler, R.R., Greenlee, W.J., Irvin, J.D. et al. (1996). *J. Med. Chem.* 39: 625–656. (d) Duncia, J.V., Chiu, A.T., Carini, D.J. et al. (1990). *J. Med. Chem.* 33: 1312–1329.
9 Carini, J.D., Duncia, J.V., Aldrich, P.E. et al. (1991). *J. Med. Chem.* 34: 2525–2547.
10 Meyers, A.I. and Mihelich, E.D. (1975). *J. Am. Chem. Soc.* 97: 7383–7385.
11 Negishi, E., King, O.A., and Okukado, N. (1994). *J. Org. Chem.* 42: 1821–1823.
12 Larsen, R.D., King, A.O., Chen, C.Y. et al. (1994). *J. Org. Chem.* 59: 6391–6394.
13 (a) Seki, M. and Nagahama, M. (2011). *J. Org. Chem.* 76: 10198–10206. (b) Seki, M. (2011). *ACS Catal.* 1: 607–610. (c) Seki, M. (2014). *Synthesis* 46: 3249–3255. (d) Seki, M. (2016). *Org. Process Res. Dev.* 20: 867–877. (e) Seki, M. (2015). *Synthesis* 47: 2985–2990. (f) Seki, M. (2014). *ACS Catal.* 4: 4047–4050. (g) Seki, M. (2012). *Synthesis* 44: 3231–3237. (h) Seki, M. (2014). *RSC Adv.* 4: 29131–29133. (i) Seki, M. *Synthesis* 47: 1423–1435.
14 (a) Diers, E., Kumar, N.Y.P., Mejuch, T., Marek, I., and Ackermann, L. (2013). *Tetrahedron* 69: 4445–4453. (b) Diers, E., Dissertation. Georg-August-Universität Göttingen (2013). Ruthenium-catalyzed synthesis of biaryls through C-H Bond functionalizations. (c) Ackermann, L. (2015). *Org. Process Res. Dev.* 19: 260–269.
15 Vergeer, M. and Stroes, E.S. (2009). *Am. J. Cardiol.* 104: 32–38.
16 O'Neill, E., Sparrow, C.P., Chen, Y. et al. (2007). *Clin. Lipidol.* 1: 367.
17 (a) Ali, A., Napolitano, J. M., Deng, Q. et al. (2006). CETP inhibitors, PCT Int. Appl. WO 2006014413. (b) Ali, A., Napolitano, J.M., Deng, Q. et al. (2007). CETP inhibitors, PCT Int. Appl. WO 2006014357; (c) Ali, A. and Sinclair, P.J. (2007). PCT Int. Appl. WO 2007041494; (d) Ali, A., Sinclair, P.J., and Taylor, G.E. (2007). PCT Int. Appl. WO 2007081570; (e) Ali, A., Lu, Z., Sinclair, P.J. et al. (2007). PCT Int. Appl. WO 2007079186.
18 Ouellet, S.G., Roy, A., Molinaro, C. et al. (2011). *J. Org. Chem.* 76: 1436–1439.
19 (a) Ackermann, L., Vicente, R., and Althammer, A. (2008). *Org. Lett.* 10: 2299–2302. (b) Ackermann, L., Vicente, R., Potukuchi, H.K., and Pirovano, V. (2010). *Org. Lett.* 12: 5032–5035.
20 Miller, R.A. and Cote, A.S. (2007) Process for synthesizing a CETP inhibitor, PCT Int. Appl. WO 2007/005572.
21 ACS Green Chemistry Institute® Pharmaceutical Roundtable. Solvent Selection Guide: Version 2.0 March 21, 2011. Retrieved [September 16, 2019] from http://www.acs.org/content/acs/en/greenchemistry/research-innovation/tools-for-green-chemistry.html.
22 Crow, J.M. (2008). *Chem. World* 5: 62–65.
23 Narayan, S., Muldoon, J., Finn, M.G. et al. (2005). *Angew. Chem. Int. Ed.* 44: 3275–3279.

24 (a) Nishikata, T., Abela, A.R., and Lipshutz, H.B. (2010). *Angew. Chem. Int. Ed.* 49: 781–784. (b) Nishikata, T. and Lipshutz, H.B. (2009). *Org. Lett.* 11: 2377–2379.

25 (a) Ackermann, L. and Vicente, R. (2009). *Org. Lett.* 11: 4922–4925. (b) Ackermann, L. (2005). *Org. Lett.* 7: 3123–3125.

26 (a) Arockiam, P., Poirier, V., Fischmeister, C. et al. (2009). *Green Chem.* 11: 1871–1875. (b) Arockiam, P.B., Fischmeister, C., Bruneau, C., and Dixneuf, P.H. (2010). *Angew. Chem., Int. Ed.* 49: 6629–6632. (c) Li, B., Bheeter, C.B., Darcel, C., and Dixneuf, P.H. (2011). *ACS Catal.* 1: 1221–1224. (d) Dixneuf, P.H. (2012). *Tetrahedron* 68: 5179–5184.

27 Koohang, A.A. and Mehta, A. (2014). Ruthenium-catalyzed synthesis of biaryl compounds in water, PCT application WO/US2014/059281 and US 2016/0263565A1.

28 Mehta, A., Saha, B., Koohang, A.A., and Chorghade, M.S. (2018). *Org. Process Res. Dev.* 22: 1119–1130.

29 Murai, S., Kakiuchi, F., Sekine, S. et al. (1993). *Nature* 366: 529–531.

30 (a) Lanke, V., Bettadapur, K.R., and Prabhu, K.R. (2016). *Org. Lett.* 18: 5496–5499. (b) Rouquet, G. and Chatani, N. (2013). *Chem. Sci.* 4: 2201–2208. (c) Gao, R. and Yi, C.S. (2010). *J. Org. Chem.* 75: 3144–3146. (d) Rit, R.K., Ghosh, K., Mandal, R., and Sahoo, A.K. (2016). *J. Org. Chem.* 81: 8552–8560.

31 (a) Kakiuchi, F., Tanaka, Y., Sato, T. et al. (1995). *Chem. Lett.* 24: 679–680; (b) Moore, E.J., Pretzer, W.R., O'Connell, T.J. et al. (1992). *J. Am. Chem. Soc.* 114: 5888–5890; (c) Szewczyk, J.W., Zuckerman, R.L., Bergman, R.G., and Ellman, J.A. (2001). *Angew. Chem. Int. Ed.* 40 (1): 216–219; (d) Imotoa, S., Uemuraa, T., Kakiuchib, F., and Chatani, N. (2007). *Synlett.*: 170–172; (e) Naoto, C., Shuhei, Y., Taku, A. et al. (2002). *J. Org. Chem.* 67: 7557–7560; (f) Trost, B.M., Imi, K., and Davies, I.W. (1995). *J. Am. Chem. Soc.* 117: 5371–5372; (g) Kochi, T., Tazawa, A., Honda, K., and Kakiuchi, F. (2011). *Chem. Lett.* 40: 1018–1020.

32 Onodera, G., Imajima, H., Yamanashi, M. et al. (2004). *Organometallics* 23: 5841–5848.

33 Ueyama, T., Mochida, S., Fukutani, T. et al. (2011). *Org. Lett.* 13: 706–708.

34 Korvorapun, K., Kaplaneris, N., Rogge, T. et al. (2018). *ACS Catal.* 8: 886–892.

35 (a) Germay, O., Kumar, N., and Thomas, E.J. (2001). *Tetrahedron Lett.* 42: 4969–4974. (b) Barks, J.M., Knight, D.W., Seaman, C.J., and Weingarten, G.G. (1994). *Tetrahedron Lett.* 35: 7259–7262. (c) Kang, S.H., Hwang, T.S., Kim, W.J., and Lim, J.K. (1990). *Tetrahedron Lett.* 31: 5917–5921. (d) Kang, S.H., Hwang, T.S., Kim, W.J., and Lim, J.K. (1991). *Tetrahedron Lett.* 32: 4015–4018. (e) Ley, S.V. and Lygo, B. (1982). *Tetrahedron Lett.* 23: 4625–4628. (f) Brussani, G., Ley, S.V., Wright, J.L., and Williams, D.J. (1986). *J. Chem. Soc., Perkin Trans.* 1: 303–307. (g) Mihelich, D. and Hite, G.A. (1992). *J. Am. Chem. Soc.* 114: 7318–7319.

36 Kondo, T., Tsunawaki, F., Sato, R. et al. (2003). *Chem. Lett.* 32: 24–25.

37 (a) Trost, B.M. and Rhee, Y.H. (2002). *J. Am. Chem. Soc.* 124: 2528–2533. (b) Trost, B.M. and Young, H.R. (2004). *Org. Lett.* 6: 4311–4313. (c) Zacuto, M.J., Tomita, D., Pirzada, Z., and Xu, F. (2010). *Org. Lett.* 12: 684–687. (d) Xu, F., Zacuto, M.J., Kohmura, Y. et al. (2014). *Org. Lett.* 16: 5422–5425. (e) Mateu, N., Kidd, S.L., Kalash, L. et al. (2018). *Chem. Eur. J.* 24: 13681–13687.

(f) Trost, B.M., Malhotra, S., Mino, T., and Rajapaksa, N.S. (2008). *Chem. Eur. J.* 14: 7648–7657. (g) Chung, J.Y.L., Scott, J.P., Anderson, C. et al. (2015). *Org. Process Res. Dev.* 19: 1760–1768. (h) Feng, K., Mary M., Kohmura, Y., Sladicka, T., Rosen, J.D., and Zacuto, M.J., Process for preparing Chiral Dipeptidyl Peptidase -IV Inhibitor Intermediates, US Patent 2009/187028; (i) Arroyo, I., Krueger, D., Chen, P., Moment, A., Biftu, T., Sheen, F., and Zhang, Y., Novel Crystalline forms of a dipeptidyl peptidase-IV inhibitor, WO2013/3249.

38 Kondo, T., Okada, T., and Mitsudo, T. (2002). *J. Am. Chem. Soc.* 124: 186–187.

39 Kondo, T., Okada, T., Suzuki, T., and Mitusdo, T. (2001). *J. Organomet. Chem.* 622: 149–154.

40 Trost, B.M., Maulide, N., and Livingston, R.C. (2008). *J. Am. Chem. Soc.* 130: 16502–16503.

41 (a) Seiller, B., Bruneau, C., and Dixneuf, P.H. (1994). *J. Chem. Soc. Chem. Commun.*: 493–494. (b) Seiller, B., Bruneau, C., and Dixneuf, P.H. (1995). *Tetrahedron* 51: 13089–13102.

42 García-Garrido, S.E., Francos, J., Cadierno, V. et al. (2011). *Chem. Sus. Chem.* 4: 104–111.

43 (a) Chinnagolla, R. and Jeganmohan, M. (2012). *Chem. Comm.* 48: 2030–2032. (b) Qiu, Y., Tian, C., Massignan, L. et al. (2018). *Angew. Chem. Int. Ed.* 57: 5818–5822.

44 Ackermann, L. and Pospech, J. (2011). *Org. Lett.* 13: 4153–4155.

45 (a) Eicher, T. and Hauptmann, S. (2003). *The Chemistry of Heterocycles.* Wiley-VCH: Weinheim. (b) Joule, J.A. and Mills, K. (2000). *Heterocyclic Chemistry.* Oxford: Blackwell Science Ltd. (c) Ackermann, L. and Lygin, A.V. (2012). *Org. Lett.* 14: 764–767. (d) Xu, F., Li, Y.J., Huang, C., and Xu, H.C. (2018). *ACS Cat.* 8: 3820–3824. (e) Liao, Y., Wei, T., Yan, T., and Cai, M. (2017). *Tetrahedron* 73: 1238–1246.

46 (a) Saxton, J.E. (1995). *Nat. Prod. Rep.* 12: 385. (b) Guo, L.W. and Zhou, Y.L. (1993). *Phytochemistry* 34: 563. (c) Zhang, H. and Curran, D.P. (2011). *J. Am. Chem. Soc.* 133: 10376. (d) Padubidri, V. and Anand, E. (2006). *Textbook of Obstetrics.* BI Publications Pvt Ltd. (e) Li, B., Ma, J., Wang, N.C. et al. (2012). *Org. Lett.* 14: 736–739. (f) Kotha, S. and Gunta, R. (2017). *J. Org. Chem.* 82: 8527–8535.

47 (a) Ackermann, L., Lygin, A.V., and Hofmann, N. (2011). *Angew. Chem. Int. Ed.* 50: 6379–6382. (b) Ackermann, L. and Fenner, S. (2011). *Org. Lett.* 13: 6548–6551. (c) Allu, S. and Swamy, K.C.K. (2014). *J. Org. Chem.* 79: 3963–3972.

48 Yang, F. and Ackermann, L. (2014). *J. Org. Chem.* 79: 12070–12082.

49 (a) Yedage, S.L. and Bhanage, B.M. (2016). *Green Chem.* 18: 5635–5642. (b) Chinnagolla, R., Pimparkar, S., and Jeganmohan, M. (2012). *Org. Lett.* 14: 3032–3035.

50 Ackermann, L., Pospech, J., Graczyk, K., and Rauch, K. (2012). *Org. Lett.* 14: 930–933.

8

Cross-couplings in Water – A Better Way to Assemble New Bonds

Tharique N. Ansari[1], Fabrice Gallou[2], and Sachin Handa[1]

[1] *University of Louisville, Department of Chemistry, 2320 S. Brook St., Louisville, KY 40292, USA*
[2] *Novartis Pharma AG, Chemical & Analytical Development, Basel, 4056, Switzerland*

8.1 Introduction

Palladium-catalyzed cross-couplings are considered as the most valuable transformations to construct the C—C and C—heteroatom bonds [1]. These transformations are easily reproducible at variable scales and on molecules with complex architecture, which led rampant advancements in the arena of total synthesis, materials for electronic applications as well as in nanoscience, and related areas [2–6]. Based on the statistical data, cross-coupling reactions are ranked as one of the most applied methods in organic chemistry because of their extensive practical use, especially on the industrial scale [7–9].

A very recent development on cross-couplings includes focusing on sustainability and environmental issues [10]. From these perspectives, significant advancements have been made, i.e. development of alternate reaction media to replace toxic organic solvents [11], design of powerful catalysts for enhanced catalytic activity [12, 13], discovery of recyclable catalysts to sustain the accessible supply of palladium or other metals involved in the catalytic pathways [14], and lowering the reaction temperature with the development of sustainable and affordable technologies.

Micellar catalysis is one of the emerging technologies with the efforts to address environmental and sustainability issues on practical cross-coupling chemistry [15]. The technology is inspired by Nature to conduct organic chemistry in water, enabled by the hydrophobic effect exerted by the self-aggregated amphiphilic molecules. Increased localized concentrations of substrates and catalysts make reaction rate remarkably faster at ambient temperature and result in exquisite selectivities [15]. Oftentimes, nonionic surfactant exhibits superior activity for such reactions than the ionic surfactant. Key parameters determining the superiority of the micellar couplings over couplings in organic solvents include a selection of appropriate amphiphile that complements the lipophilicity of substrates, shape, and size of the resulting micelles, compatibility of the catalyst with the amphiphile, a sequence of addition of reaction components, etc. These parameters in micellar chemistry are derived as a result of critical analysis of the existing

Organometallic Chemistry in Industry: A Practical Approach, First Edition.
Edited by Thomas J. Colacot and Carin C.C. Johansson Seechurn.

plethora of reports [16–19]. Because the scope of this topic is too broad, this chapter only focuses on the detailed practice and procedure of micellar catalysis with selected examples of Suzuki–Miyaura, Heck, Negishi, Buchwald–Hartwig amination, and Miyaura borylation and α-arylation reactions.

In the following sections, fundamental knowledge of micellar cross-couplings and the applications of micellar catalysis in the pharmaceutical industry are provided.

8.2 Transition Metal Catalysis in Organic Solvents vs Micellar Catalysis

Conventional organic synthesis involves the use of organic solvents, which serve two purposes: (i) solubilization of reactants and catalyst [20] and (ii) interaction of solvent with the reaction components to lower the transition-state energy [21]. Therefore, the selection of the solvent for a particular reaction pathway plays a crucial role to achieve the optimal results. Depending on the type of transformation, either a polar or a nonpolar solvent is selected. The solvent's boiling point and vapor pressure are crucial parameters, which are always considered before reaction setup, especially for those reactions requiring elevated temperatures. The nature of the reaction pathway and the optimal concentration of reactants dictate reaction kinetics, while the catalyst governs chemical events involved in the bond-breaking and bond-forming processes [20]. After completion of the reaction, inorganic or water-soluble by-products are removed by water wash followed by product extraction with a common organic solvent to obtain a crude product. Finally, the product is preferably purified by recrystallization. These major operations produce enormous amounts of organic as well as aqueous waste. To circumvent this situation, application of micellar technology in organic synthesis is a good option (Figure 8.1). To better understand how micellar chemistry (chemistry in water) works, first it is necessary to understand its significant advantages and drawbacks, which are highlighted below in a step-by-step manner.

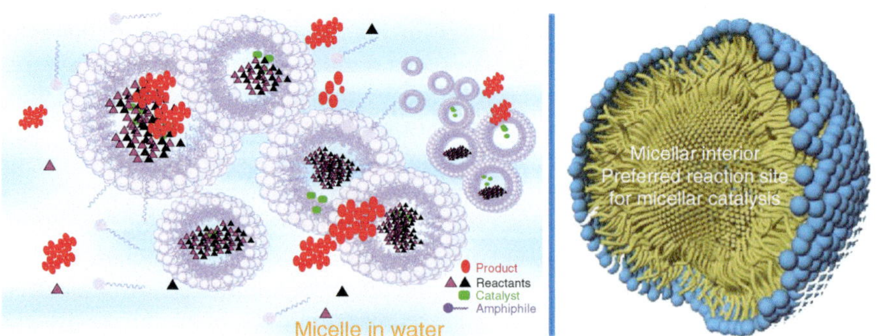

Figure 8.1 Pictorial overview of micellar catalysis.

8.2.1 Micellization

In the micellar catalysis, an aqueous micellar solution of an amphiphile (surfactant) is used as a solvent [15]. Thus, the choice of amphiphile is a key reaction variable. Upon dissolution, amphiphilic molecules self-aggregate to form micelles of variable sizes. Size of micelles depends on the nature of amphiphiles, such as the chain length of the hydrophilic and hydrophobic component and the size of the linker [15]. Longer chain length preferentially forms larger micelles. These micelles are dynamic in nature, and there is always an exchange of all components (individual amphiphile, reactants, catalyst, and product) between different micelles (Figure 8.2). Oftentimes, smaller micelles fuse to form larger micelles, which happens continuously. The purity of amphiphile and water are two critical parameters to reproduce chemistry at a large scale. Generally, high purity of amphiphile and electrolyte-free water is beneficial to make surfactant solution. Distilled water is recommended in order to bring reliable outcome. Nonionic amphiphiles are predominantly superior to ionic amphiphiles because they do not react with coupling partners and the catalyst, which could otherwise lead to side reactions.

Micellization is a crucial factor in harnessing the beneficial effect of this technology. Therefore, any variable that obliterates micellization could adversely affect the outcome. In the micellar catalysis, c. 2–3 wt% amphiphile is dissolved in water, which is far above the critical micellar concentration. Such a high concentration of surfactant is required to ensure micellization in the presence of reaction partners, catalyst, and product. More than 3 wt% surfactant concentration is oftentimes detrimental because it reduces the effective concentration of reactants in the micellar cavities, which slows the reaction rate as well as the exchange process, and exhibits higher viscosity [22–24]. However, there are some exceptions when a higher concentration of surfactant is required [25]. Many nonionic amphiphiles are now commercially available and have more extensive applications in organic synthesis [15]. The structures of commonly used amphiphiles in the micellar catalysis are provided in Figure 8.3.

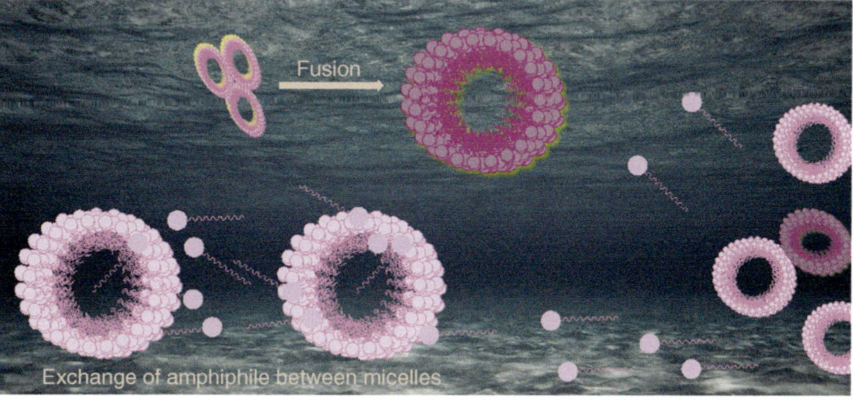

Figure 8.2 Exchange of amphiphiles between micelles.

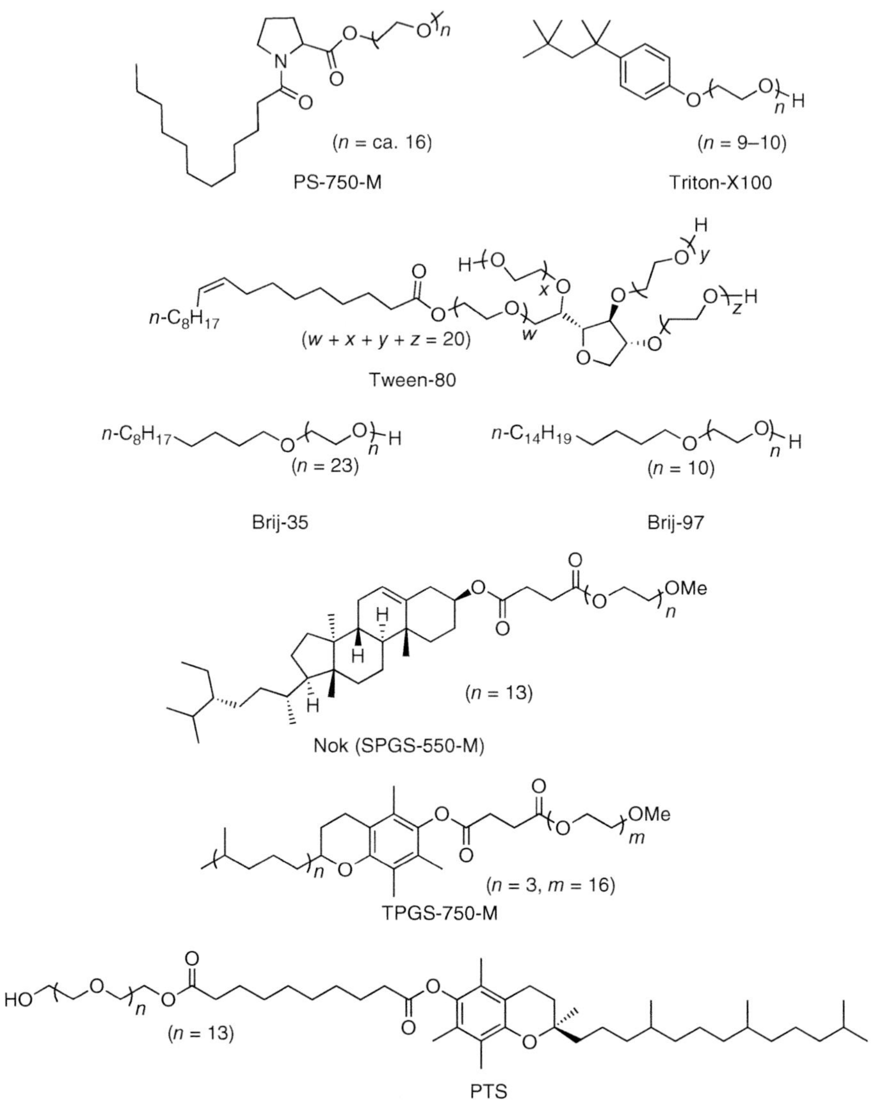

(n = ca. 16)
PS-750-M
(n = 9–10)
Triton-X100
n-C8H17
(w + x + y + z = 20)
Tween-80
n-C8H17
(n = 23)
Brij-35
n-C14H19
(n = 10)
Brij-97
OMe
(n = 13)
Nok (SPGS-550-M)
OMe
(n = 3, m = 16)
TPGS-750-M
HO
(n = 13)
PTS

Figure 8.3 Structures of commonly used nonionic amphiphiles.

8.2.2 Surfactant Solution – A Highly Organized Reaction Medium to Enhance Reaction Rate

An aqueous surfactant solution containing reactants, additives, and catalysts is a highly organized array of reaction sites (Figure 8.1). The reactant and catalyst concentration is relatively higher inside the micelles compared to the exterior, which enhances the reaction rate and enables the chemistry at the ambient temperature without side reactions. Compared to the use of organic solvents, transformations are relatively cleaner and faster in the micellar media, albeit with the low catalyst loading [26]. Because the micelles are dynamic in nature, during the exchange of amphiphiles between two or more micelles, reactants also exchange while the catalyst remains within the same micelle. Therefore, the sequence of addition is highly essential in most of the times. For example, if a catalyst is dissolved in a surfactant solution before addition of other reaction components, the probability of its uniform distribution among all the micelles is much higher compared to the situation when other components are added prior. In the latter situation when the micellar interior is already occupied with the reactants and additives, the probability of the catalyst to accommodate smaller micelles is much higher, which are more dynamic compared to the one with larger size.

8.2.3 Reaction Temperature

Transformations that require elevated temperature for desired chemoselectivity can be achieved at ambient temperature with micellar technology. The upper limit of temperature is ca. 60 °C [26]. As the temperature increases, exchange of reactants between different micelles also increases, which allow less time for the reactants to stay in the micelles to interact with the catalyst, especially when the catalyst is designed to have more lipophilicity to remain within the micelle. Whenever the elevated temperature is required for any transformation, higher weight percent of surfactant is desirable; otherwise, it may deteriorate micellization. Reaction temperature close to the boiling point of water (>80 °C) adversely affects the micellar catalysis, and under such conditions, it can be considered as on water catalysis. The surfactant may act as an additive to enhance the solubility of reactants through individual amphiphile interaction with the reactant molecules (Figure 8.4). More impurities tend to form under such conditions. Therefore, as a precaution, if the reaction is smoothly proceeding under room temperature or with milder heating, it is recommended not to overheat the reaction mixture for wrongly anticipating faster reaction rate.

8.2.4 Size of Micelles

Particle size is a very crucial factor for the successful implementation of this technology. Too small (50 nm) or too large (>500 nm) sized micelles of aqueous surfactant (without any reactant) are not very effective and beneficial. In general, other than the nature of micelles and additives, particle size expands upon

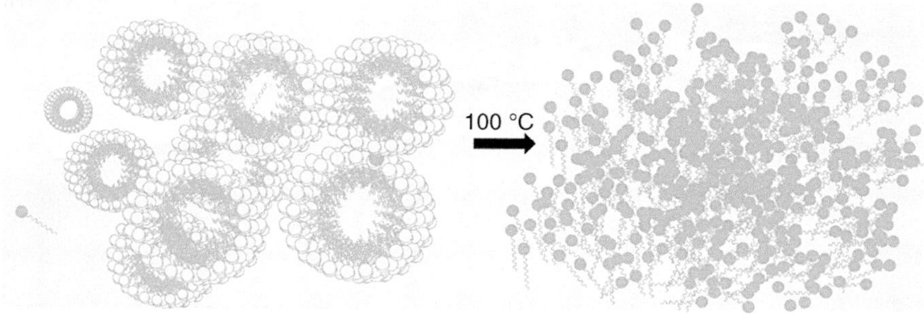

Figure 8.4 Demicellization at higher temperature.

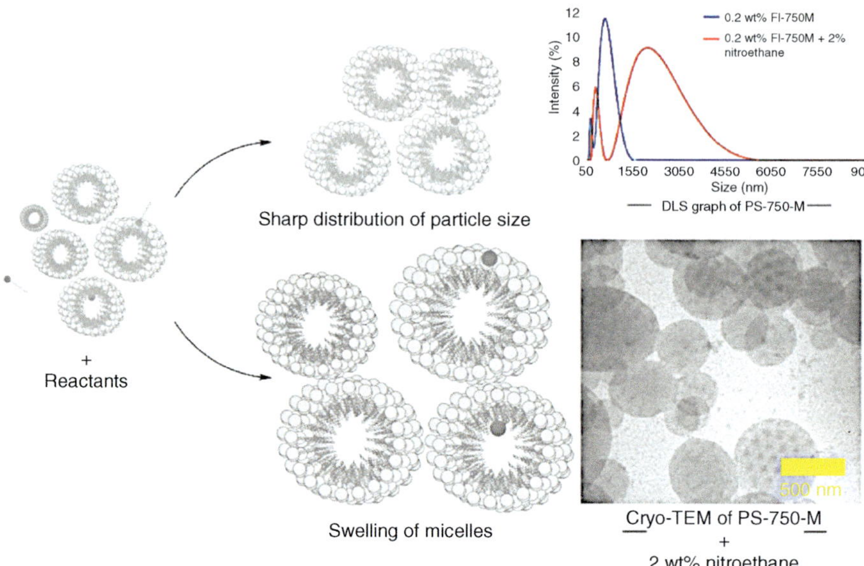

Figure 8.5 Effect of the reactant on the micellar size and distribution.

addition of miscible liquid reactants (Figure 8.5) [19]. Much larger particle size hinders the effective exchange of reactants between different micelles. This also causes retention of product in the micelles for a more extended period, resulting in slower reaction rate and difficulties in the product extraction from the reaction mixture. Selecting a surfactant that provides micelles of c. 100 nm size is typically ideal for most of the challenging palladium-catalyzed cross-coupling reactions. In general, a micellar solution is composed of micelles of different sizes, and this size distribution curve becomes sharper upon addition of reaction component [19, 27]. Upon addition of coupling partners, average particle size also exponentially increases [27]. This allows for the accommodation of higher concentration of coupling partners within the micellar core, the site where chemistry happens.

8.2.5 Nature of Catalyst

Because the micellar interior is relatively lipophilic compared to the exterior, the polarity of the catalyst plays a crucial role in determining the success of this methodology. Depending on the type of the surfactant, the polarity of micellar interior varies, e.g. the micellar interior of D/L-α-Tocopherol methoxypolyethylene glycol succinate (TPGS)-750-M and Tween is more lipophilic than that of PS-750-M [15]. Matching the polarity of the micellar interior with the catalyst could lead to the powerful catalytic system. Along the same lines, a highly lipophilic HandaPhos-Pd species in SPGS-550-M was reported as a highly efficient catalyst system, so that palladium catalysis can be done at the ppm level [12]. The reason behind the success of this catalyst system is the greater binding of catalyst in the micellar interior. The longer the time spent by the catalyst in the micellar interior means the more catalysis takes place, and therefore, less catalyst is needed for this purpose. With this approach, Suzuki–Miyaura and Sonogashira couplings have been recently reported [12, 28]. This particular approach for many other reactions is discussed in more detail in Section 8.3 [19, 27].

8.2.6 Increasing the Efficiency in Micellar Catalysis

Not only the factors listed above but many other factors contribute to enhance the efficiency in micellar catalysis. Sometimes, coupling partners employed in the catalytic process are highly crystalline, and it is impossible for those partners to get solubilized and enter the micelle. They are merely suspended in the water or settled at the bottom. Such circumstances lead to poor catalytic efficiency. There are two ways to solve this problem, i.e. (i) grinding the crystalline material to powder so that it could be soluble in the micellar media or (ii) adding 5–10% of a cosolvent, such as tetrahydrofuran (THF), toluene, and acetone [18, 29, 30]. Notably, the addition of a cosolvent causes dissolution of reactants as well as the expansion of micellar size, which may slow down the exchange process required for effective catalysis (Figure 8.6). Preferably, use of cosolvent can be avoided if the reaction smoothly proceeds without it. Alternatively, the use of an appropriate amphiphile can easily resolve the issue of solubility, e.g. an amphiphile PS-750-M is primarily designed to conduct cross-couplings between polar coupling partners [27]. Notably, the cosolvent could also solubilize the resulting desired product, which otherwise could have been crystallized in the reaction mixture and isolated by simple filtration [31].

Mild heating to carry out the reaction causes evaporation of volatile reaction components, which could result in poor reaction yield [31]. Readers may think to use a condenser to avoid such evaporation. However, this decreases the reaction efficiency because dissolution and uniform distribution of condensed material are not as easy in a micellar medium as in organic solvents. To solve this problem, an appropriate cosolvent having a boiling point very close to the volatile material must be selected. Pentane, acetone, and diethyl ether are preferred volatile cosolvents. Use of cosolvents in the reaction mixture containing volatile components allow the formation of layers of vapors of pentane or ether in the empty headspace, which increases the pressure above the reaction mixture and prevents

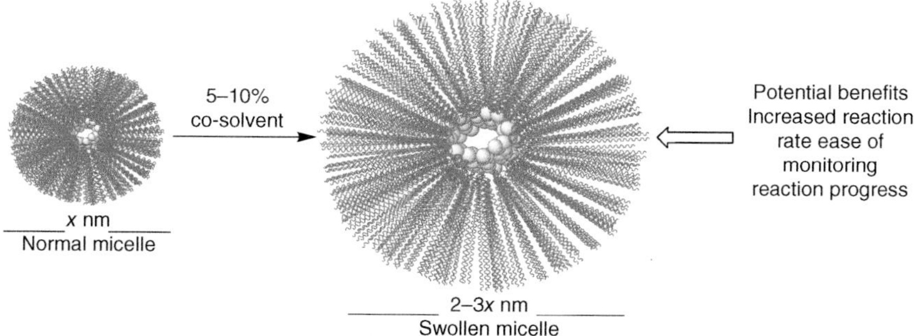

Figure 8.6 Expansion of micelle size upon addition of cosolvent.

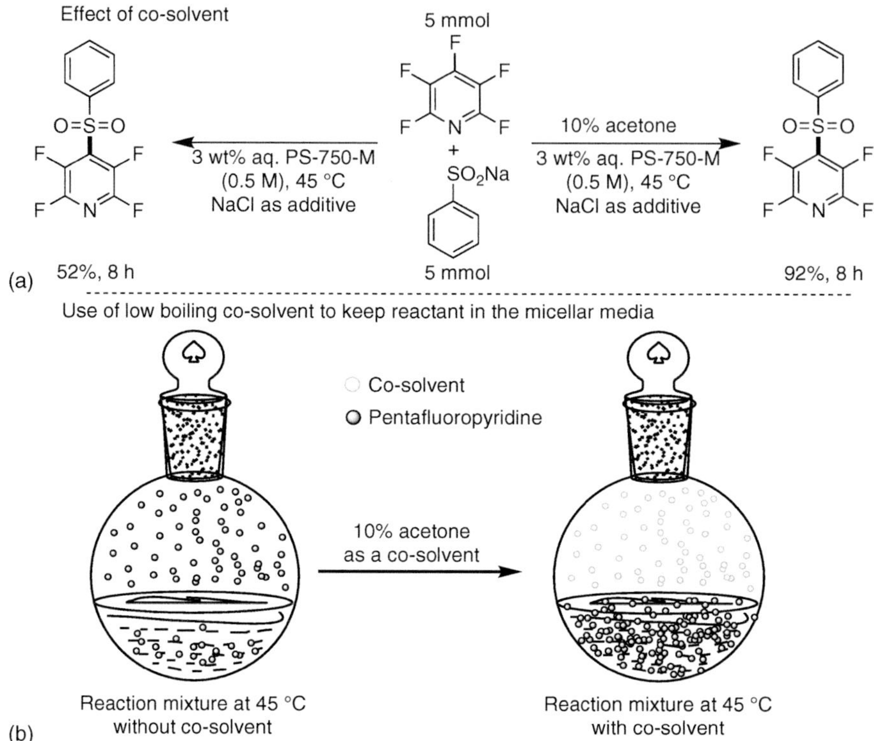

Figure 8.7 Co-solvent technique when a reaction mixture contains volatile substrate.

vaporization of volatile reaction partners. Notably, a cosolvent must have at least 5–10 °C higher boiling point than the reactants. For example, pentafluoropyridine is highly volatile, and its reaction with arylsulfinate salt does not proceed smoothly because of its high volatility [31]. It just covers the empty headspace of a tightly sealed container and the reaction occurs only at the surface where its vapors reversibly diffuse into the micelles for a reaction to proceed (Figure 8.7). Upon use of acetone as a cosolvent, the empty headspace is covered by acetone vapors, which ensures the presence of all pentafluoropyridine molecules in the micellar media. About 92% of isolated yield of the desired sulfone was obtained when acetone was used as a cosolvent.

8.2.7 Order of Addition

Catalytic reactions in the micellar media require careful attention to the order of addition, especially, when a very low amount of catalyst is used, i.e. 500–1000 ppm. Addition of the catalyst to the micellar media as the first component makes a big difference in the yield as well as reaction time. Addition of catalyst in the form of a solution using a minimal amount of cosolvent is preferred. Upon addition of the catalyst solution, the reaction mixture must be allowed to stir at room temperature for about five minutes (Figure 8.8). This

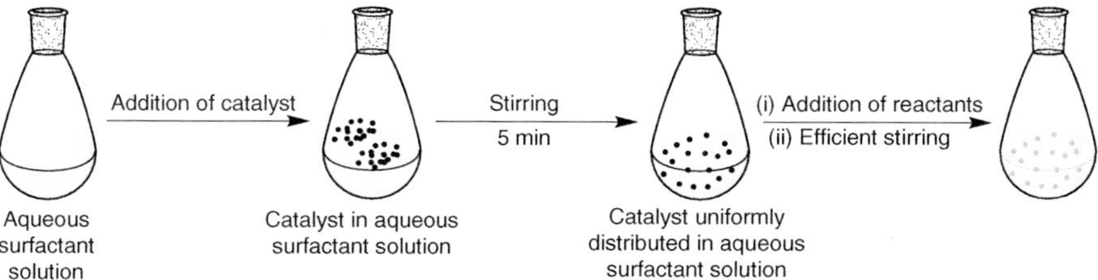

Figure 8.8 General sequence of addition. Notably, uniform and effective stirring is required from the beginning until product extraction or filtration. If the reaction is air-sensitive, an aqueous solution of surfactant should be purged with argon for a few minutes before use.

allows the catalyst to distribute in the micelles uniformly. Addition of other components must be made afterward as per reaction requirement.

8.2.8 Product Precipitation or Extraction

Depending on the solubility of the product in the surfactant solution (reaction medium) and polarity of the cosolvent used, the product either crystallizes in the surfactant solution or needs extraction with a minimal amount of the appropriate organic solvent. For monitoring the progress of the reaction, analyte must be withdrawn while the reaction mixture is stirred. The analyte must then be diluted with water and EtOAc. This process breaks the micelles and allows the product and unconsumed starting material to get dissolved in the organic solvent for accurate analysis.

After completion of the reaction, if the product appears as a solid, it could be filtered over a frit and washed with water to remove any residual surfactant. In the event of the opposite scenario, if the product is not crystallized, EtOAc (equal to the amounts of surfactant solution) should be added, and the mixture was stirred for a few minutes before centrifugation. This rule applies to reactions at all scales. Separation of organic layer after centrifugation extracts most of the product in the organic phase. If the residual product is still in the surfactant solution, it must be diluted with water and needs another extraction with a minimum amount of organic solvent.

8.2.9 Trace Metal in the Product

One of the many advantages of micellar catalysis is the very low trace metal impurities in the product (Figure 8.9). The reasons behind the low trace metal in the product are the high association of the catalyst with the micellar interior, binding of metal nanoparticles with the polyethylene glycol (PEG) chains, and upfront low catalyst loading. As a caution, dichloromethane must be avoided as a solvent for extraction because it breaks micelles and allow the metal catalyst in the organic layer.

By carefully following all points mentioned above, switching to micellar catalysis can give significant benefits such as lower reaction time, lower energy consumption, and cleaner reactions, thereby minimizing organic waste with the reduction of cost.

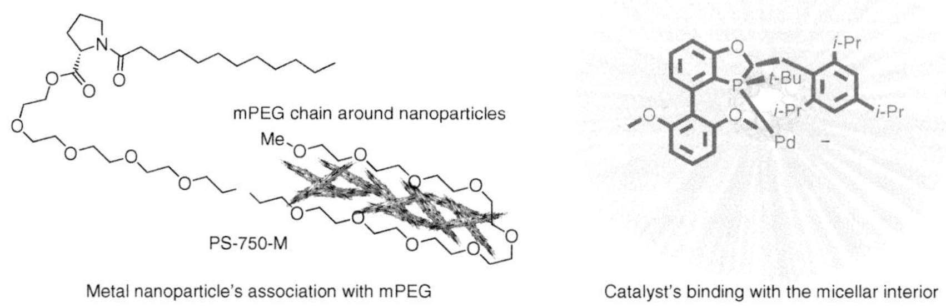

Metal nanoparticle's association with mPEG

Catalyst's binding with the micellar interior

Figure 8.9 Catalyst and micelle association.

8.3 Highly Valuable Reactions in Water

In this section, we will briefly discuss cross-couplings in micellar media followed by their applications in the pharmaceutical industry.

8.3.1 Suzuki–Miyaura Couplings

Since the first report on the topic of palladium-catalyzed cross-coupling by Suzuki and Miyaura, the cross-coupling chemistry has flourished as the most potent transformation to achieve biaryl compounds [32, 33]. The broad applications raised its popularity among different industries [7–9]. Many reports described novel ways to accomplish this transformation sustainably especially via micellar catalysis [11]. Along these lines, Lipshutz and coworkers have tremendously contributed to this area [12, 15]. In the first report, authors explored nonionic surfactant polyoxyethanyl α-tocopheryl sebacate (PTS) as the standard reaction media, although similar conversions were observed in Triton-X100, TPGS-1000, and Brij-30. Pd(dtbpf)Cl_2 (2 mol%) as an efficient catalyst [34] and triethylamine (3.0 equiv) as a base were used to obtain products derived from a variety of coupling partners (Scheme 8.1) [35]. Notably, transformations were

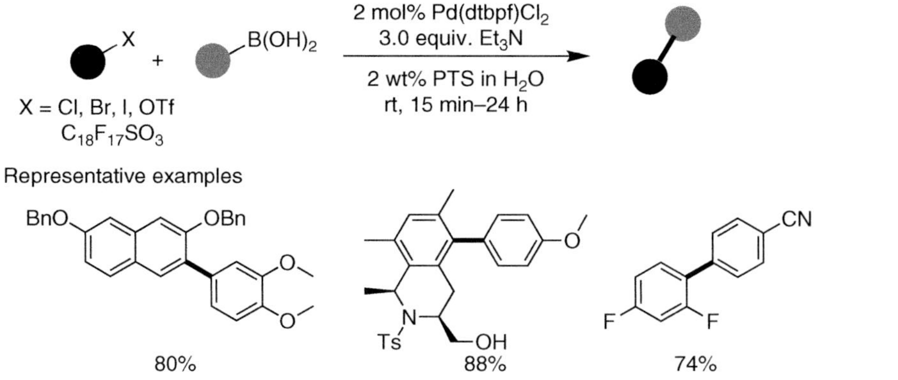

Scheme 8.1 Suzuki–Miyaura couplings in the micelles of PTS.

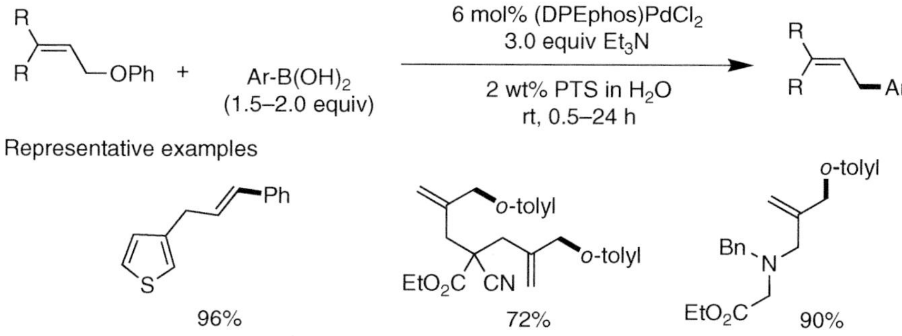

Scheme 8.2 Coupling of allylic ethers assisted by micelles of PTS.

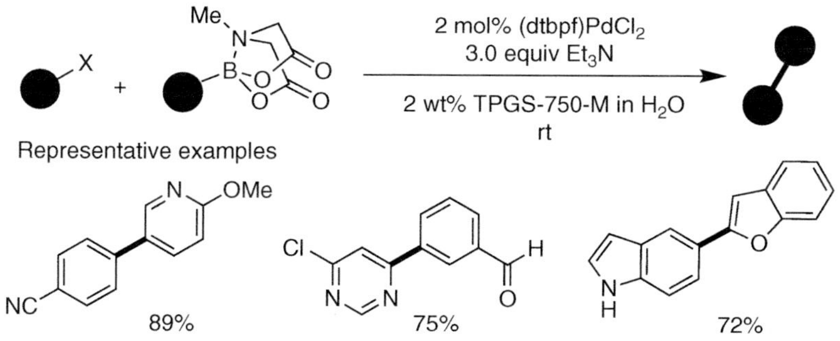

Scheme 8.3 Suzuki–Miyaura couplings – no use of organic solvents at any stage.

achieved at room temperature. Good-to-excellent reactivity was reported with boronic acids and trifluoroborate nucleophiles.

Couplings of boronic acids with challenging functionalized allylic ethers have also been documented [36]. For such systems, the catalyst containing a chelating bidentate ligand, i.e. Pd(DPEPhos)Cl_2, provides better activity, presumably because of stabilization of the allyl complex by chelate, which is formed during the catalytic cycle. Catalyst loading (2 mol%), 3.0 equiv base, and 2 wt% PTS were optimal. This reaction protocol was found to be effective for diverse substrate scope (Scheme 8.2).

In 2013, a complete organic solvent-free Suzuki–Miyaura coupling of *N*-methyliminodiacetic acid (MIDA) boronate esters with aryl halides was reported [37]. This green technology enabled couplings of MIDA boronate ester with (hetero)aryl halide in water at room temperature (Scheme 8.3). The optimized procedure utilizes 2 mol% Pd(dtbpf)Cl_2 as a catalyst with 3.0 equiv of Et_3N as a base in the nanomicelles of 2.0 wt% aqueous TPGS-750-M. This methodology was explored on a broad substrate scope with a focus on functional group tolerance. The method was also employed on a gram-scale reaction where the reaction was carried out in water and the product was isolated via simple filtration.

In 2015, micelle-enabled nanonickel-catalyzed Suzuki–Miyaura (SM) couplings were also explored (Scheme 8.4) [38]. This process was unique because of prior catalyst activation with MeMgBr and order of addition. Addition

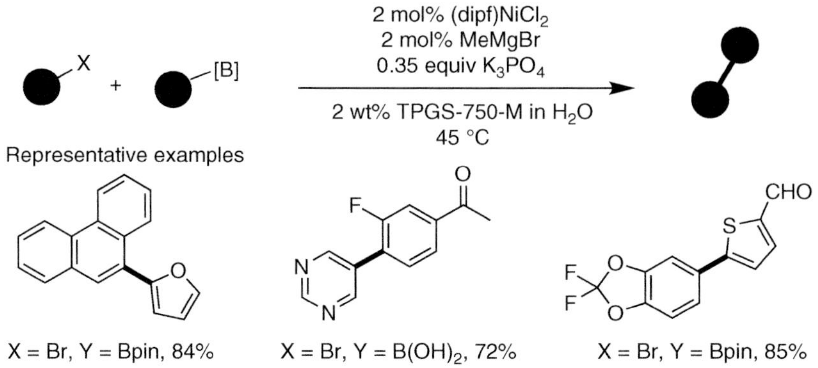

2 mol% (dipf)$NiCl_2$
2 mol% MeMgBr
0.35 equiv K_3PO_4
2 wt% TPGS-750-M in H_2O
45 °C

Representative examples

X = Br, Y = Bpin, 84%   X = Br, Y = $B(OH)_2$, 72%   X = Br, Y = Bpin, 85%

Scheme 8.4 Nano-nickel technology for cross-couplings.

of aryl halide to the active catalyst was a first key step, which stabilizes the catalyst after oxidative addition. Addition of nucleophile and base before the addition of aryl halide was found to be detrimental because of the replacement of ligand by a nucleophile, which immediately generates ligand-free Ni(II) species. Notably, in this methodology, excess of coupling partners was not required, and only 0.35 equiv of K_3PO_4 as a base was used. This methodology was explored on a variety of combinations, i.e. aryl–aryl, aryl–heteroaryl, and heteroaryl–heteroaryl. Broad substrate scope and excellent functional group tolerance were documented while reaction yields were moderate to excellent.

Catalysis using copper and ppm levels of palladium toward selective Suzuki–Miyaura couplings exclusive to aryl iodides was also reported in micellar media (Scheme 8.5)[39]. Unlike palladium, copper-catalyzed reactions start with transmetallation by aryl boronic acid. The scope of such a process can be easily affected by the nature of nucleophile and oxygen. The reaction proceeds with the events caused by both metals. Coupling occurs selectively with the iodo without touching the bromo functional group in the same substrate and without formation of diarylated by-product.

The P–N ligand used in the catalysis undergoes oxidation to form catalytically active copper complex P(O)–N-Cu. The $Cu(OTf)_2$ was the key for oxidation of P–N ligand because couplings only proceeded with the use of $Cu(OTf)_2$ while no catalysis was observed when other Cu sources such as $Cu(OAc)_2$, CuI, CuBr, etc., were used. After several control experiments, it was hypothesized that in the presence of air and triflate counterion, the P–N ligand oxidizes. As a critical finding, before oxidation of the P–N ligand, use of 1.0 equiv of TfOH followed by the $Cu(OAc)_2$ generated an active catalyst containing an oxidized form of ligand. Notably, only 150–200 ppm $Pd(OAc)_2$ was needed, which assists oxidative addition at the Cu(I) center.

Another advancement in this area is the development of "HandaPhos" ligand in 2016 by Handa and Lipshutz [12]. This highly lipophilic oxaphosphole ligand enabled cross-couplings of highly functionalized molecules using ppm levels of palladium under micellar conditions (Scheme 8.6). The superior catalytic activity of this ligand under micellar conditions is completely attributed to the

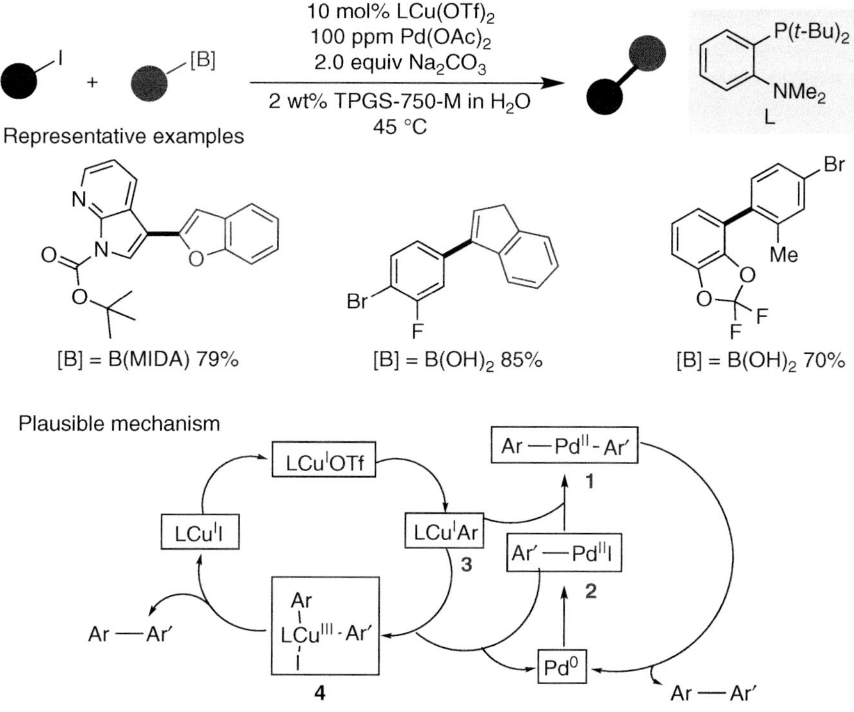

Scheme 8.5 Selective Suzuki–Miyaura couplings using copper and ppm levels of palladium.

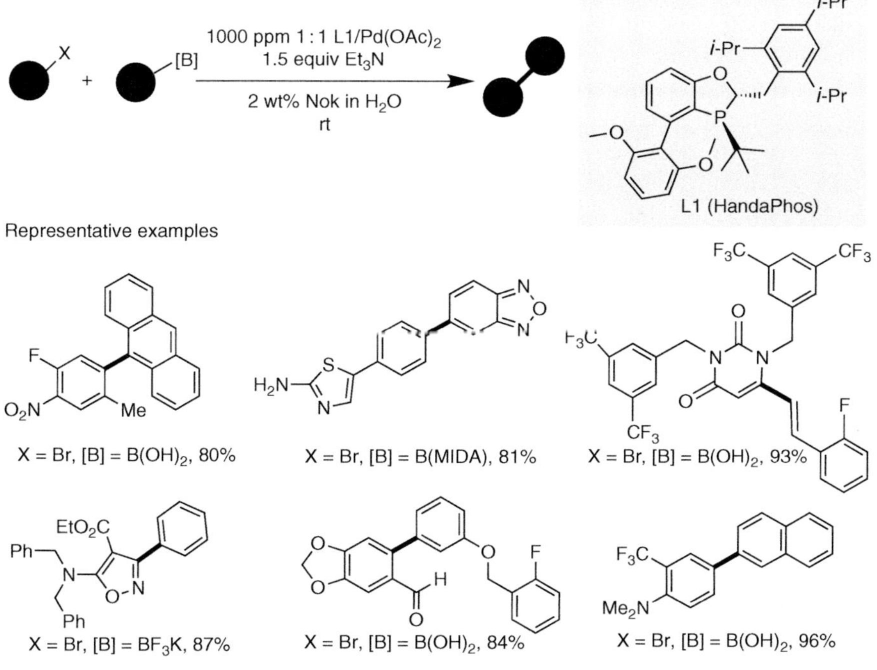

Scheme 8.6 HandaPhos technology for cross-couplings with ppm palladium.

ligand design, which includes (i) the presence of three isopropyl substituents on the phenyl ring to impart lipophilicity for a better binding of a catalyst with the micellar interior; (ii) the presence of electron-donating groups on the phenyl ring constitute an electron-rich phosphine to enhance the rate of oxidative addition; (iii) phosphorus as a donor atom in the cyclic ring to minimize the catalytically inactive conformations; and (iv) the presence of t-butyl group on the phosphorus atom and flexible triisopropylphenyl ring for faster reductive eliminations. Optimized reaction conditions include precomplexation between $Pd(OAc)_2$ and HandaPhos ligand in a 1 : 1 ratio in toluene, addition of coupling partners in approximately 1 : 1 ratio, 1.5 equiv of Et_3N as a base, and 2 wt% aqueous solution of Nok surfactant. Notably, only up to 1000 ppm of palladium is required in each transformation. The optimized conditions were highly effective for couplings of substrates containing a wide variety of functional groups and sensitive protecting groups. The lower level of palladium loading in the catalytic reactions results in an extremely low amount of trace metal impurities in the product, which is a key feature of this methodology. Inductively coupled plasma mass spectrometry (ICP-MS) analysis of the representative compounds revealed very low palladium content in the products, i.e. up to 9 ppm. The reaction medium containing catalyst can be at least recycled four more times without the loss of catalytic activity.

Handa and Lipshutz also documented a technology using nanoparticles of iron-containing ppm levels of palladium for cross-couplings of highly functionalized moieties in 2015 [40]. Nanoparticles derived from $FeCl_3$ containing ppm levels of palladium and SPhos as a ligand were explored as a highly efficient catalyst for Suzuki and Sonogashira couplings in water. This remarkable protocol only utilized 350–400 ppm of $Pd(OAc)_2$. Notably, low-purity commercially available $FeCl_3$ used in this methodology inherently contained ppm levels of palladium. Therefore, no additional palladium was required to obtain an active catalyst. The broad range of substrate scope revealed the strength of this protocol both regarding functional group tolerance and reaction yields (Scheme 8.7).

Very recently, Handa and coworkers reported highly challenging cross-couplings of unactivated (iso)quinolines [27]. Amphiphile PS-750-M with less lipophilic interior enables accommodation of relatively polar bromo (iso)quinoline as a coupling partner and Colacot's π-allyl palladium catalyst for efficient couplings (Scheme 8.8). Optimized reaction conditions provided good-to-excellent yields up to a gram scale with high selectivity and functional group tolerance. Control studies revealed the long-term stability of catalyst in PS-750-M. Both the catalyst and micellar reaction medium are recyclable. The behavior of the nanomicelles has been elucidated with dynamic light scattering (DLS) and cryo-transmission electron microscopy (TEM) measurements, and mechanistic investigations have revealed the reversible binding of quinoline nitrogen with palladium that competitively inhibits reaction rate. Notably, nanomicelles of PS-750-M played a crucial role in the catalysis. A DLS experiment revealed the existence of micelles with ca. 107 nm average diameter, and upon addition of 2-bromoquinoline (0.01 M) to the surfactant solution, particle size significantly increased, which was indicative of substrate accommodation in the micelles.

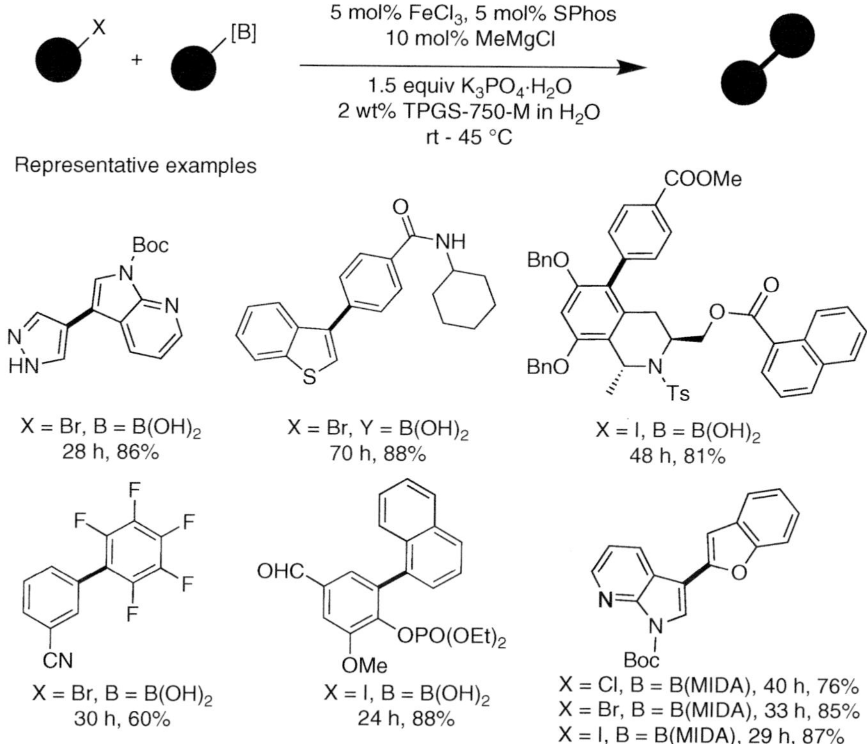

Scheme 8.7 Iron ppm palladium technology for cross-couplings.

8.3.2 Heck Couplings

Since the initial report by Heck and Mizoroki, a plethora of reports have appeared in the literature [41–43]. Among these, Jeffery's report entitled "Heck-type reactions in the water" was the first that completely avoided organic solvents as a reaction medium [44]. The approach incorporated $R_4N^+Br^-$ as a phase transfer catalyst to serve as an organic inner-core to accommodate the coupling partners. Later, Bumagin and coworkers improved this method by eliminating the phase transfer catalyst [45], which was further advanced by Sengupta [46]. In the reported methodology, reaction proceeds on water and needs elevated temperature, which also suffers limited substrate scope because of the insolubility of substrates.

In 2008, Lipshutz and coworkers adopted a micellar strategy for Heck couplings on a wide variety of lipophilic substrates [25]. In this method, micelles derived from Triton X-100 and PTS enable couplings in water under mild conditions. Reaction conditions include the use of commercially available 2 mol% (dtbpf)$PdCl_2$ as a catalyst and Et_3N as a base (Scheme 8.9a). Notably, a higher concentration of surfactant, i.e. 15 wt%, was optimal. The role of surfactant was further investigated with a series of control experiments, which includes the use of neat Et_3N or pH = 10, an aqueous buffer solution, or dimethylformamide

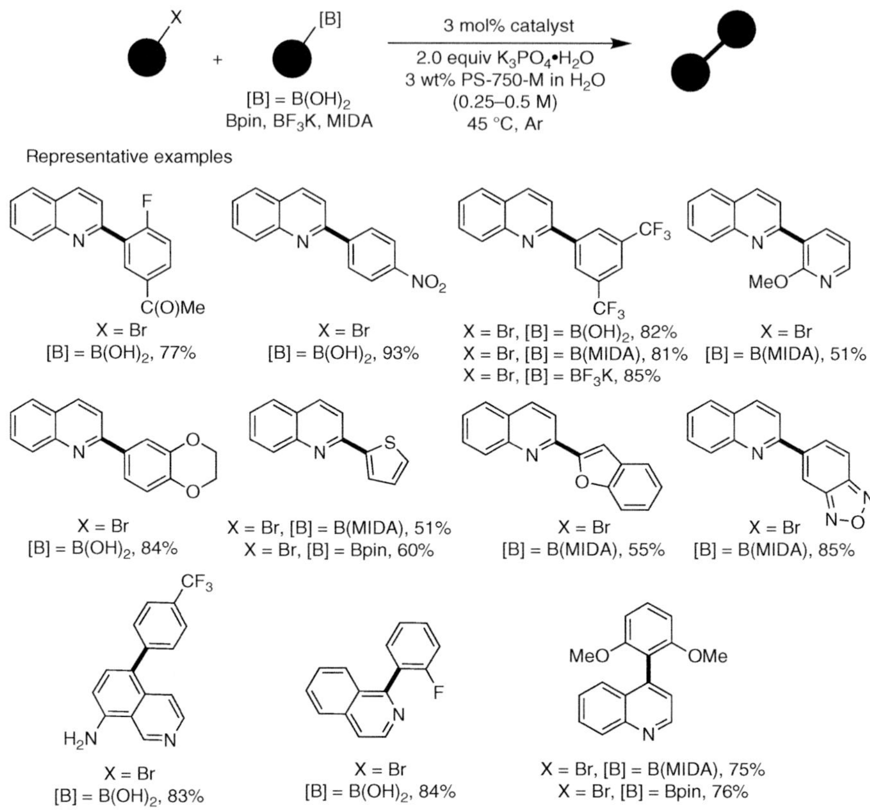

Scheme 8.8 Micelle-enabled couplings of unactivated (iso)quinoline systems.

(DMF)/Et_3N as the reaction medium. In all these experiments, poor conversions were detected, which reveals the role of surfactant as the reaction medium. The methodology was successful in carrying out the couplings of highly lipophilic substrates, which were previously not achieved at a high temperature in water. Instead PTS, Triton X-100 enabled superior conversion for more sterically hindered substrates. Under these conditions, styrenes exhibited more reactivity than their corresponding acrylates. Although the substrate scope proved the strength of this protocol about functional group tolerance, regioselectivity, and reaction yield, the primary concern was the use of a high concentration of surfactant. Comments on surfactant removal from the product were missing in this report.

Later, in 2011, the authors suggested that by improving the ionic strength of the medium, it is possible to lower the surfactant concentration to 5 wt%. The addition of 3M NaCl can improve the ionic strength [47]. Notwithstanding, the activity of Pd(0) catalyst containing monodentate electron-rich phosphines was superior to the catalyst comprising bidentate ligand, i.e. (dtbpf)$PdCl_2$ (Scheme 8.9b). Later, with the development of designer surfactant TPGS-750-M, the PTS was replaced under the same conditions [17]. Although the micellar technology

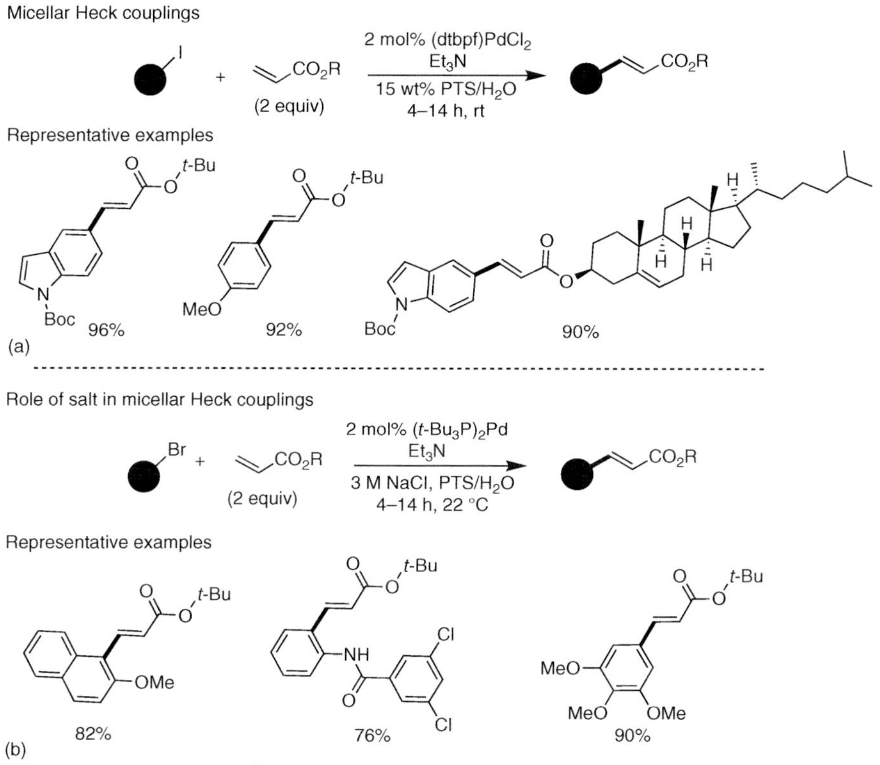

Scheme 8.9 Micellar Heck couplings.

was able to achieve traditional Heck reactions, reductive and asymmetric intramolecular Heck reactions remain a challenge with the micellar approach.

8.3.3 Negishi Couplings

Transformations involving alkyl zinc reagents in protic media are rarely explored in synthetic organic chemistry. Reports from the Wolinsky and coworkers [48], Luche and coworker [49], and Fleming et al. [50] groups suggested the possibility of such transformations. However, substrate scope and functional group tolerance were minimal. With the micellar technology, Krasovskiy and Lipshutz reported the first aqueous micellar Negishi-type couplings in 2 wt% aqueous PTS [51]. The remarkable feature of this protocol was an *in situ* generation of alkyl zinc reagents and their subsequent use in couplings (Scheme 8.10a). Critical components of optimized reaction conditions include (i) use of 5.0 equiv of tetramethylethylenediamine (TMEDA), which activates the Zn surface, allows it to insert to the C—X bond, and enhances the stability of alkyl zinc species; (ii) the $(Amphos)_2$ $PdCl_2$ as a catalyst; and (iii) aqueous PTS as a reaction medium, which also assists the generation of alkyl zinc reagents. Broad substrate scope, excellent functional group tolerance, and moderate-to-excellent yields were obtained. Although arylation of primary halides proceeded smoothly, their secondary

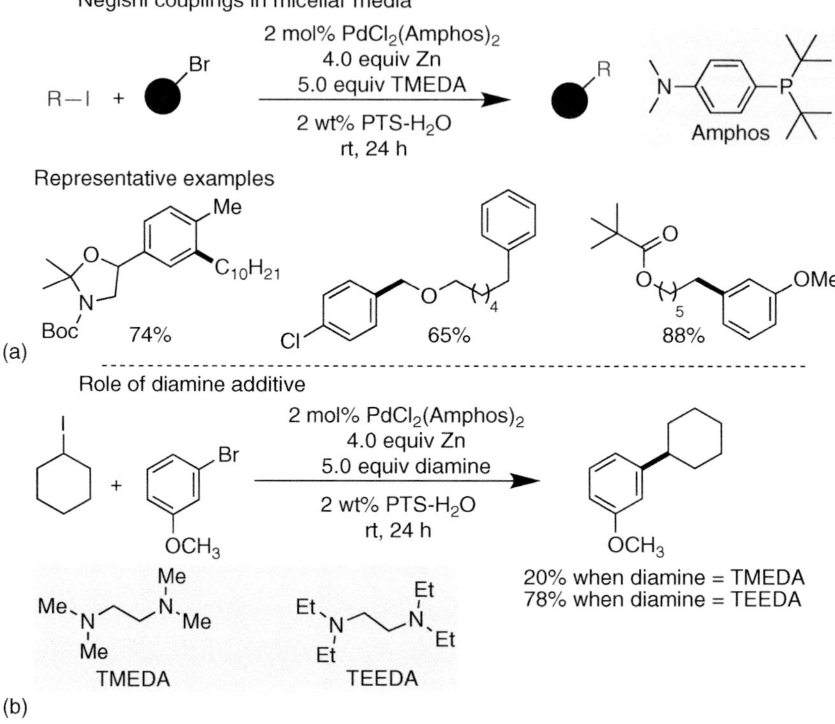

Scheme 8.10 Micelle-assisted Negishi couplings.

counterparts gave poor conversion (<20%) while secondary alkyl iodide was completely consumed. Thus, the faster reactivity of secondary alkyl iodides with zinc must be controlled, which otherwise results in protio-quenching and form the corresponding alkane. The tuning of reactivity of secondary alkyl iodides was achieved by modifying the steric bulk of diamine ligand (TMEDA). A switch from TMEDA to *N,N,N′,N′*-tetraethylethylenediamine (TEEDA) gave satisfactory conversions (Scheme 8.10b). This modification broadened the scope of this methodology. However, the arylation of tertiary alkyl iodide was not possible with this approach, which may be because of the problematic insertion of tertiary alkyl iodides to the zinc.

The authors conducted various control experiments to ensure the micellar effect, which also includes the use of neat THF as reaction media with and without alkyl zinc reagents. All these experiments confirmed the importance of aqueous PTS system. In a separate report, applications of micellar technology for the coupling of alkenyl halides with alkyl bromides were also reported [52]. The double-bond configuration remains unaffected during these transformations. When a mixture of *E* and *Z* alkenyl halides was subjected to the catalysis, the *E*/*Z* ratio slightly improved. Reaction kinetics also evidenced that the E isomer reacts faster than the *Z*. The report also mentioned the possibility of isomerization of *Z* alkenyl systems to the corresponding *E* alkenes. Reactions smoothly proceeded with the alkyl iodides and the aryl bromide as coupling partners.

This green technology was further extended to achieve sp^3–sp^2 couplings between alkyl halides with heteroaromatic halides, benzylic, and alkenyl halides (Scheme 8.11a) [53].

With the methods mentioned above, reactions do not proceed with the alkyl bromides. To address this issue, Lipshutz and coworkers developed an alternate method for sp^2–sp^3 couplings between alkyl and aryl bromides under micellar conditions [54]. In the modified process, the PTS-H_2O system was replaced with 4 wt% aqueous Brij-30 (Scheme 8.11b). The superiority of Brij system is because of larger micellar size (particle size: Brij-30 = 110 nm, PTS = 25 nm), which could accommodate more coupling partners for efficient catalysis and avoid homocoupling and dehalogenation. Another exciting feature associated with this report is the low catalyst loading (0.5 mol%) and broader substrate scope. Aromatics or heteroaromatics possessing saturated heterocyclic residues are found in the wide variety of active pharmaceutical ingredients (APIs), and with this technology, such moieties can be obtained in one step. Notably, this methodology is challenging because of possible side reactions, including the β-hydride elimination that could result in a few other by-products (Scheme 8.11c). To obtain the desired product, fine-tuning of electronic and steric properties of the catalyst is frequently required.

With the refined methodology, Buchwald and coworkers introduced a newly designed ligand VPhos [55] as an efficient ligand for these couplings (Scheme 8.11d). In the micelles of octanoic acid/sodium octanoate, XPhos exhibited superior activity over CPhos, RuPhos, and SPhos. However, product isomerization was the primary issue. SPhos provided moderate conversion but very high selectivity. With CPhos, lower conversion was detected, but no product isomerization was observed.

From the structure–activity relationship of ligands, the authors judiciously designed a ligand VPhos, with the optimal steric and electronics. Notably, a very low amount of alkyl halide was required, i.e. 1.5 equiv. However, 5 mol% of precatalyst, 2.5 equiv of zinc, and 3.0 equiv of TMEDA were required for optimal catalytic activity. The methodology was successful in carrying out the sp^2–sp^3 couplings between a wide range of (hetero)aryl halides with protected bromo cyclic amines.

8.3.4 C–H Arylations

Since the past two decades, a vast number of reports have appeared in the literature [56–58]. For this reaction class, palladium is still preferred. Some of the sustainability features in this reaction class are still missing. Lipshutz and coworkers reported the room temperature and micelle-mediated ortho-monoarylation of anilides via C–H activation pathway (Scheme 8.12) [59]. The reaction protocol utilizes 10 mol% $Pd(OAc)_2$, 2.0 equiv of $Ag(OAc)_2$ as an oxidant to regenerate the active palladium species after each catalytic cycle, and 5.0 equiv of HBF_4. Brij-35 was the best amphiphile for this transformation. High functional group tolerance was demonstrated with moderate-to-excellent yield.

In 2015, using polysorbate-based surfactant Tween-20, Ren and coworkers reported ortho-monoarylation of benzoic acids via C–H activation [60]. The

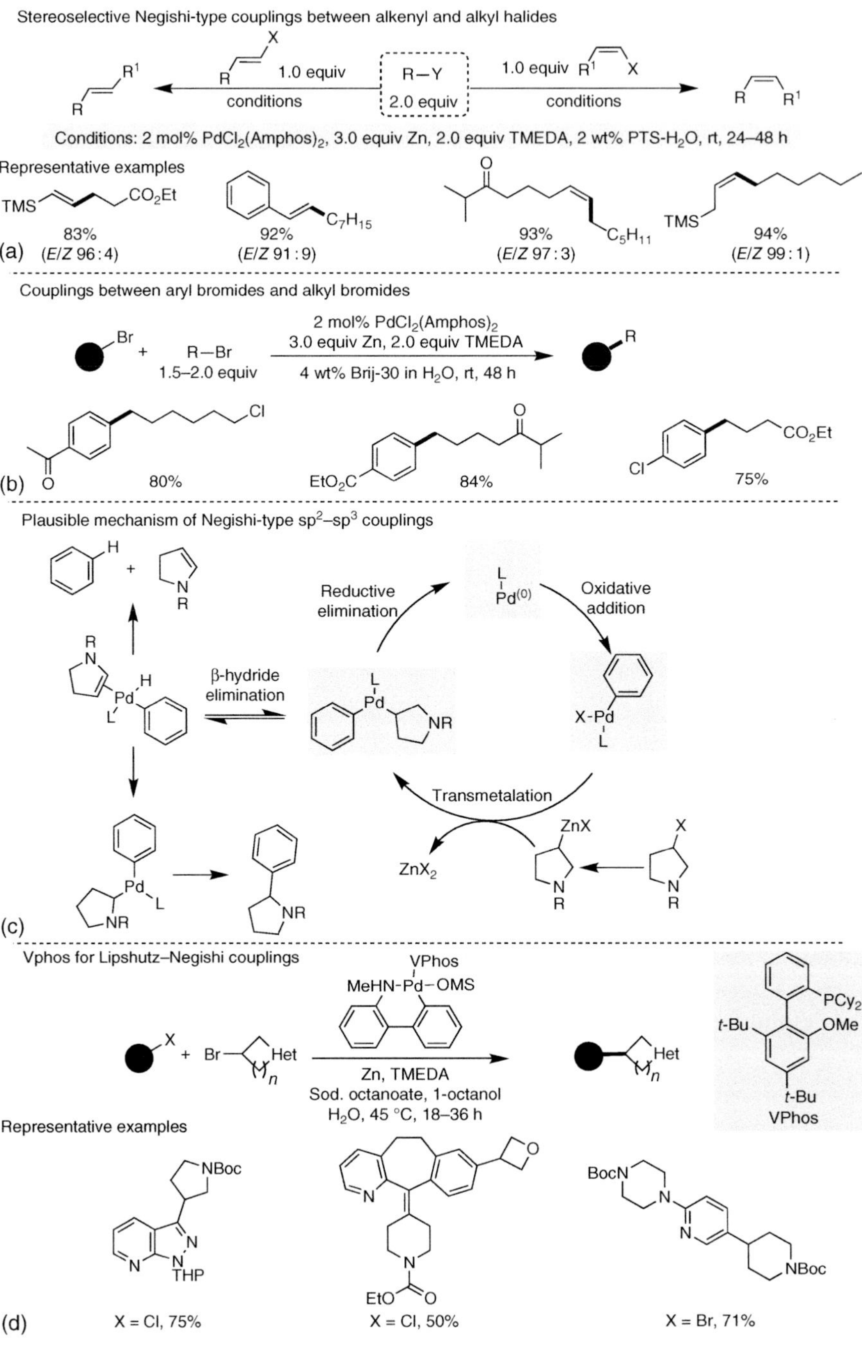

Scheme 8.11 Advancements in the micellar Negishi couplings.

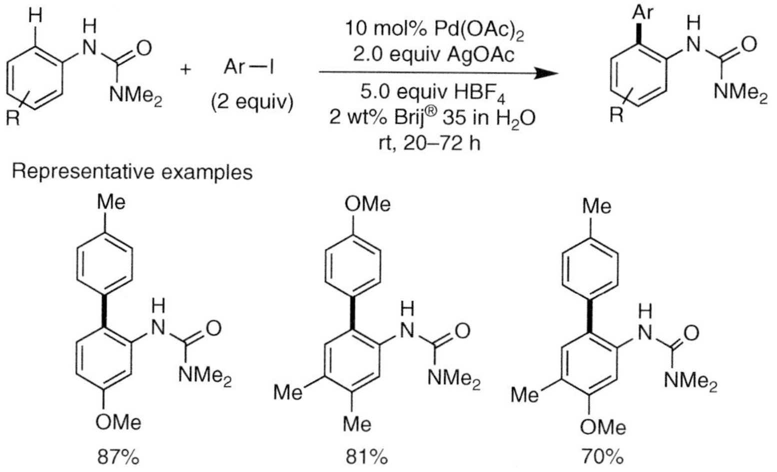

Scheme 8.12 C–H activations in micellar media.

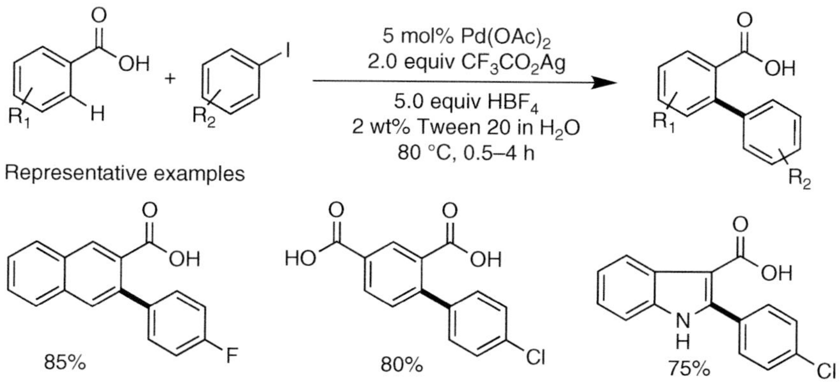

Scheme 8.13 C–H activation of benzoic acids.

catalytic cycle operates via Pd^{II}/Pd^{IV}. The procedure is exclusive for aryl iodides and highly selective for ortho-monoarylation. Arylation of indoles at the C-2 position is also accomplished but in low yield. Reactions need elevated temperature (Scheme 8.13).

Lipshutz and coworkers later adopted the strategy developed by Fujiwara for couplings of anilides with acrylates to produce corresponding cinnamate derivatives. Such reactions are normally conducted in acidic media under harsh conditions, whereas Lipshutz strategy eliminates the use of stoichiometric acids. The optimized procedure includes cationic palladium catalyst, i.e. 10 mol% $[Pd(MeCN)_4](BF_4)_2$, 1.0 equiv of benzoquinone, and 2.0 equiv of $AgNO_3$ (Scheme 8.14) [61].

Ren and coworkers also reported an elegant C–H arylation at the C-2 position of indoles, benzofurans, and benzothiophenes using aryl iodides, under micellar conditions at room temperature [62]. The protocol can be extended to both protecting group-free and methylated indoles. The indole arylation reactions

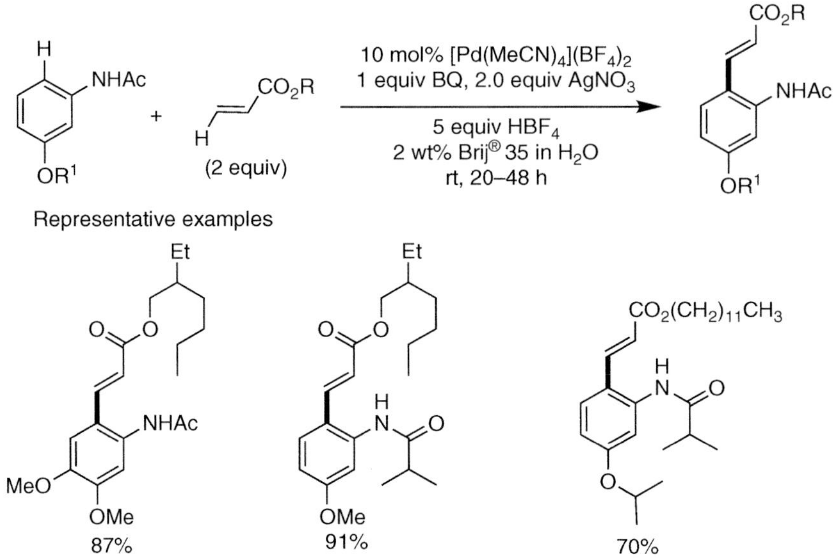

Scheme 8.14 Synthesis of cinnamate esters via micellar catalysis.

were considerably faster compared to benzofuran and benzothiophene systems (Scheme 8.15a). Although reaction conditions were mild compared to previously published reports, only aryl iodides display reactivity.

In 2016, Chemists from Novartis reported a micelle-assisted regioselective C-2 C–H arylation in thiophene systems (Scheme 8.15b) [63]. The reaction conditions

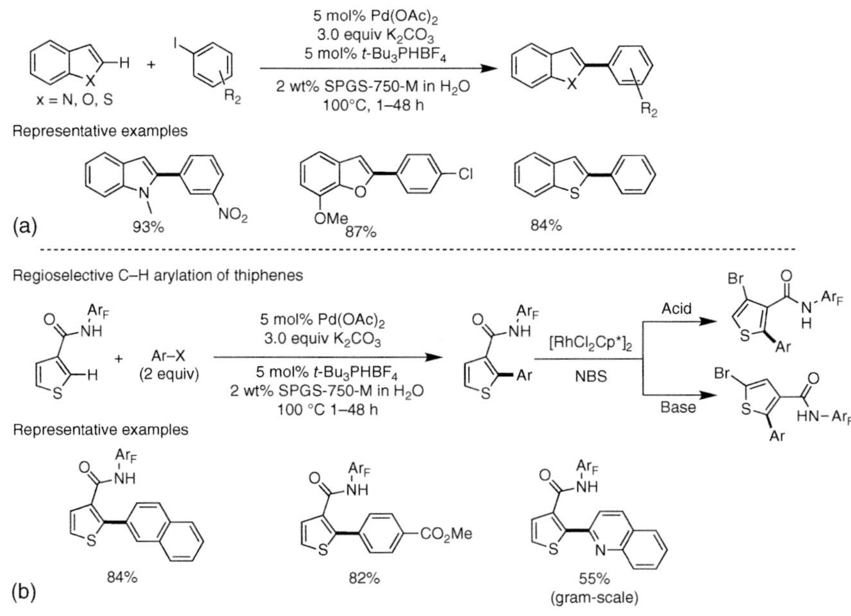

Scheme 8.15 C–H arylation of heterocycles.

were optimized using 3-perfluorotoluimide as a benchmark substrate. 5 mol% $Pd(OAc)_2$ and 5 mol% tri-*tert*-butyl phosphine were used in optimal catalyst loading. Catalytic reactions were conducted in aqueous SPGS-750. Optimized conditions were explored on 15 different thiophene systems, which displayed good reactivity. Depending on the reaction conditions, resulting C-2-arylated thiophenes could be further selectively brominated at the C-4 or C-5 position. This methodology was also applied on multigram-scale reactions.

8.3.5 Aminations

The transition-metal-catalyzed C–N bond formation via cross-coupling chemistry is a powerful technology to access aromatic amines [5]. Initial investigations by Migita and Yagupolskii postulated the possibility of transition-metal-catalyzed arylation of amines. In 1995, the two independent reports from Buchwald and coworkers [64] and Hartwig and coworker [65] set the foundation for modern methods of aminations [66–69].

With the many advancements on this topic, only fewer reports focus on general and sustainable protocols that address green chemistry principles. In 2009, Lipshutz and coworkers reported a series of protocols for the catalytic amination of aryl halides [70], allylic alcohols [71], and allylic ethers [72] under micellar conditions. For the amination of aryl halides, PTS surfactant was found to be the best (Scheme 8.16a). The optimized reaction condition utilized $[(\pi\text{-allyl})PdCl]_2$ (0.5 mol%) and Taksago's ligand di-*tert*-butyl(2,2-diphenyl-1-methyl-1-cyclopropyl) [cBRID-P] (2 mol%). The reaction temperature and equivalents of the base were significantly reduced while harnessing the hydrophobic effect of the micelle. With the 1-bromo-4-methylnaphthalene and indoline as benchmark coupling partners, the authors demonstrated the role of a lipophilic base on the reaction rate. However, its effect on micelle size was completely missing. When an inorganic base such as potassium hydroxide (KOH) was used, only 72% conversion was detected after 44 hours. When triethylamine was used, the conversion was improved to 79%. The rate was considerably enhanced when more lipophilic bases were used. With the use of potassium butoxide (*t*-BuOK), $KOSiMe_3$, and KOH-triisopropylsilane (TIPS-OH), complete conversions were observed within 6, 3, and 0.5 hours, respectively. These observations evidenced that lipophilic bases tend to enhance the reaction rate. This strategy was further explored for the installation of amine equivalents (carbamate, sulfonamide, urea, etc.) onto the (hetero)aryl ring (Scheme 8.16b) [73]. In this case, the authors replaced PTS with TPGS-750-M. Notably, the use of KBr as an additive significantly enhanced the reaction rate. The *E*-factor for this process was calculated as 2.7 (Scheme 8.16). In 2014, Schmitt and coworkers further extended this methodology to amination of more challenging heterocycles. However, high catalyst loading was required [74, 75].

The methyl formate introduces lability to the hydroxyl group and assists the generation of π-allyl species, which could further participate in regioselective amination in the micellar media [71]. Using this strategy, facile access to allylic amines is documented via a highly regioselective pathway. The reaction pathway

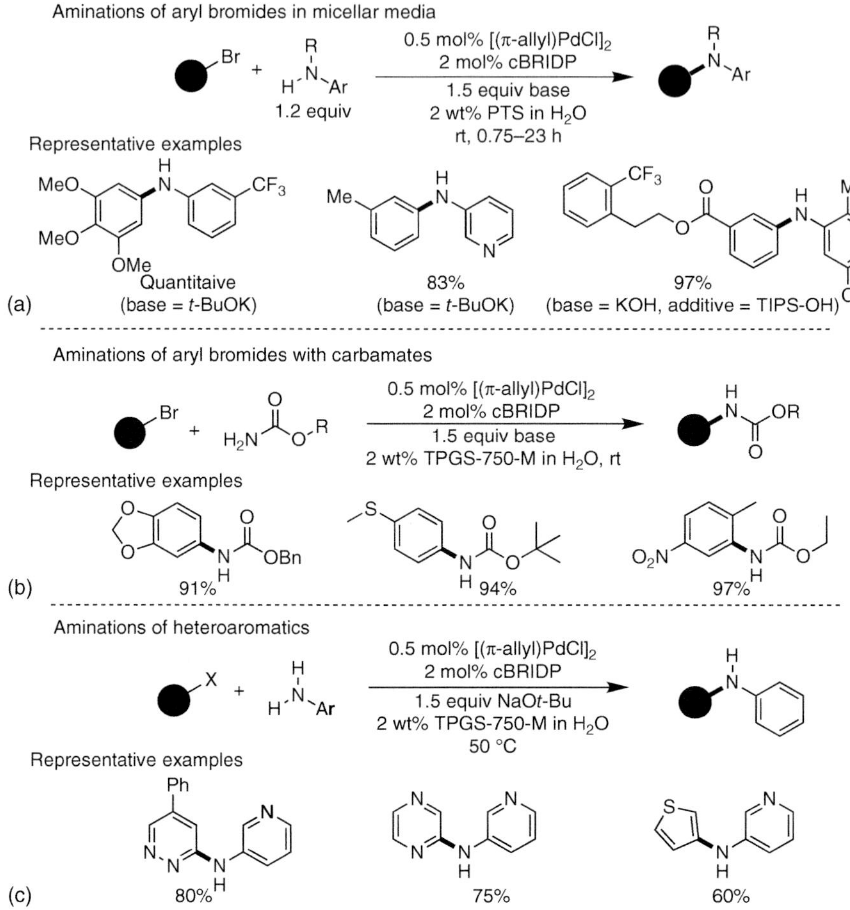

Scheme 8.16 Aminations in micellar media.

is highly selective to the formation of *E* isomer. This methodology was further utilized to obtain antifungal agent Naftifine (Scheme 8.17a).

In the same year, Lipshutz and coworkers reported similar aminations of allylic phenyl ethers under micellar conditions at room temperature [72].

Nonetheless, using $[Pd(allyl)Cl]_2$ (0.5 mol%) and DPE-Phos as a ligand (1 mol%), K_2CO_3 (1.5 equiv) as a base, and methyl formate (4.0 equiv) as an additive led to successful aminations in good-to-excellent yields (Scheme 8.17b). The reaction pathway was dictated by steric and electronic effects, favoring the less hindered terminal site, which results in very high $E:Z$ ratios. In 2014, Kumar and coworkers reported a nonligated copper nanoparticle as a powerful catalyst for the *N*-(hetero)arylation of amines and azoles under micellar conditions at room temperature (Scheme 8.17c) [76]. The scope of this work was showcased for the synthesis of tryptanthrin and fragments of many bioactive molecules, such as imatinib, nilotinib, and oxcarbazepine. The optimized reaction condition includes 2 wt% TPGS 750-M and NaOH (3.0 equiv) as a base. Although many

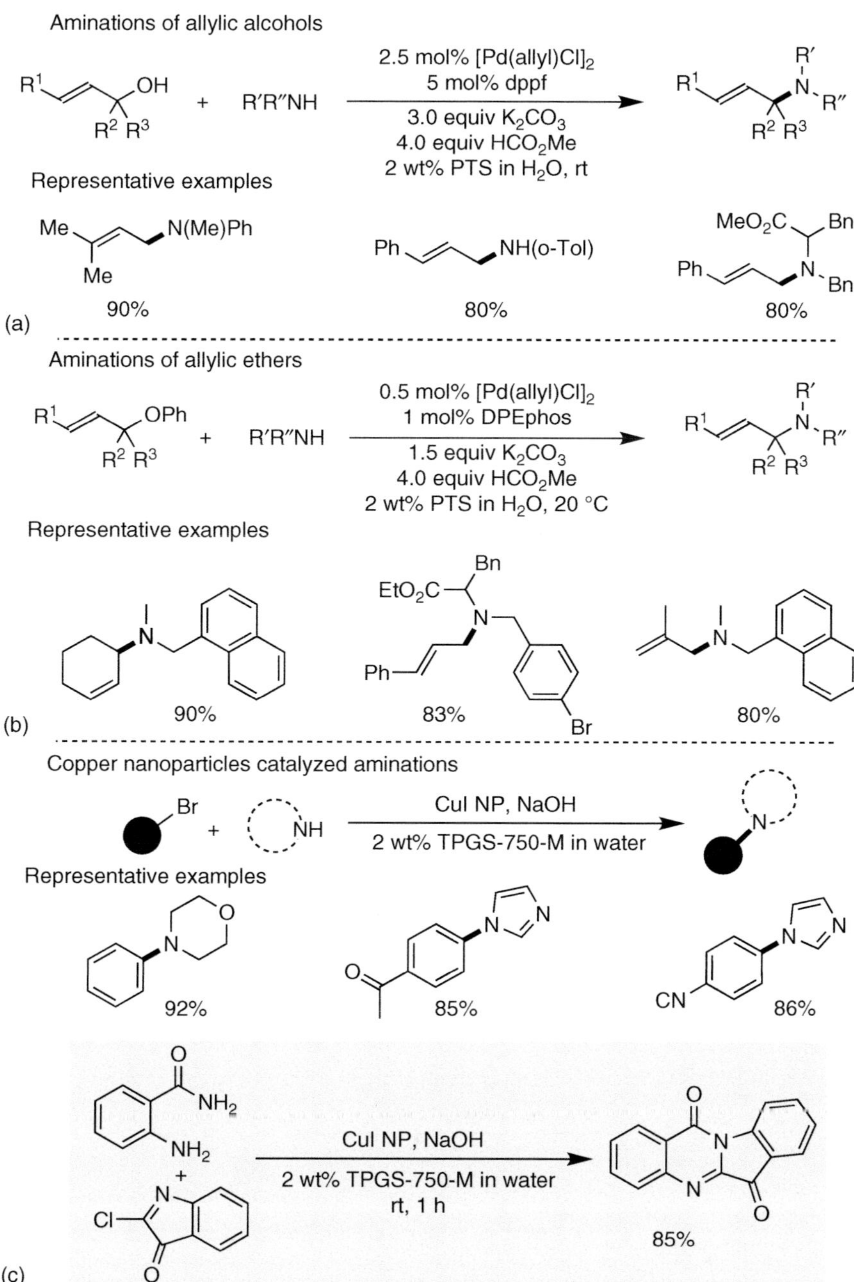

Scheme 8.17 A broader scope of micellar aminations.

reports are available on the sustainable catalytic aminations, a truly general, sustainable, and recyclable catalytic system remains a challenge.

8.3.6 Borylation

C—X (X = Cl, Br, and I) bond can be easily transformed to the corresponding C—B bond via formation of a Grignard reagent followed by its quenching with electrophilic boron source. Significant drawbacks of this method include limited functional group tolerance, use of the air-sensitive reagent, and difficult scale-up. Therefore, a catalytic method for borylation is always preferred. Miyaura reported a palladium-catalyzed process where aryl halides are coupled with B_2pin_2 in the presence of weak base KOAc in dimethyl sulfoxide (DMSO) at 80 °C [77]. Furstner, Buchwald, and Marder further improved this transformation [78–80]. In all these reports, the primary focus was to lower the reaction temperature, reduce the catalyst loading, and broaden the scope. However, all this work was done in organic solvents. Till 2010, there was no report on borylation in water or micellar media.

In 2010, Lipshutz and coworkers adopted a micellar technology for this transformation, which assisted the reaction to occur at room temperature [81]. The method was successful for borylation of various aryl bromides with a focus on functional group tolerance (Scheme 8.18). Functional groups such as esters, ketones, α,β-unsaturated esters, and free amines were well tolerated. In this report, borylation of 4-bromoanisole was achieved with a 1000 ppm catalyst (turn over number [TON] of 1000).

8.3.7 Arylation of Nitro Compounds

Arylation of sp^3-nitro compounds typically requires polar-aprotic solvents such as 1,4-dioxane or *N*-methyl-2-pyrrolidone (NMP) [82]. Toxicity associated with these solvents, anhydrous conditions needed for successful reactions, high catalyst loading, and elevated temperature are the significant drawbacks of traditional methodology. Very recently, Handa and coworkers developed an environmentally benign proline-based nonionic amphiphile PS-750-M, which when dissolved in

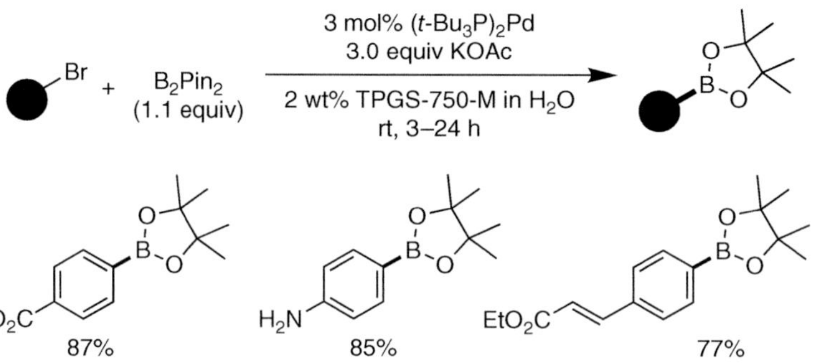

Br + B_2Pin_2 (1.1 equiv) —[3 mol% $(t\text{-}Bu_3P)_2Pd$, 3.0 equiv KOAc; 2 wt% TPGS-750-M in H_2O, rt, 3–24 h]→ Bpin product

MeO_2C 87%; H_2N 85%; EtO_2C 77%

Scheme 8.18 Borylation at room temperature.

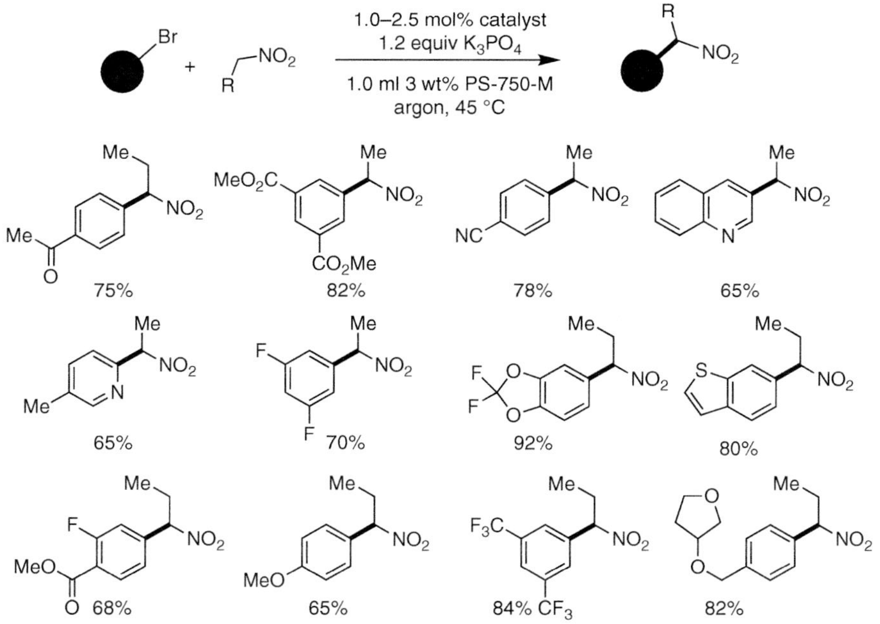

Scheme 8.19 Arylation of nitroalkane in micellar media.

water, self-aggregate to form micelles with a relatively polar interior compared to the already existing nonionic amphiphiles [19]. The resulting micelles swell with the addition of nitroalkanes, enabling the micelle to accommodate the catalyst and coupling partners in high concentration so that transformations were possible under mild conditions. With this approach, the authors reported sp^2–sp^3 couplings of nitroalkanes with aryl/heteroaryl halides in water under mild conditions (Scheme 8.19). The rational design of PS-750-M enables amphiphile to mimic polar aprotic solvents such as DMF, 1,4 dioxane, and NMP.

Broad substrate scope and high functional group tolerance were demonstrated with critical examples. Good-to-excellent yields were obtained on a broad range of substrates. Notably, both aryl bromide and chloride underwent complete conversions, whereas the conversions were incomplete for aryl iodides. The optimized procedure includes the use of [(*t*-BuXPhos)Pd(allyl)]OTf (1–2.5 mol%) as a catalyst, K_3PO_4 (1.2 equiv) as a base, and 3 wt% PS-750-M as a surfactant solution. The metal complex generated by treating *t*-BuXPhos with $Pd(OAc)_2$ or Pd_2dba_3 was not highly catalytically active, suggesting the active role of allyl group as a ligand in the catalytic cycle, which was further confirmed by control experiments. The reaction medium can be at least recycled four additional times and the calculated *E*-factor was 5.3. This methodology was also reproducible on a gram-scale reaction.

8.3.8 Adoption of Micellar Technology by Pharmaceutical Industry

So far, micellar technology is mainly adopted by Novartis Pharmaceuticals. Gallou and coworkers first reported the astounding translation of this technology to a multistep kilogram-scale process for the synthesis of an undisclosed API [83]. In the published report, the authors evaluated the scalability of key reactions using micellar technology. The key reactions mainly involve Pd-catalyzed cross-coupling, S_NAr, amide couplings, and protection–deprotection in the micellar medium. The authors also provided a detailed comparison of economics between the traditional synthesis in organic solvents and in micellar media, which reveals the unique aspects of micellar catalysis.

The authors did not disclose the identity of the final targeted compound, reaction intermediates, and some of the details about reaction conditions because of the protection of intellectual property. According to the published report, the synthesis of targeted API starts with the S_NAr reaction between highly functionalized aromatic amine and highly functionalized aryl halide (Scheme 8.20). The transformation leads to form a new C—N bond between trisubstituted heteroaromatic ester **1** and a nucleophilic amine **2** bearing a free hydroxyl group. Notably, this transformation to obtain reaction intermediate **3** requires an elevated temperature (60 °C) when conducted in organic solvents (*i*-PrOH/toluene). However, reaction cleanly proceeded in micellar medium (2 wt% TPGS-750-M) at room temperature. At the same time, a possible side reaction O-arylation is minimized by rigorous optimization, which circumvented the protecting group chemistry.

In this report, the authors did not disclose full optimization. The reported isolated yield of **3** is 75% in TPGS-750-M. Although the yield is slightly less compared to the reaction in organic solvents (*i*-PrOH/toluene), purification was simplified. Upon a Suzuki–Miyaura reaction between **3** and boronic acid **4**, a new C—C bond was constructed to obtain compound **5**. For the construction of this particular bond, the reaction in organic solvent requires 85 °C, while the reaction in TPGS-750-M cleanly proceeded at 40 °C. Without isolating the reaction intermediate **5**, *in situ* hydrolysis of the ester provided carboxylic acid **6**. Again, the outcome was much better when TPGS-750-M was used as a reaction medium. Next, amide coupling between carboxylic acid **6** and amine **7** was performed. Comparison of the reaction outcome was also reported, i.e. a reaction in CH_3CN vs TPGS-750-M/H_2O. Reaction yield for the formation of **8** was 80% when TPGS was used as a reaction medium while 76% in CH_3CN. 12% side product **9** was obtained when the reaction medium was CH_3CN. No such side product was observed in the reaction medium TPGS-750-M. This may be because of the requirement of elevated temperature in a reaction in CH_3CN. Finally, deprotection of a protecting group in **8** provided the API in 92% isolated yield for the final step. The overall reaction yield was 42.5% when all the transformations were conducted in an organic solvent. Overall yield in Scheme 8.20 was 45% when all the transformations were conducted in aqueous TPGS-750-M as a recyclable reaction medium. Notably, this difference in the overall yield is quite significant at kilogram-scale synthesis of API.

The first S_NAr reaction between **1** and **2** in the organic solvent system produced a minor amount of O-arylated side product, which was further purified by tedious recrystallization (details about recrystallization process was undisclosed). The same transformation in designer surfactant TPGS-750-M was selective to the N-arylation and provided an overall yield of 75% of the desired product. The advantage of the micelle-enabled process was the extra-purity of

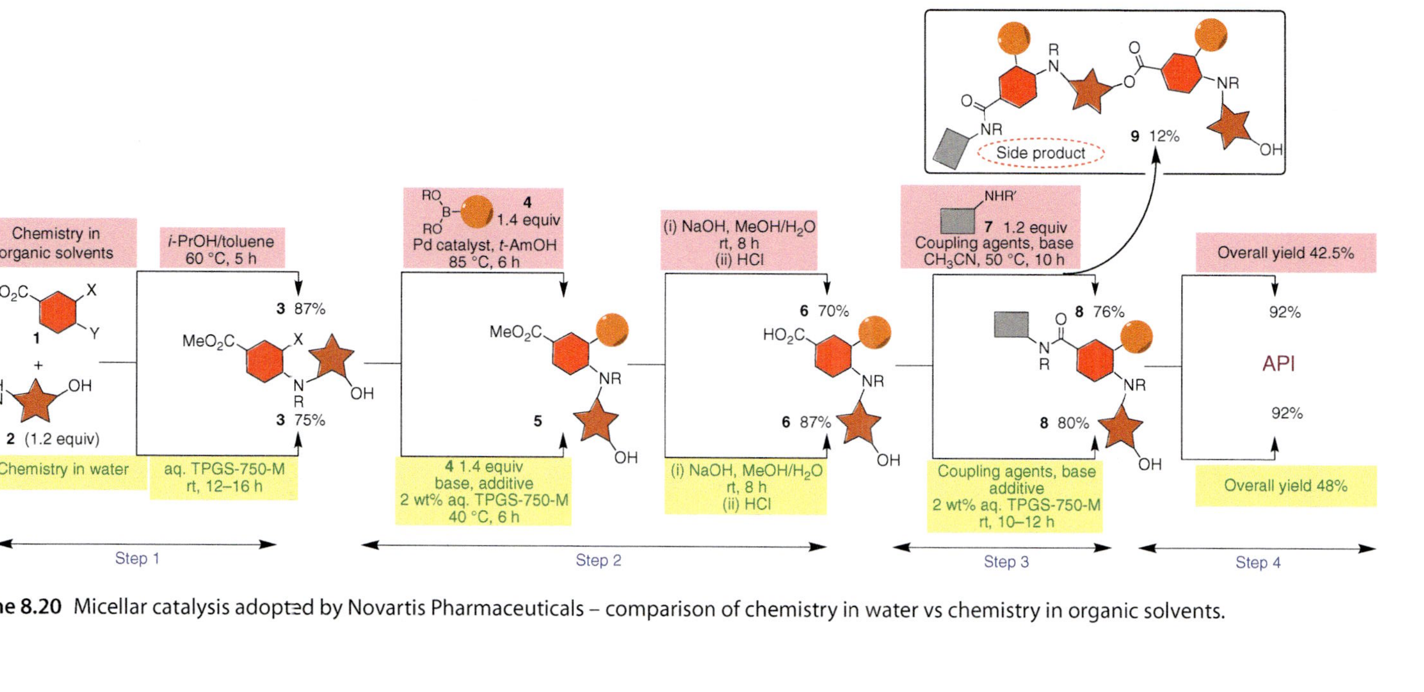

Scheme 8.20 Micellar catalysis adopted by Novartis Pharmaceuticals – comparison of chemistry in water vs chemistry in organic solvents.

Table 8.1 Effect of reaction medium on the outcome.

Step	In organic solvent	In aq. TPGS-750-M
Yield for formation of **3** (S_NAr reaction)	87%	75%
Yield for one-pot SM coupling and hydrolysis to obtain **6**	70%	87%
Yield for amide coupling to obtain **8**	76%	80%
Yield for final step to obtain API	92%	92%
Overall yield	42.5%	48%
Overall PMI	238	161
Overall PMI change	—	−32%
Overall PMI water change	—	−8%
Overall PMI solvent change	—	−48%
Overall PMI organic mass change	—	−52%

crude product. Extraction using isopropyl acetate followed by precipitation with heptane provided a pure product, which was directly used in the next step, i.e. SM coupling reaction. SM coupling in organic solvent requires an excess of boronate ester and higher palladium loading, which significantly affects the cost. These conditions demanded exhaustive purification procedures of the crude mixture and rigorous process to remove the traces of palladium from the product, while reactions in the micellar medium inherently addressed all these issues. From the thorough comparison by the authors, it can be concluded that the scope of micelle-enabled processes is beyond the replacement of toxic organic solvents. The additional benefits associated with such methods include reduced process mass intensity (PMI) value, improved reaction yields, high purity of API with reduced metal contents, and high cost efficiency (see Table 8.1 and Figure 8.10).

In another report published by Boehringer Ingelheim, the role of the reaction medium was demonstrated [84]. The authors synthesized an 11-β-HSD1 inhibitor on a multigram scale (Scheme 8.21). Excellent reaction yield was obtained using aqueous Triton X-100 as a reaction medium (although this is a

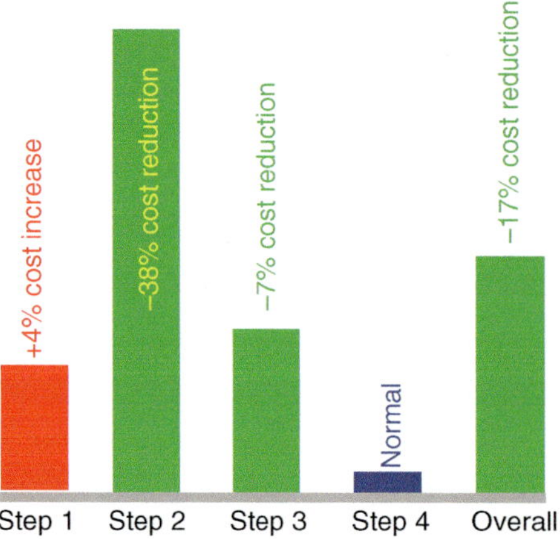

Figure 8.10 Effect of reaction medium on the cost reduction.

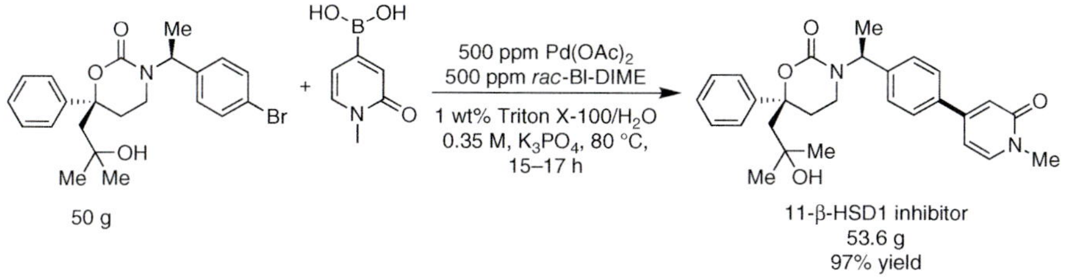

Scheme 8.21 Use of micellar media for the synthesis of bioactive molecule (by Boehringer Ingelheim).

toxic surfactant). Notably, only 500 ppm of palladium catalyst was needed for this particular reaction. However, a high reaction temperature was required for a clean transformation.

For this book chapter, some of the unpublished results are also generously provided by Novartis Pharmaceuticals (Scheme 8.22). In the example of cross-coupling given below, both the bromopyridyl and boronate ester suffered from limited stability. Bromopyridyl system was prone to debromination. The boronate ester was especially thermolabile and prone to rapid protodeboronation. An obvious alternative was to utilize the more stable MIDA boronate, but the additional expense that occurred with this change was severely detrimental to such a strategy. Researchers' attention, therefore, turned to micellar catalysis that enables the reaction under very mild conditions. At that time, standard optimization revealed that the preformed $PdCl_2$(dtbpf) catalyst was superior to all other catalytic systems evaluated and could be minimized to 0.2 mol% at most, when used in combination with triethylamine as the base (1.5 equiv), at 40 °C in aqueous solution of TPGS-750-M (2 wt% in water) with acetone as a cosolvent (20% volume relative to water). An additional benefit of such a low catalyst loading was the rapid depletion upon simple treatment and the already integrated crystallizations along the synthetic sequence. Of paramount importance was the slow addition of the least stable reaction component

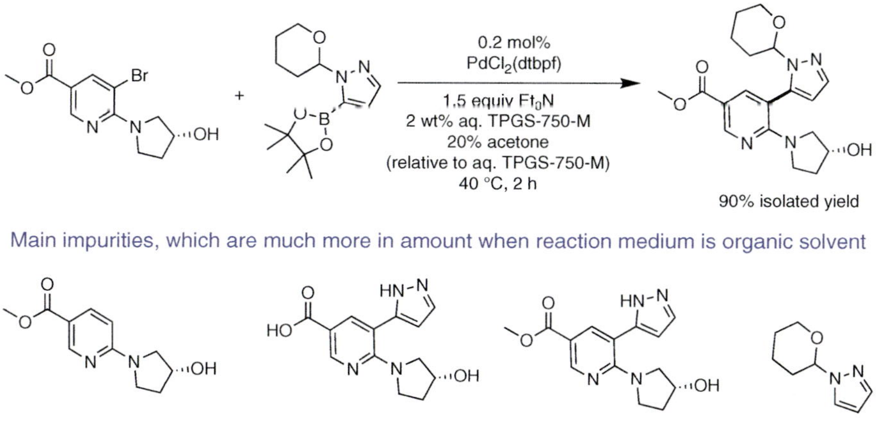

Main impurities, which are much more in amount when reaction medium is organic solvent

Scheme 8.22 Highly challenging Suzuki–Miyaura coupling accomplished in the micellar medium for better economics.

boronate ester. The latter was added slowly within two hours to a premixed solution of the bromopyridine, the catalytic system, and the base. This slow addition minimizes accumulation of this poorly stable chemical entity, and hence, its excess to no more than 5% (1.05 equiv in total) was required. The reaction was complete within an additional hour and resulted into c. 95% chemical yield, 90% isolated yield, and a purity >98.0% (desbromopyridine as the main impurity that would not carry over to the subsequent steps). From a mass efficiency, the overall PMI [82] for this complex cross-coupling was calculated to 14.2 kg/kg product (4.9 and 6.2, respectively, for PMI water and solvent), compared to the fully optimized previous process in organic solvents that showed the following performance: 23.7, 10.3, and 9.4, respectively, for the overall PMI, PMI water, and PMI solvent. The economics of the process obviously decreased significantly as a result of these improvements.

8.4 Conclusions

In concluding remarks, micellar catalysis should not only be envisioned as an alternate to organic solvents but it also has many more exciting features than anticipated to date. However, a wealth of fundamental information is still missing to further grow this field into the areas of asymmetric catalysis, photoredox chemistry, and electrocatalysis. Many more asymmetric transformations can be achieved by leveraging hydrophobic effect. Many questions are still unanswered, especially when micellar catalysis is associated with the metal nanoparticles. The current status in terms of use of this technology by the industry is so far limited to Novartis Pharmaceuticals, AbbVie, and Boehringer Ingelheim. This may be because of a lack of sufficient literature and awareness about the practical skills needed to reproduce this chemistry at kilogram scales in the hands of chemists trained in doing chemistry in organic solvents. To overcome this barrier, future onsite workshops organized by the researchers who already explored the benefits of this technology could encourage and convince the broader community of its usage. This will not only reduce the cost but also help to reduce the waste as well as consumption of precious metal catalysts.

Biography

Tharique N. Ansari completed his integrated M.S. degree in the field of interdisciplinary sciences from the Mahatma Gandhi University (MGU), Kerala. Under the guidance of Prof. I. Ibnusaud at MGU, he carried research on the synthesis of enantiopure nitrogen heterocycles using chiralpool approach. In Fall 2017, under the guidance of Prof. Sachin Handa, he started his graduate studies in the Department of Chemistry at the University of Louisville.

His research primarily focuses on the development of new and highly selective reaction pathways mediated by nanocatalysts.

Fabrice Gallou received his Ph.D. degree from the Ohio State University in the field of total synthesis of natural products. After Ph.D., he started his career as a process chemist at Boehringer Ingelheim and subsequently moved to the chemical development group at Novartis Pharmaceuticals, where he is now responsible for global scientific activities worldwide, overseeing development and implementation of practical and economical chemical processes for large-scale production of APIs, and leads cross-functional innovation projects.

Sachin Handa was raised in Patti, a small town of Punjab (India), where he completed his undergraduate studies. He received a Ph.D. degree from Oklahoma State University where his research focus was the development of nonracemic acyclicdiaminocarbenes for asymmetric gold catalysis. Upon completion of postdoctoral training under the guidance of Prof. Bruce H. Lipshutz at UC Santa Barbara, he started his independent career at the University of Louisville in Fall 2016. His research laboratory focuses on the development of sustainable catalysts and reaction pathways for chemistry in water and photoredox catalysis.

References

1 Johansson Seechurn, C.C.C., Kitching, M.O., Colacot, T.J., and Snieckus, V. (2012). *Angew. Chem. Int. Ed.* 51: 5062.

2 Nicolaou, K.C., Bulger, P.G., and Sarlah, D. (2005). *Angew. Chem. Int. Ed.* 44: 4442.

3 Biffis, A., Centomo, P., Del Zotto, A., and Zecca, M. (2018). *Chem. Rev.* 118: 2249.

4 O'Brien, H.M., Manzotti, M., Abrams, R.D. et al. (2018). *Nat. Catal.* 1: 429.

5 Ruiz-Castillo, P. and Buchwald, S.L. (2016). *Chem. Rev.* 116: 12564.

6 Hazari, N., Melvin, P.R., and Beromi, M.M. (2017). *Nat. Rev. Chem.* 1: 25.

7 Colacot, T. (2015). *New Trends in Cross-coupling; Catalysis Series*. The Royal Society of Chemistry.

8 Magano, J. and Dunetz, J.R. (2015). *Transition Metal-catalyzed Couplings in Process Chemistry: Case Studies From the Pharmaceutical Industry*. Wiley.

9 Brown, D.G. and Boström, J. (2016). *J. Med. Chem.* 59: 4443.

10 Bihani, M., Ansari, T.N., Smith, J.D., and Handa, S. (2018). *Curr. Opin. Green Sustainable Chem.* 11: 45.

11 La Sorella, G., Strukul, G., and Scarso, A. (2015). *Green Chem.* 17: 644.
12 Handa, S., Andersson, M.P., Gallou, F. et al. (2016). *Angew. Chem. Int. Ed.* 55: 4914.
13 Roy, D. and Uozumi, Y. (2017). *Adv. Synth. Catal.* 360: 602.
14 Molnár, Á. (2011). *Chem. Rev.* 111: 2251.
15 Lipshutz, B.H., Ghorai, S., and Cortes-Clerget, M. (2018). *Chem. Eur. J.* 24: 6672.
16 Lipshutz, B.H. (2018). *Curr. Opin. Green Sustainable Chem.* 11: 1.
17 Lipshutz, B.H., Ghorai, S., Abela, A.R. et al. (2011). *J. Org. Chem.* 11: 4379.
18 Gabriel, C.M., Lee, N.R., Bigorne, F. et al. (2017). *Org. Lett.* 19: 194.
19 Brals, J., Smith, J.D., Ibrahim, F. et al. (2017). *ACS Catal.* 7: 7245.
20 Temkin, O.N. (2012). *Homogeneous Catalysis with Metal Complexes*. Wiley.
21 Andersson, M.P., Gallou, F., Klumphu, P. et al. (2018). *Chem. Eur. J.* 24: 6778.
22 Handa, S., Fennewald, J.C., and Lipshutz, B.H. (2014). *Angew. Chem. Int. Ed.* 53: 3432.
23 Handa, S., Lippincott, D.J., Aue, D.H., and Lipshutz, B.H. (2014). *Angew. Chem. Int. Ed.* 53: 10658.
24 Klumphu, P., Desfeux, C., Zhang, Y. et al. (2017). *Chem. Sci.* 8: 6354.
25 Lipshutz, B.H. and Taft, B.R. (2008). *Org. Lett.* 10: 1329.
26 Lipshutz, B.H. and Ghorai, S. (2014). *Green Chem.* 16: 3660.
27 Handa, S., Ibrahim, F., Ansari, T.N., and Gallou, F. (2018). *ChemCatChem* 10: 4229.
28 Handa, S., Smith, J.D., Zhang, Y. et al. (2018). *Org. Lett.* 20: 542.
29 Gallou, F., Parmentier, G.M., and Zhou, J. (2016). *Org. Process Res. Dev.* 20: 1388.
30 Serrano-Luginbühl, S., Ruiz-Mirazo, K., Ostaszewski, R. et al. (2018). *Nat. Rev. Chem.* 2: 306.
31 Smith, J.D., Ansari, T.N., Andersson, M.P. et al. (2018). *Green Chem.* 20: 1784.
32 Miyaura, N. and Suzuki, A. (1979). *Chem. Commun.*: 866.
33 Miyaura, N. and Suzuki, A. (1995). *Chem. Rev.* 95: 2457.
34 Grasa, G.A. and Colacot, T.J. (2007). *Org. Lett.* 9: 5489.
35 Lipshutz, B.H., Petersen, T.B., and Abela, A.R. (2008). *Org. Lett.* 10: 1333.
36 Nishikata, T. and Lipshutz, B.H. (2009). *J. Am. Chem. Soc.* 131: 12103.
37 Isley, N.A., Gallou, F., and Lipshutz, B.H. (2013). *J. Am. Chem. Soc.* 135: 17707.
38 Handa, S., Slack, E.D., and Lipshutz, B.H. (2015). *Angew. Chem. Int. Ed.* 54: 11994.
39 Handa, S., Smith, J.D., Hageman, M.S. et al. (2016). *ACS Catal.* 6: 8179.
40 Handa, S., Wang, Y., Gallou, F., and Lipshutz, B.H. (2015). *Science* 349: 1087.
41 Dounay, A.B. and Overman, L.E. (2003). *Chem. Rev.* 103: 2945.
42 Beletskaya, I.P. and Cheprakov, A.V. (2000). *Chem. Rev.* 100: 3009.
43 Dieck, H.A. and Heck, R.F. (1974). *J. Am. Chem. Soc.* 96: 1133.
44 Jeffery, T. (1994). *Tetrahedron Lett.* 35: 3051.
45 Bumagin, N.A., Bykov, V.V., Sukhomlinova, L.I. et al. (1995). *J. Organomet. Chem.* 486: 259.

46 Bhattacharya, S., Srivastava, A., and Sengupta, S. (2005). *Tetrahedron Lett.* 46: 8809.
47 Lipshutz, B.H., Ghorai, S., Leong, W.W.Y. et al. (2011). *J. Org. Chem.* 76: 5061.
48 Killinger, T.A., Boughton, N.A., Runge, T.A., and Wolinsky, J. (1977). *J. Organomet. Chem.* 124: 131.
49 Petrier, C. and Luche, J.L. (1985). *J. Org. Chem.* 50: 910.
50 Fleming, F.F., Gudipati, S., and Aitken, J.A. (2007). *J. Org. Chem.* 72: 6961.
51 Krasovskiy, A., Duplais, C., and Lipshutz, B.H. (2009). *J. Am. Chem. Soc.* 131: 15592.
52 Krasovskiy, A., Duplais, C., and Lipshutz, B.H. (2010). *Org. Lett.* 12: 4742.
53 Krasovskaya, V., Krasovskiy, A., Bhattacharjya, A., and Lipshutz, B.H. (2011). *Chem. Commun.* 47: 5717.
54 Duplais, C., Krasovskiy, A., and Lipshutz, B.H. (2011). *Organometallics* 30: 6090.
55 Bhonde, V.R., O'Neill, B.T., and Buchwald, S.L. (2016). *Angew. Chem. Int. Ed.* 55: 1849.
56 Yi, H., Zhang, G., Wang, H. et al. (2017). *Chem. Rev.* 117: 9016.
57 Lyons, T. and Sanford, M. (2010). *Chem. Rev.*: 1147.
58 Liu, C., Yuan, J., Gao, M. et al. (2015). *Chem. Rev.* 115: 12138.
59 Nishikata, T., Abela, A.R., and Lipshutz, B.H. (2010). *Angew. Chem. Int. Ed.* 49: 781.
60 Xu, Z., Yang, T., Lin, X. et al. (2015). *Tetrahedron Lett.* 56: 475.
61 Nishikata, T. and Lipshutz, B.H. (2010). *Org. Lett.* 12: 1972.
62 Xu, Z., Xu, Y., Lu, H. et al. (2015). *Tetrahedron* 71: 2616.
63 Daniels, M.H., Armand, J.R., and Tan, K.L. (2016). *Org. Lett.* 18: 3310.
64 Guram, A.S., Rennels, R.A., and Buchwald, S.L. (1995). *Angew. Chem. Int. Ed. Engl.* 34: 1348.
65 Louie, J. and Hartwig, J.F. (1995). *Tetrahedron Lett.* 36: 3609.
66 Surry, D.S. and Buchwald, S.L. (2008). *Angew. Chem. Int. Ed.* 47: 6338.
67 Park, N.H., Vinogradova, E.V., Surry, D.S., and Buchwald, S.L. (2015). *Angew. Chem. Int. Ed.* 54: 8259.
68 Lavoie, C.M., Macqueen, P.M., Rotta-Loria, N.L. et al. (2016). *Nat. Commun.* 7: 1.
69 Bariwal, J. and Van der Eycken, E. (2013). *Chem. Soc. Rev.* 42: 9283.
70 Lipshutz, B.H., Chung, D.W., and Rich, B. (2009). *Adv. Synth. Catal.* 351: 1717.
71 Nishikata, T. and Lipshutz, B.H. (2009). *Org. Lett.* 11: 2377.
72 Nishikata, T. and Lipshutz, B.H. (2009). *Chem. Commun.*: 6472.
73 Isley, N.A., Dobarco, S., and Lipshutz, B.H. (2014). *Green Chem.* 16: 1480.
74 Wagner, P., Bollenbach, M., Doebelin, C. et al. (2014). *Green Chem.* 16: 4170.
75 Salomé, C., Wagner, P., Bollenbach, M. et al. (2014). *Tetrahedron* 70: 3413.
76 Kumar, A. and Bishnoi, A.K. (2015). *RSC Adv.* 5: 20516.
77 Ishiyama, T., Murata, M., and Miyaura, N. (1995). *J. Org. Chem.* 60: 7508.
78 Fürstner, A. and Seidel, G. (2002). *Org. Lett.* 4: 541.

79 Billingsley, K.L., Barder, T.E., and Buchwald, S.L. (2007). *Angew. Chem. Int. Ed.* 46: 5359.
80 Kleeberg, C., Dang, L., Lin, Z., and Marder, T.B. (2009). *Angew. Chem. Int. Ed.* 48: 5350.
81 Lipshutz, B.H., Moser, R., and Voigtritter, K.R. (2010). *Isr. J. Chem.* 50: 691.
82 VanGelder, K.F. and Kozlowski, M.C. (2015). *Org. Lett.* 17: 5748.
83 Gallou, F., Isley, N.A., Ganic, A. et al. (2016). *Green Chem.* 18: 14–19.
84 Patel, N.D., Rivalti, D., Buono, F.G. et al. (2017). *Asian J. Org. Chem.* 6: 1285.

9

Aspects of Homogeneous Hydrogenation from Industrial Research

Stephen Roseblade

Johnson Matthey, 28 Cambridge Science Park, Milton Road, Cambridge CB4 0FP, UK

9.1 Homogeneous Hydrogenation: A Brief Introduction

Homogeneous catalytic hydrogenation is a reaction of fundamental importance in the chemical industry as evidenced by excellent reviews and book chapters on this topic [1–3] as well as the award of the Nobel Prize to Noyori and Knowles in 2001 [4, 5]. These chemical transformations typically occur under relatively mild reaction conditions, potentially leading to high chemical yields with very low levels of side product formation because of high selectivity.

A homogeneous catalyst typically contains a transition metal, preferably a precious metal, with a specific ligand. Ligands are used to tune reactivity and selectivity, often in subtle ways that are not yet fully understood. Although very high enantioselectivity cannot be achieved in an asymmetric hydrogenation in all cases, it can often be possible to further increase the enantiomeric excess of the reaction product by a simple purification technique such as recrystallization. In favorable cases, high turnover numbers are obtained, leading to low catalyst loadings, which are necessary for a successful commercialization leading to sustainability and affordability in the long term. An example is metolachlor, the active ingredient of the grass herbicide Dual. A stereoselective process based on an iridium catalyst, which was still generally underdeveloped at the time, was developed to achieve an extremely high catalytic activity of turnover number (TON) = 1 000 000 and turnover frequencies (TOF) > 200 000 h^{-1} (Figure 9.1). The ligand effect was crucial in this reaction and this process is currently being used on a scale of more than 10 000 tons per year [6, 7].

Minimization of catalyst loadings is particularly important not only to achieve a cost-effective process but also to minimize the residual amount of metal to be removed at the end of the reaction. For pharmaceutical applications, there are guidelines for the amount of metals in the active pharmaceutical ingredients (API), which go down to parts per million (PPM) levels. With homogeneous hydrogenation becoming an established tool in industry and with the progress of new, complex APIs with catalytic steps toward commercialization, a recent trend is for industrial researchers to focus more on the problem of availability

Organometallic Chemistry in Industry: A Practical Approach, First Edition.
Edited by Thomas J. Colacot and Carin C.C. Johansson Seechurn.

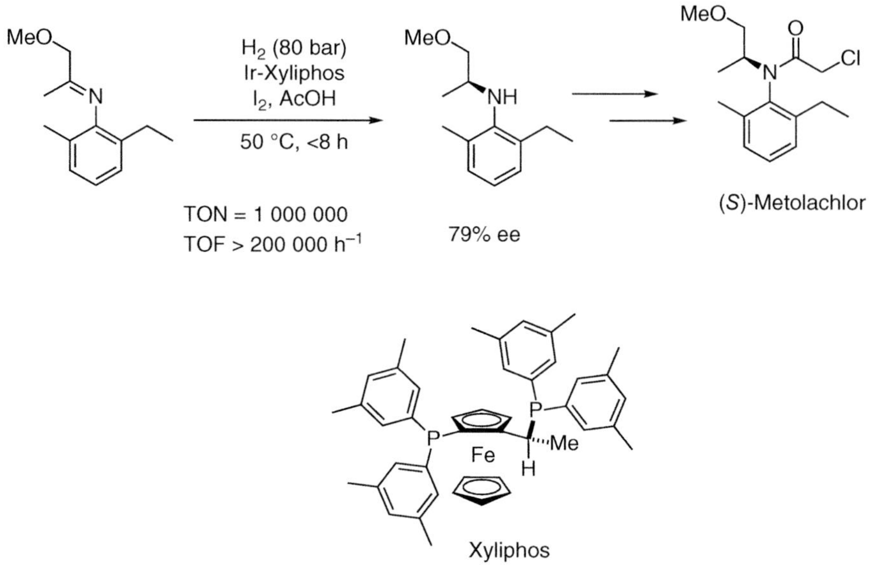

MeO N
H_2 (80 bar)
Ir-Xyliphos
I_2, AcOH
50 °C, <8 h
TON = 1 000 000
TOF > 200 000 h^{-1}
MeO NH
79% ee
MeO N O Cl
(*S*)-Metolachlor

P Fe P Me H
Xyliphos

Figure 9.1 Asymmetric Ir-catalyzed imine hydrogenation in the (S)-metolachlor process.

on scale of chiral phosphine ligands. In fact, only a small number of these ligands have been made on multikilogram scale suitable to support the large-scale commercial production of chiral intermediates. This is of course one of the factors to be considered while selecting a suitable catalyst for a transformation. Although ligands commercially produced at larger scale are the obvious first choice, there are cases of low-volume intermediates that can accept the use of a technically superior, specialized catalyst.

One advantage of hydrogenation is that it is a simple addition reaction that does not in itself produce waste by-products and hence is environmentally safe and can be considered as a sustainable technology. During hydrogenation, excess hydrogen gas is normally used, which is vented out safely after the completion of the reaction: use of hydrogen gas as a reagent is practically straightforward and economical. Homogeneous hydrogenation can often be carried out successfully in a range of organic solvents, depending on the substrate solubility, to achieve useful concentrations. Suitable solvents may include relatively benign organic solvents such as alcohols (MeOH, EtOH, and 2-PrOH). The scope of homogeneous hydrogenation is vast, with excellent substrate scope, functional group tolerance, and high levels of chemo- and enantioselectivity attainable.

9.2 Catalyst Selection by Effective Screening Approaches

The field of homogeneous hydrogenation has enjoyed intense research efforts over many years, and this has resulted in the development of a wide variety of organometallic complexes with diverse ligands. For many reactions, a meaningful prediction can be made about which type of catalyst system will be most appropriate: factors to consider include the metal, the ligand, and reaction conditions such as solvent, additives, hydrogen pressure, and temperature. However, such recommendations typically work for model substrates and they rarely predict with real accuracy for a particular or unique substrate such as the functionalized molecules found in the life science industries. Therefore, a knowledge-based screening approach, in conjunction with high-throughput screening, is a way to identify a catalytic system for a particular transformation from a practical point of view.

Both preformed catalysts, which are often organometallic complexes, or *in situ*-generated catalysts, in which a metal precursor is combined with a ligand to form the catalyst within the reaction mixture, are tested frequently. The use of preformed catalysts can have advantages over *in situ* systems. One advantage of preformed catalysts is that they are synthesized, isolated, and characterized fully; hence, their purity and identify are under control and side products related to catalyst impurities can be minimized. This can also make the scalability of the process easier. Activation processes for preformed catalysts are also understood and can be controlled, such as hydrogenation of a cyclooctadiene ligand in the [Rh(cod)(P^P]BF_4 catalysts [8] or thermolysis of a [Ru-(BINAP)(*p*-cymene)Cl]Cl complex by dissociation of the *p*-cymene. The invention of BINAP and related ligands, together with their use in combination with ruthenium as the reactive metal center, was an important advance in making asymmetric hydrogenation a truly practical methodology for large-scale asymmetric synthesis [4, 9]. Preformed catalysts of the type [Ru-(P^P)(*p*-cymene)Cl]Cl or [Ru-(P^P)$(OAc)_2$] are easy to screen and the reaction temperature needed for the reaction to proceed well enough is often in the region of 60–80 °C. This promotes catalyst activation and good reaction rates.

With an *in situ* catalyst system, the nature of the metal–ligand complex is not clearly understood at the beginning of the reaction and this can lead to uncertainties; for example, if the rate of catalyst formation for a certain transition metal precursor varies significantly with different ligands. One example of a catalytic system found by screening preformed catalysts is discussed in more detail in a separate section, in which catalytic asymmetric hydrogenation was used as a key step in a novel synthesis of *aliskiren*.

Nevertheless, some *in situ* catalytic systems work very well to deliver high enantioselectivity and catalytic turnover and experiments can be carried out to screen chiral Ru catalysts through *in situ* catalyst formation. Common precursors for these experiments are [Ru(*p*-cymene)Cl_2]$_2$, [Ru(benzene)Cl_2]$_2$, and [Ru(Me-allyl)(cod)], which can be reacted with the appropriate chiral ligand to form a catalyst *in situ*. This mixture, which is often uncharacterized during screening, is used in the hydrogenation reaction. Iridium-bisphosphine catalysts for imine hydrogenation are good examples of *in situ*-formed systems that can be highly efficient. The precursor used is normally [Ir(cod)Cl]$_2$, which is mixed with the ligand, substrate, solvent, and additive to form a catalytic system that is able to deliver excellent results in certain cases [6]. One example of imine hydrogenation using an *in situ*-formed catalyst uses an [Ir(cod)Cl]$_2$/PPhos/H_3PO_4 system for asymmetric hydrogenation of a functionalized dihydroisoquinoline,

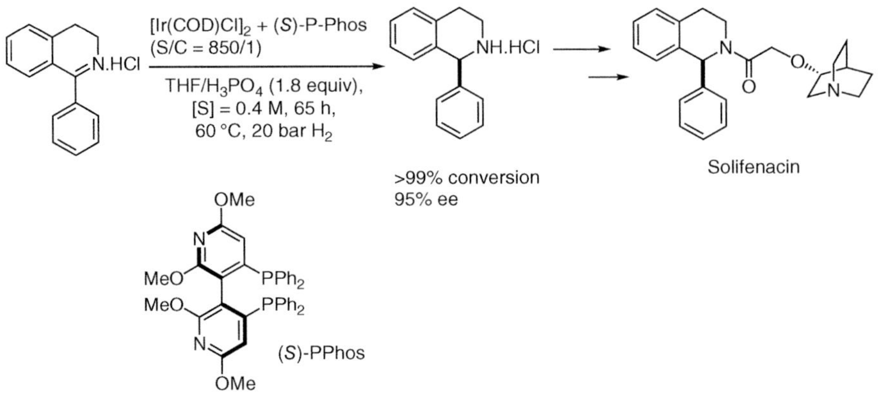
[Ir(COD)Cl]2 + (S)-P-Phos
(S/C = 850/1)
THF/H3PO4 (1.8 equiv),
[S] = 0.4 M, 65 h,
60 °C, 20 bar H2
N.HCl
NH.HCl
>99% conversion
95% ee
Solifenacin
OMe
MeO
MeO
PPh2
PPh2
(S)-PPhos

Figure 9.2 Asymmetric hydrogenation of a cyclic imine en route to solifenacin.

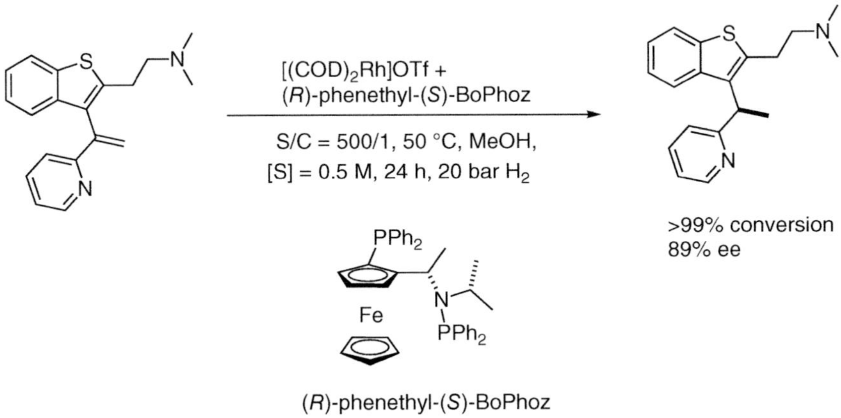
[(COD)2Rh]OTf +
(R)-phenethyl-(S)-BoPhoz
S/C = 500/1, 50 °C, MeOH,
[S] = 0.5 M, 24 h, 20 bar H2
>99% conversion
89% ee
PPh2
Fe
PPh2
(R)-phenethyl-(S)-BoPhoz

Figure 9.3 Asymmetric hydrogenation of a complex olefin substrate.

en route to Solifenacin (Figure 9.2). The work undertaken to find this system was carried out at Johnson Matthey and the screening approach is detailed in the published work, as well as some information about reaction scale-up in Parr autoclaves and workup/product isolation [10].

Screening programs can also include the use of *in situ*-generated rhodium catalysts using a range of commercially available precursors such as $[Rh(cod)Cl]_2$, $[Rh(nbd)Cl]_2$, $[Rh(cod)_2]BF_4$, and $[Rh(CO)_2(acac)]$. The example shown in the scheme below (Figure 9.3) highlights an *in situ*-generated catalytic system using a BoPhoz derivative, which was discovered at Johnson Matthey (JM) during a catalyst screening program for the asymmetric hydrogenation of a highly substituted heteroaromatic olefin [11].

The success of screening work and the collection of meaningful scientific data rely on having a reliable analytical method for timely analysis of reaction samples. This is most often performed using liquid and/or gas chromatography (GC) that gives information on conversion to product, side product formation, and

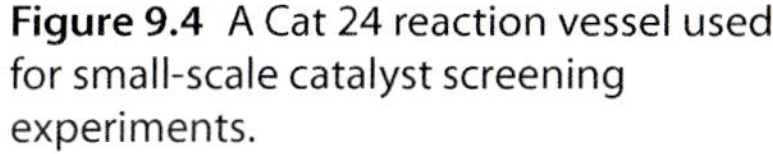

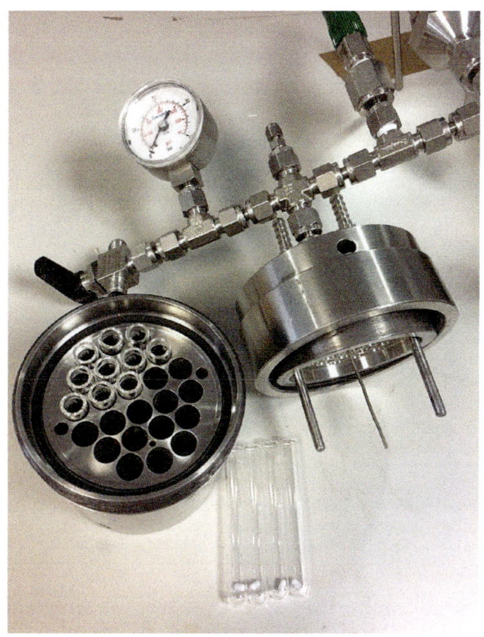

Figure 9.4 A Cat 24 reaction vessel used for small-scale catalyst screening experiments.

enantioselectivity. For liquid chromatography, normal and reverse-phase methods exist, which use a variety of columns and stationary phases. These can often be used in parallel, and in combination with mass spectrometry, as a powerful tool to generate experimental data. Normally, direct analysis of the crude reaction mixture with minimal workup is performed to minimize variability and ensure fair comparisons between experiments.

The practical equipment used for catalyst screening can affect the number and type of experiment(s) that are performed, and which piece of equipment is used will depend on the laboratory in question. In academic laboratories, testing might be carried out simply in an autoclave that is able to hold a few glass vials charged with small magnetic stirrer bars and a very small reaction volume. In our laboratory at Johnson Matthey, the HEL Cat 24 is used for screening: this consists of a metal block in which 24 small glass tubes can be stirred magnetically and heated under hydrogen in parallel (Figure 9.4). Control of temperature and stirring speed is possible through a thermocouple and a heated stirrer hot plate. Hydrogen pressure can also be controlled and measured through a valve and gauge. The HEL Cat24 is a relatively simple and effective piece of equipment for catalyst screening in homogeneous hydrogenation applications.

The Biotage Endeavour has also proved extremely useful and robust in our laboratory as it can be conveniently used for catalyst screening studies as well as reaction optimization at a slightly larger scale. Eight reaction vials of 8 ml capacity may be charged, which are used with overhead stirring to allow excellent mixing of the gas and liquid interface (Figure 9.5). Reaction temperature and pressure can be controlled individually, which makes the equipment useful for optimization experiments as well as catalyst screening. The reaction wells can be purged with nitrogen gas before addition of solvent through injection ports,

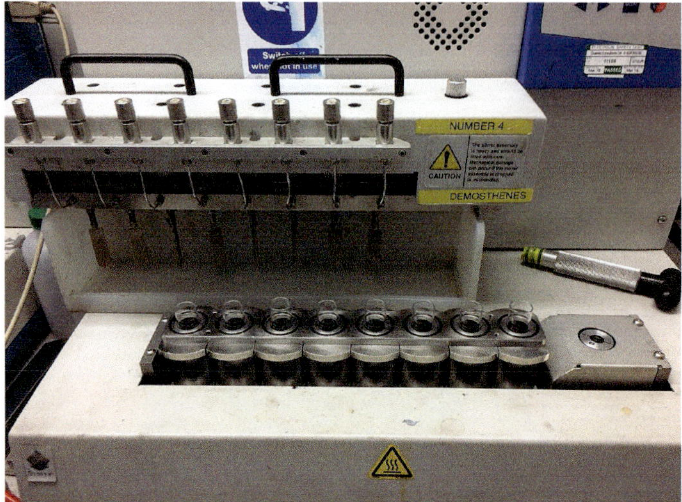

Figure 9.5 A Biotage Endeavour used for screening of reaction conditions and optimization.

allowing hydrogenation reactions to be performed under inert conditions. Different reaction solvents can also be used as each of the reaction wells is sealed and isolated from each other. The stirring speed can be varied, allowing the effect of this parameter on the reaction to be determined. Hydrogen uptake curves can be measured for each of the reaction wells, allowing kinetic data to be obtained.

9.3 Considerations for Reaction Scale-up

Homogeneous hydrogenation, including asymmetric hydrogenation, can often be scaled-up from small-scale screening experiments to multigram and kilogram reactions in a straightforward manner. Once a catalytic system has been identified in the screening experiments, further work will be needed to assess how easily the hydrogenation can be scaled-up. This can be done using standalone autoclaves of which various sizes are available. As an example, reaction in a 25–50 ml autoclave might be first attempted, and if this proceeds well, giving similar results to the screening experiments, then further scale-up might be possible to a 1 l reactor or larger volume (Figure 9.6).

One factor to consider in ensuring a successful reaction scale-up is that homogeneous hydrogenation may be sensitive to oxygen in the reaction solvents, which means degassing of reaction solvents can be needed. Homogeneous hydrogenation is a two-phase reaction mixture in which hydrogen gas must dissolve in the reaction solution to drive the reaction, which means that factors such as hydrogen pressure, gas solubility, and stirring speeds/dynamics might be important [12]. H_2 solubility could be a factor in solvent choice but is not the only factor that determines the choice of solvent; solubility in MeOH is much worse than in heptane, for instance, but the polarity of the solvent plays a role in the catalytic cycle [13].

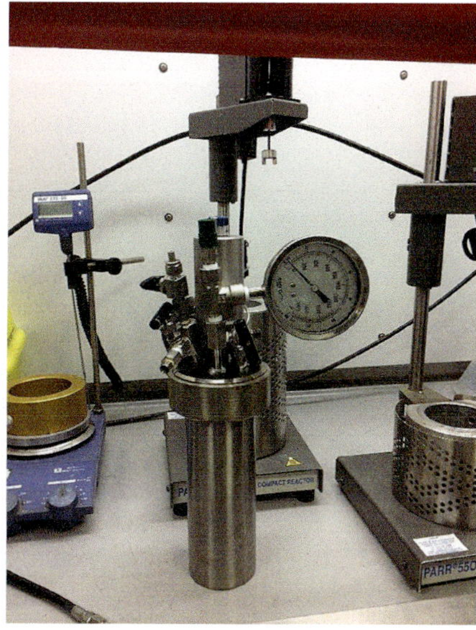

Figure 9.6 Stand-alone autoclaves used for preliminary reaction scale-up.

Inadequate stirring of the hydrogenation reaction could lead to a reaction mixture in which the amount of hydrogen dissolved in the solution is not enough to fully saturate the solution, so this could affect the reaction rate and selectivity. High stirring rates, together with adequate baffling or the use of entrainment impellers, should normally be used to enhance reaction rates and prevent catalyst deactivation. The effect of stirring rates on the hydrogenation reaction can be determined experimentally at small scale during optimization before scale-up is attempted and mass transfer effects can also be studied to ensure an efficient process [14].

Hydrogen gas is obviously consumed during the hydrogenation reaction, and this can lead to a pressure drop during the reaction if the system is closed. The effect of hydrogen pressure should be investigated during reaction optimization so that this factor is understood before moving to larger scale. It can be the case that the hydrogen is completely consumed from the reaction headspace during hydrogenation, which means the catalyst is present under "hydrogen-starvation" conditions. Some catalytic systems can survive such conditions and are quite robust because of this, which means when hydrogen is reintroduced to the reaction mixture, the hydrogenation reaction proceeds as expected. Such an experiment should be performed during optimization to explore robustness of the system, toward mechanical failure of the reactor, for example.

It is possible that a hydrogenation reactor or autoclave overheats in the initial stages of the reaction if the temperature is not controlled well enough, partly because of the hydrogenation reaction being exothermic. This can lead to the reaction taking place at a higher temperature than desired, which could result in increased side products, reduced selectivity of the catalytic reaction, or catalyst deactivation. Therefore, the optimal temperature should be established during

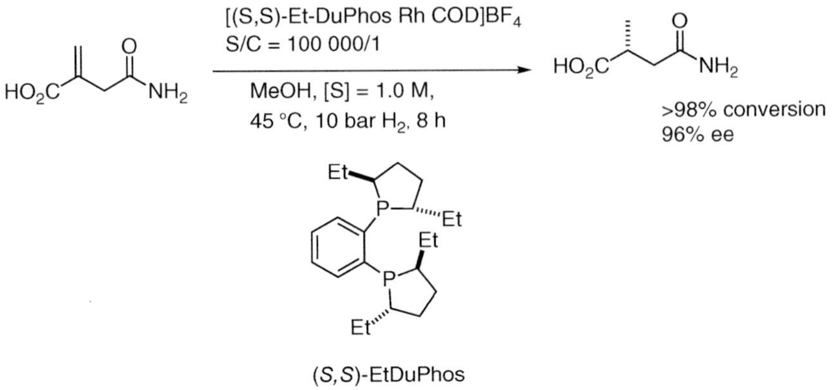

Figure 9.7 Efficient Rh-catalyzed asymmetric hydrogenation of an unsaturated carboxylic acid.

smaller scale optimization experiments and then carefully controlled during scale-up.

Substrate purity can have a strong impact on the catalyzed reaction, and a pure substrate free from impurities is much preferred. The nature of catalyst poisons and how they act can be in many ways and include organic residues from reactions such as Ac_2O or AcOH and halide anions such as chlorides and bromides. In the example shown in Figure 9.7, excellent results in terms of catalyst loading and reaction rates are obtained if halide-free substrate is used. When chloride ion contamination is present, inferior catalytic performance is observed [15].

A second example of an asymmetric hydrogenation using a cationic rhodium complexes, which are affected by impurities in the substrate, is shown in Figure 9.8. With a highly pure material, obtained by column chromatography, a TON of 1000 was attained but at larger scale using material that was not as pure TON was lower (300) [16].

Contamination of the hydrogenation vessel must be avoided to perform a hydrogenation reaction reproducibly. For example, an asymmetric hydrogenation reaction performed in a reaction vessel contaminated with residues of heterogeneous catalyst can result in a hydrogenation reaction proceeding with a

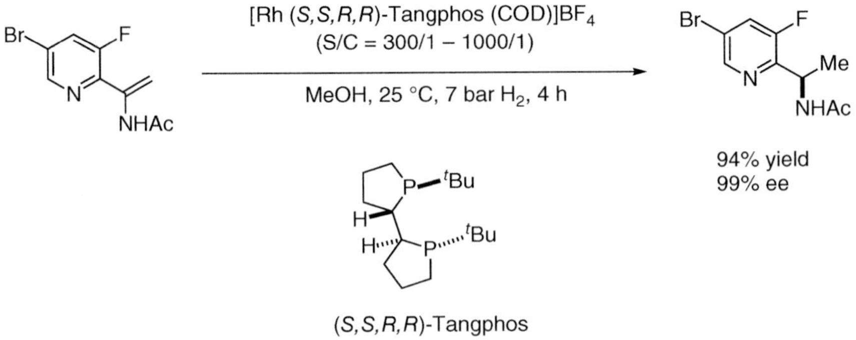

Figure 9.8 Rh-catalyzed asymmetric hydrogenation of a pyridine-substituted enamide.

decreased enantioselectivity. This can be a problem with reactions that are quite facile, making it a real issue for processes suited to industrial application at scale. Therefore, careful cleaning of the reactor should be carried out before use.

Hydrogen gas has a number of properties that make it hazardous: it is a flammable gas that is colorless, odorless, and tasteless. It is lighter than air with a specific gravity of 0.06. It is asphyxiant if released in high concentrations and may be ignited by sparks, heat, or even by releasing gas from a cylinder under pressure. Hydrogen undergoes a reverse Joule–Thomson effect, which means it warms slightly when released from a cylinder. It has a high flammability range in air (4–75%). Hydrogen diffuses rapidly and may leak from systems that may be gas tight for other gases. Because of these hazards, hydrogen cylinder valves must never be rapidly opened and closed. Earthing lines and equipment where there is the possibility of electrostatic discharge should be considered. Excellent ventilation by extraction through fume cupboards is needed, with the ability to remove hydrogen gas in the event of a leak or in the case of a sudden release from the vessel following the rupture of a bursting disk. Leaks from reactors, valves, regulators, cylinders, and lines must be avoided, normally by replacing defective equipment. Further information can be found through the British Compressed Gas Association.

9.4 Notes on Additive Effects

The effect of additives on catalytic asymmetric reactions has been reviewed [17]. Halide anions can have a strong effect on the use of asymmetric hydrogenation at scale, and ammonium chloride has been found to have a positive impact on asymmetric hydrogenation using a chiral rhodium–bisphosphine complex: although usually the presence of halides is negative, as they modify the catalyst from cationic to a less reactive halide coordinated one [15], in the example shown below, an unexpected result was observed when the source of reaction variability was traced back to the contents (or lack of) of ammonium chloride in the hydrogenation feedstock (Figure 9.9) [18].

In another example shown below, $Zn(OTf)_2$ was found to have a very strong positive effect on the hydrogenation of a tetrasubstituted enone substrate (Figure 9.10) [19]. An *in situ* catalyst was used, formed from a cationic Rh-precursor, and a Josiphos ligand. The addition of $Zn(OTf)_2$ increased the turnover frequency and reduced epimerization in the hydrogenation product.

Application of this system to use in continuous flow has also been reported [20]. The use of continuous flow for homogeneous hydrogenation is an area of active research and development. Examples in the literature include the use of tube-in-tube reactors [21], the use of ionic liquids and supercritical CO_2 as reaction phases [22, 23], and solid-supported or heterogenized catalysts [24, 25].

Noyori's group has shown that the Lewis acidic additive, $B(O^iPr)_3$, is beneficial for ketone hydrogenation using heteroaromatic substrates able to coordinate to the catalytically active metal center (Figure 9.11) [26].

$LiBF_4$ has been used in combination with a preformed Ru-BINAP catalyst in the asymmetric hydrogenation of a dehydro-β-amino acid [27]. The additive was

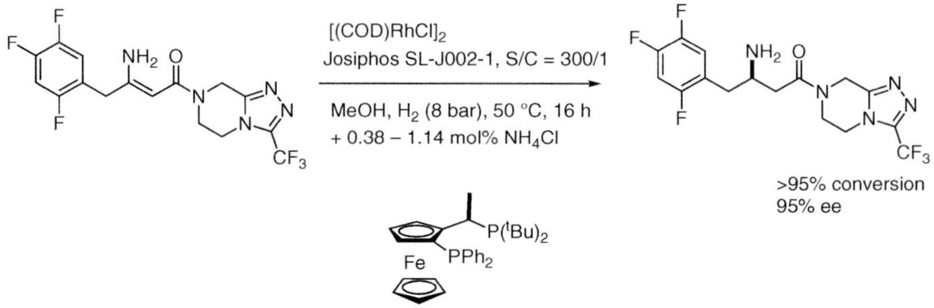

$[(COD)RhCl]_2$
Josiphos SL-J002-1, S/C = 300/1
MeOH, H_2 (8 bar), 50 °C, 16 h
+ 0.38 – 1.14 mol% NH_4Cl

>95% conversion
95% ee

$P(^tBu)_2$
PPh_2
Fe

Figure 9.9 Asymmetric hydrogenation using a rhodium catalyst with ammonium chloride as an additive.

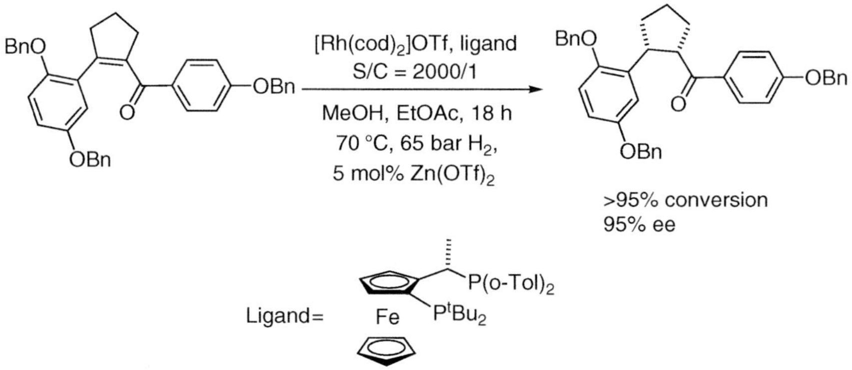

$[Rh(cod)_2]OTf$, ligand
S/C = 2000/1
MeOH, EtOAc, 18 h
70 °C, 65 bar H_2,
5 mol% $Zn(OTf)_2$

>95% conversion
95% ee

Ligand= Fe $P(o\text{-}Tol)_2$ P^tBu_2

Figure 9.10 Asymmetric hydrogenation of an enone with zinc triflate as an additive.

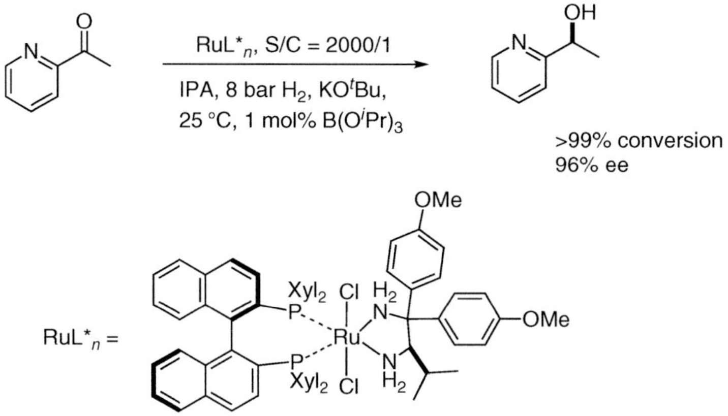

RuL^*_n, S/C = 2000/1
IPA, 8 bar H_2, KO^tBu,
25 °C, 1 mol% $B(O^iPr)_3$

>99% conversion
96% ee

RuL^*_n =

Figure 9.11 Ru-catalyzed asymmetric hydrogenation of a pyridyl ketone in the presence of triisopropylborate.

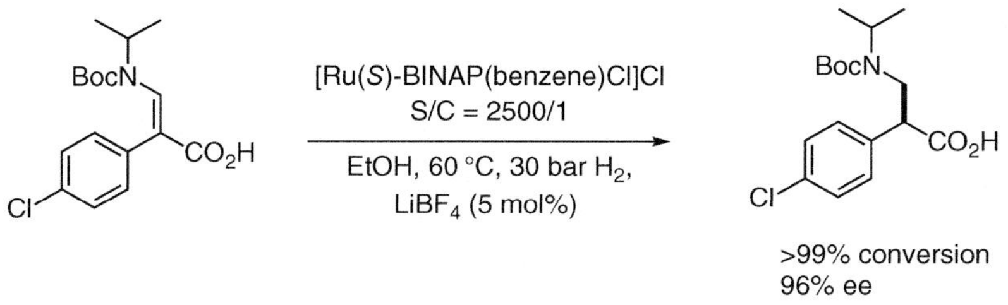

Figure 9.12 Ru-catalyzed asymmetric hydrogenation of a dehydro beta-amino acid derivative with $LiBF_4$ as an additive.

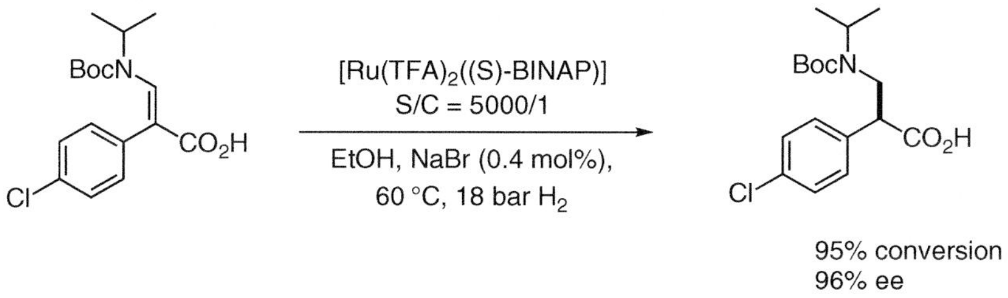

Figure 9.13 An alternative Ru-BINAP-based catalytic system using NaBr as an additive.

used in catalytic amount and was found to increase reaction rate allowing hydrogenation at lower catalyst loading and decreased hydrogen pressure (Figure 9.12).

In further work carried out at Roche, the system was reinvestigated. A Ru-BINAP system was again identified as giving the optimal performance in EtOH as the reaction solvent. An *in situ* catalytic system was formed using $[Ru(COD)(TFA)_2]$ as the precursor and BINAP. NaBr was found to lead to improved performance giving reproducible results at a catalyst loading of 5000/1 and lower [28]. As in a previous example from Merck, in which NH_4Cl was found to aid a Rh-Josiphos-catalyzed asymmetric hydrogenation for sitagliptin production [18], this additive effect was discovered because of variations in reaction performance caused by an impurity in the substrate (Figure 9.13).

The above examples show that additives of different types can have significant effects on the reaction being investigated, although the mode of action may not be very well understood in some cases. Screening additives which might behave as acids, bases and Lewis acids can be a useful exercise and sometimes unexpected findings can be made. To some extent such discoveries do still rely on intuition, open-mindedness and careful experimental work.

9.5 A Novel Approach to Aliskiren Using Asymmetric Hydrogenation as a Key Step

Aliskiren **1** is a well-known renin inhibitor used for treatment of high blood pressure and was developed originally by Novartis (see Figure 9.15 for structure, **1**). It has a complex structure that incorporates four stereogenic centers and is used in the clinic at a relatively high dosage (150 mg/d). Its chemical development has been challenging, but a variety of approaches have been successfully

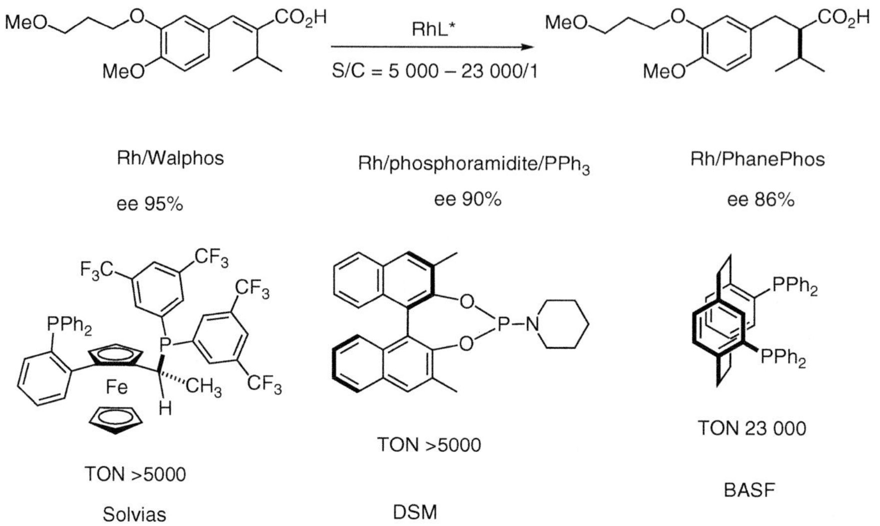

RhL*

S/C = 5 000 – 23 000/1

Rh/Walphos | Rh/phosphoramidite/PPh_3 | Rh/PanePhos

ee 95% | ee 90% | ee 86%

TON >5000 | TON >5000 | TON 23 000

Solvias | DSM | BASF

Figure 9.14 Rh-catalyzed asymmetric hydrogenation of an unsaturated acid for Aliskiren production.

carried out, which include an asymmetric hydrogenation of an unsaturated acid (Figure 9.14). This process has been performed with three different catalysts, developed at DSM, Solvias, and BASF. BASF were able to use Phanephos, which gave the lowest ee of the three but gave the highest activity, as the acid can be recrystallized to increase enantiomeric excess [3].

Chemessentia became interested in developing a non-intellectual property (IP) infringing route to the generic drug and designed a novel approach in which asymmetric hydrogenation was also a key step, although as part of a different synthetic strategy to the previous routes. Chemessentia worked with JM on the hydrogenation of a tetra-substituted enol-acetate **4**, which would eventually be converted to a 1,2 amino alcohol derivative **2** by way of a Curtius rearrangement using a chiral carboxylic acid **3** [29]. Diastereoselective hydrogenation using a variety of chiral and achiral homogeneous catalysts was investigated using several unsaturated ester and unsaturated acid substrates, which were investigated consecutively during this collaboration (Figure 9.15).

Experiments were initially performed using the unsaturated ester substrates **5** and **6** shown below (Figure 9.16), which both incorporate a rather unreactive tetra-substituted C=C double bond. Screening experiments showed that conversion at relatively high catalyst loadings could be achieved using highly active rhodium bis-phosphine catalysts such as with [Rh(*S*)-Phanephos(cod)]BF_4. This catalyst was shown to give the best performance in terms of conversion and stereoselectivity. Two ester substrates were tested: one bearing an oxazolidinone auxiliary **5** and the other a benzyl ester functional group **6** and a multigram hydrogenation was carried out in a Parr autoclave using the benzyl ester **6** (Figure 9.16).

In an effort to improve the performance of the asymmetric hydrogenation reaction, a carboxylic acid substrate **7** was prepared at Chemessentia and used in screening experiments at Johnson Matthey (Figure 9.17). Using

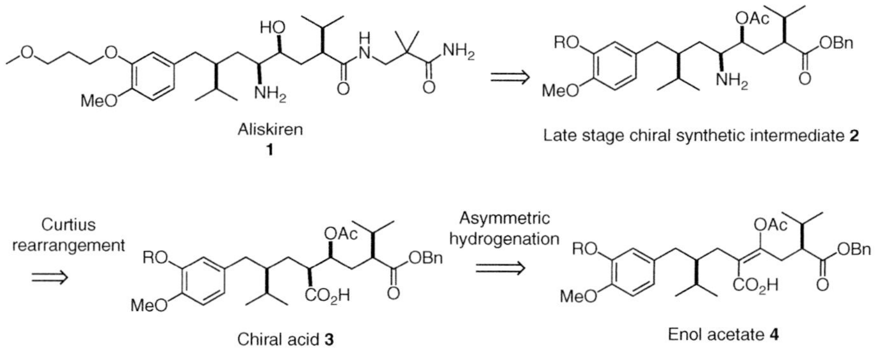

Aliskiren
1

Late stage chiral synthetic intermediate **2**

Curtius rearrangement

Asymmetric hydrogenation

Chiral acid **3**

Enol acetate **4**

Figure 9.15 Retrosynthesis of Aliskiren via Curtius rearrangement and asymmetric hydrogenation.

5

[Rh-(*S*)-Phanephos (cod)]BF_4 (2 mol%)
DCE, 80 °C, 30 bar H_2, 48 h, 1 equiv $HBF_4.Et_2O$

77% yield, dr = 4 : 1

(*S*)-Phanephos

6

[Rh-(*S*)-Phanephos (cod)]BF_4 (5 mol%)
DCE, 60 °C, 30 bar H_2, 18 h

91% conversion,
9% side-product
dr = 8 : 92

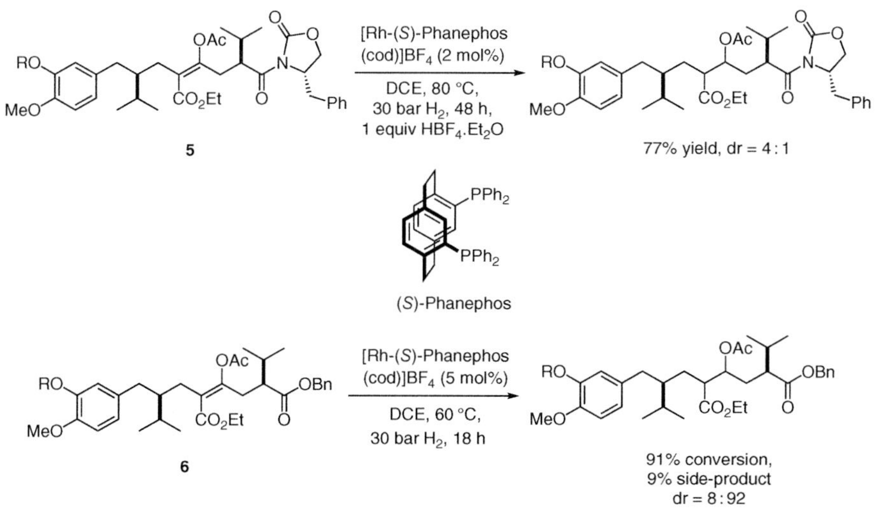

Figure 9.16 Rhodium-Phanephos-catalyzed asymmetric hydrogenation of synthetic intermediates **5** and **6**.

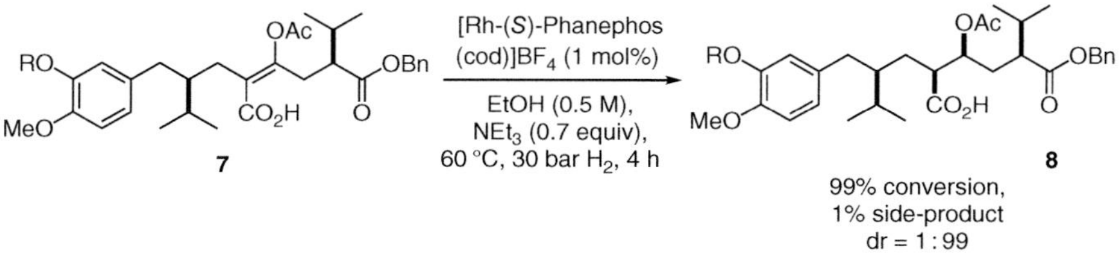

7

[Rh-(*S*)-Phanephos (cod)]BF_4 (1 mol%)
EtOH (0.5 M), NEt_3 (0.7 equiv), 60 °C, 30 bar H_2, 4 h

8

99% conversion,
1% side-product
dr = 1 : 99

Figure 9.17 Rhodium-Phanephos-catalyzed asymmetric hydrogenation of unsaturated carboxylic acid intermediate **7** from screening experiments.

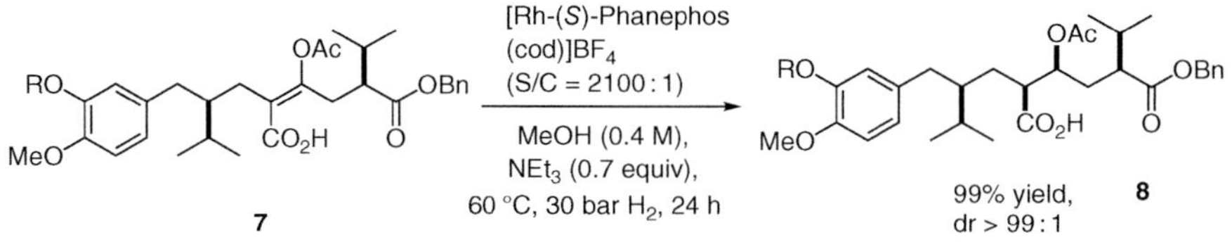

[Rh-(*S*)-Phanephos(cod)]BF_4 (S/C = 2100 : 1)
MeOH (0.4 M), NEt_3 (0.7 equiv), 60 °C, 30 bar H_2, 24 h
7 → 8: 99% yield, dr > 99 : 1

Figure 9.18 Rhodium-Phanephos-catalyzed asymmetric hydrogenation at lower catalyst loading.

this substrate, much better results were obtained, particularly in the presence of base that was added to deprotonate the carboxylic acid. The use of [Rh-(*S*)-Phanephos(cod)]BF_4 was found to give the desired diastereoisomer **8** in 98% de with the correct relative stereochemistry (Figure 9.17). Other catalysts, such as the achiral [Rh(dipfc)(cod)]BF_4, were tested but gave lower stereoselectivity and product purity.

Optimization work before scale-up of the hydrogenation of **7** was carried out but has not been reported in detail. However, the system described for use at 1-l scale is actually very similar to that found in the initial experiments (MeOH, 0.7 equiv NEt_3, 60 °C, 30 bar H_2). Catalyst loadings of >2000/1 are reported when the solvent and hydrogenation reactor are degassed carefully with nitrogen (Figure 9.18).

9.6 Efficient Chemoselective Aldehyde Hydrogenation

Reduction of carbonyl compounds to alcohols is an important transformation for industrial processes [30]. Large-scale reductions have been performed using stoichiometric reducing agents such as sodium borohydride under various reaction conditions. Catalytic hydrogenation presents a potential alternative, simplifying reaction processes and reducing the amount of waste generated [31–34]. Solvent-free catalytic processes are of great interest in reducing the amount of waste produced in large-scale processes, helping to reduce costs and environmental impacts as well as improving efficiency [35–40]. A current trend has been the application of highly active achiral homogeneous catalysts to carbonyl reduction for industrial use, which poses an additional challenge in terms of further reduction of catalyst loadings, requiring very active catalysts [41]. For this, ruthenium complexes are used, which bear ligands imparting high activity toward certain carbonyl reductions, such as complexes **9–11** shown in Figure 9.19 [33].

Using these catalysts (**9–11**), aldehydes can be reduced in a highly efficient reaction under solvent-free conditions. For example, the hydrogenation of benzaldehyde was examined using [$RuCl_2$(dppb)(ampy)] **10**, a commercially available catalyst. A range of temperatures were tested with a catalyst loading of 2000/1, in combination with 1N KOH (2 mol%). MeOH was used as a solvent initially and at increasing substrate concentration. Temperatures of 50–90 °C were probed at 30 bar H_2 pressure. Results from these experiments show that reactions at 90 °C gave high conversion to the benzyl alcohol product and that the hydrogenation in

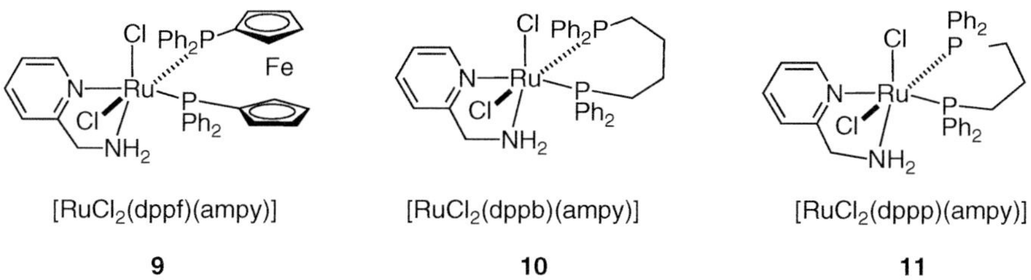

Figure 9.19 Multifunctional ruthenium-ampy catalysts from Prof. Baratta et. al.

the absence of solvent performed very well (T. Angelini and S. Roseblade, unpublished results).

A variety of different aldehydes was hydrogenated using the [$RuCl_2$(dppb)(ampy)] catalyst **10** under the conditions: S/C = 10 000/1, KOH (0.1–0.2 mol% added as a 1N solution), neat, 90 °C, 28 bar H_2, 16 hours. Hydrogen uptake was observed using the Biotage Endeavour. The reactions were analyzed by ^{1}H NMR and GC (Figure 9.20).

High chemoselectivity is obtained using substrates that bear halogen atoms on the aromatic ring: with a heterogeneous catalyst, these functional groups might be reduced by hydrogenolysis. A base is needed to activate the catalyst, but only in low levels of 0.1–0.2 mol%, which can be added as an aqueous solution of KOH. These conditions result from a straightforward and progressive optimization, in which the reaction can be performed at very low loadings of catalyst and base to avoid side product formation.

Hydrogenation of commercial grade benzaldehyde using [$RuCl_2$(dppb)(ampy)] **10** was attempted in a stainless steel Parr autoclave at a loading of S/C = 40 000/1 under the conditions: KOEt (0.125 mol%), neat, 90 °C, 25 bar H_2. Hydrogen uptake was observed and the pressure was maintained between 20 and 25 bars. Once uptake had finished, after about 24 hours, GC analysis was performed and showed >98% conversion to benzyl alcohol (Figure 9.21).

Although cheap reagents such as sodium borohydride could be used for such transformations and are probably a more common approach, there are practical benefits to using a hydrogenation reaction: essentially, no waste is generated in the hydrogenation and the reaction can be performed solvent free. Potentially laborious and inefficient workup of borohydride reactions, involving aqueous extraction of the desired product from the crude reaction mixture, is avoided. The use of a relatively inexpensive and highly active catalyst is a key to making such an approach feasible.

9.7 Closing Remarks/Summary

Homogeneous hydrogenation has long been known for its practicality in terms of allowing asymmetric synthesis to be carried out at scale. This is still true today, and as a selection of ligands and catalysts becomes commercially available at lower cost, its use should continue. New applications are being discovered

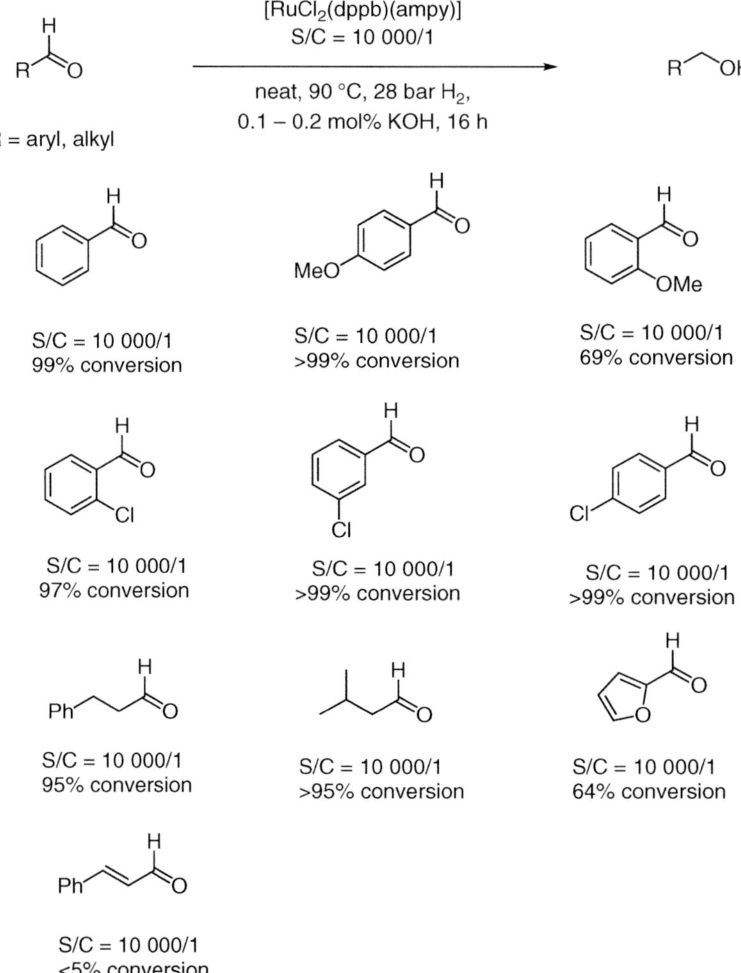

Figure 9.20 Examples of solvent-free conditions for chemoselective aldehyde hydrogenation at low catalyst loading.

with a wide variety of substrate classes, making this still a very active field of great interest to researchers. Equipment for catalyst screening and reaction analysis allows this task to be carried out in a straightforward manner, and although reaction scale-up is not always easy, it can sometimes be achieved without problems. In this regard, first-hand experiences have been shared about the practical aspects of hydrogenation reactions in general as well as findings related to impurity and/or additive effects. Asymmetric hydrogenation can be applied in new and imaginative ways and an example of a novel aliskiren synthesis was described in which unsaturated acid hydrogenation and Curtius rearrangement were combined to form a 1,2-amino alcohol functional group stereoselectively. Finally, a recent application for the sustainable synthesis of alcohols by chemoselective reduction of aldehydes has been disclosed. Again,

Ru-721, **10**
(S/C = 40 000/1)
neat, 90 °C, 20 – 25 bar H_2,
0.125 mol% KOEt

OH

>98% by GC and ^1H NMR

Cl Ph_2P N Ru P Cl Ph_2 NH_2

"Ru-721" **10**

Figure 9.21 Highly efficient and user-friendly hydrogenation of benzaldehyde on multigram scale.

this is a straightforward reaction to apply, which can be carried out using highly active, commercially available catalysts. Future developments are difficult to predict, but there is always a drive toward more efficient, cost-effective processes that are sustainable and environmentally friendly.

Biography

Stephen James Roseblade was a chemistry undergraduate at Keble College, Oxford, and graduated with a first class degree. He holds a Ph.D. degree from the University of Bristol after three years with Varinder Aggarwal's research group working on asymmetric organic synthesis. He carried out post-doctoral research in homogeneous catalysis as a Marie Curie Fellow at the CSIC in Seville and also as a research fellow with Andreas Pfaltz at the University of Basel. Since 2007 he has worked at Johnson Matthey on the Cambridge Science Park, actively engaged in catalyst research for applications in the fine chemical and pharmaceutical industry.

References

1 Ager, D.J. and DeVries, J.G. (2012). *Comprehensive Chirality*, vol. 9, 73–82. Elsevier.
2 Komiya, S., Shimizu, H., and Nagasaki, I. (2012). *Comprehensive Chirality*, vol. 9, 83–102. Elsevier.
3 Blaser, H.-U. (2012). *Applications of Transition Metal Catalysis in Drug Discovery and Development: An Industrial Perspective* (eds. M.L. Crawley and B.M. Trost), 315–342. Wiley.
4 Noyori, R. (2002). *Angew. Chem. Int. Ed.* 41: 2008–2022.
5 Knowles, W.S. (2002). *Angew. Chem. Int. Ed.* 41: 1998–2007.

6 Blaser, H.-U. (2002). *Adv. Synth. Catal.* 344 (1): 17–31.
7 Colacot, T.J. (2003). *Chem. Rev.* 103: 3101–3118.
8 Cobley, C.J., Lennon, I.C., McCague, R. et al. (2001). *Tetrahedron Lett.* 42: 7481–7483.
9 Noyori, R. and Takaya, H. (1990). *Acc. Chem. Res.* 23: 345–350.
10 Ružič, M., Pečavar, A., Prudič, D. et al. (2012). *Org. Process Res. Dev.* 16: 1293–1300.
11 Gross, T., Chou, S., Dyke, A. et al. (2012). *Tetrahedron Lett.* 53: 1025–1028.
12 Sun, Y., LeBlond, C., Wang, J. et al. (1995). *J. Am. Chem. Soc.* 117: 12647–12648.
13 Brunner, E. (1985). *J. Chem. Eng. Data* 30: 269–273.
14 Sun, Y., Landau, R.N., Wang, J. et al. (1996). *J. Am. Chem. Soc.* 118: 1348–1353.
15 Cobley, C.J., Lennon, I.C., Praquin, C. et al. (2003). *Org. Process Res. Dev.* 7 (3): 407–411.
16 O'Shea, P.D., Gauvreau, D., Gosselin, F. et al. (2009). *J. Org. Chem.* 74: 4547–4553.
17 Hong, L., Sun, W., Yang, D. et al. (2016). *Chem. Rev.* 116: 4006–4123.
18 Clausen, A.M., Dziadul, B., Cappuccio, K.L. et al. (2006). *Org. Process Res. Dev.* 10: 723–726.
19 Calvin, J.R., Frederick, M.O., Laird, D.L.T. et al. (2012). *Org. Lett.* 14 (4): 1038–1041.
20 Johnson, M.D., May, S.A., Calvin, J.R. et al. (2012). *Org. Process Res. Dev.* 16: 1017–1038.
21 Newton, S., Ley, S.V., Casas-Arcé, E., and Grainger, D. (2012). *Adv. Synth. Catal.* 354 (9): 1805–1812.
22 Hintermair, U., Franciò, G., and Leitner, W. (2013). *Chem. Eur. J.* 19 (14): 4538–4547.
23 Geier, D., Schmitz, P., Walkowiak, J. et al. (2018). *ACS Catal.* 8: 3297–3303.
24 Amara, Z., Poliakoff, M., Duque, R. et al. (2016). *Org. Process Res. Dev.* 20: 1321–1327.
25 Tsubogo, T., Ishiwata, T., and Kobayashi, S. (2013). *Angew. Chem. Int. Ed.* 52 (26): 6590–6604.
26 Ohkuma, T., Koizumi, M., Yoshida, M., and Noyori, R. (2000). *Org. Lett.* 2 (12): 1749–1751.
27 Remarchuk, T., Babu, S., Stults, J. et al. (2014). *Org. Process Res. Dev.* 18: 135–141.
28 Han, C., Savage, S., Al-Sayed, M. et al. (2017). *Org. Lett.* 19: 4806–4809.
29 Cini, E., Banfi, L., Barreca, G. et al. (2016). *Org. Process Res. Dev.* 20: 270–283.
30 Magano, J. and Dunetz, J.R. (2012). *Org. Process Res. Dev.* 16: 1156.
31 Dupau, P., Bonomo, L., and Kermorvan, L. (2013). *Angew. Chem. Int. Ed.* 52: 1–5.
32 Bonomo, L., Kermorvan, L., and Dupau, P. (2015). *ChemCatChem* 7: 907–910.
33 Baldino, S., Facchetti, S., Zanotti-Gerosa, A. et al. (2016). *ChemCatChem* 8: 2279–2288.

34 Baldino, S., Facchetti, S., Nedden, H.G. et al. (2016). *ChemCatChem* 8: 3195–3198.
35 Hermans, S., Raja, R., Thomas, J.M. et al. (2001). *Angew. Chem. Int. Ed.* 40: 1211.
36 Kimmich, B.F.M., Fagan, P.J., Hauptman, E., and Bullock, R.M. (2004). *Chem. Commun.*: 1014–1015.
37 Kimmich, B.F.M., Fagan, P.J., Hauptman, E. et al. (2005). *Organometallics* 24: 6220–6229.
38 Amenuvor, G., Makhubela, B.C.E., and Darkwa, J. (2016). *ACS Sustainable Chem. Eng.* 4: 6010–6018.
39 Al-Shaal, M.G., Hausoul, P.J.C., and Palkovits, R. (2014). *Chem. Commun.* 50: 10206–10209.
40 Selka, A., Levesque, N.A., Foucher, D. et al. (2017). *Org. Process Res. Dev.* 21: 60–64.
41 Nedden, H. and Zanotti-Gerosa, A. (2013). Emerging trends in homogeneous hydrogenation catalysis. *Speciality Chemicals Magazine*, pp. 26–29.

10

Latest Industrial Uses of Olefin Metathesis

John H. Phillips

Materia, Inc., 60 N San Gabriel Blvd., Pasadena, CA 91107, USA

10.1 Introduction

The ability to control carbon–carbon bond formation is an essential aspect of organic synthesis. In the past few decades, one of the chemical transformations that has shown great potential in chemical industries is alkene metathesis. Although metathesis has enjoyed recent success in the pharmaceutical industry, metathesis had been utilized for decades in the petroleum industry, with some of the first reports from DuPont [1] followed by patents by the standard Oil Co of Indiana [2] and publications from Phillips Petroleum describing the disproportion of propylene and ethylene [3]. These early efforts in the petroleum industry led to more discoveries in the rubber industry with patents and publications from Calderon at the Goodyear with examples of metathesis to create polybutadiene replacements [4]. Subsequently, in the early 1990s, the well-defined metathesis work were published from the labs of Schrock at MIT [5] followed by the Grubbs group at Caltech [6]. These new well-defined catalysts ushered in a new era of metathesis culminating in the 2005 Nobel Prize for Chemistry awarded to Chauvin [7], Grubbs [8], and Schrock (Figure 10.1) [9].

In the past twenty years the chemical industry has seen an increased adoption of metathesis across various sectors from pharmaceutical to materials, primarily driven by the advances in the development of well-defined metathesis catalysts, which are increasingly utilized for all types of reactions from ring-closing metathesis (RCM) to build biologically active macrocycles to ring-opening metathesis polymerization (ROMP) for commercially interesting polymers used in applications from thermal insulation to carbon fiber composites. Although there is a deep history of heterogeneous metathesis processes in industry, this review will focus only on recent industrial uses of well-defined catalysts. Although there are many excellent reviews on metathesis [10, 11], including the utilization in various industries such as pharmaceuticals [12], fine chemicals [13], and materials [14], this chapter will highlight a few of the commercial uses of well-defined metathesis catalysts with an extra emphasis on the roadblocks

Organometallic Chemistry in Industry: A Practical Approach, First Edition.
Edited by Thomas J. Colacot and Carin C.C. Johansson Seechurn.

Figure 10.1 2005 Chemistry Nobel Prize Winners: Yves Chauvin, Robert H. Grubbs, and Richard R. Schrock (left to right).

encountered at the industrial scale. This chapter will conclude with recommendations for reaction design and strategies for troubleshooting problems that are encountered in metathesis transformations.

10.2 General Information

Well-defined metathesis catalysts can be divided into two main categories based on the metal of choice: Ruthenium or nonruthenium-based catalysts. A nonruthenium catalyst was the first well-defined metathesis catalyst to be isolated as a highly reactive 14-electron molybdenum imido alkylidene species [15]. The ruthenium catalysts were discovered a couple of years later and consist of a less reactive 16-electron ruthenium carbene catalyst [6].

10.2.1 Non-ruthenium Catalysts

The first well-defined catalysts discovered by Schrock and coworkers were based on molybdenum and tungsten (Figure 10.2). These Mo- and W-based catalysts are not as commonly used in industry as the ruthenium catalysts. The slow adoption of well-defined non-ruthenium metathesis catalysts on the industrial scale is because of both catalyst availability at scale coupled with the handling and functional group tolerance. Recently, there has been some progress to improve the handling of non-ruthenium catalysts via wax coating to encase the catalyst, thus protecting it from any oxidation or degradation [16]; however, this process has yet to be used on the industrial scale. In summary, the major commercial use of non-ruthenium metathesis catalysts has been relegated to olefin disproportion and the polymerization of dicyclopentadiene (DCPD) [11], but with the recent advancements from Schrock and coworkers in making the air-stable catalysts, their use in industrial applications will hopefully grow with time.

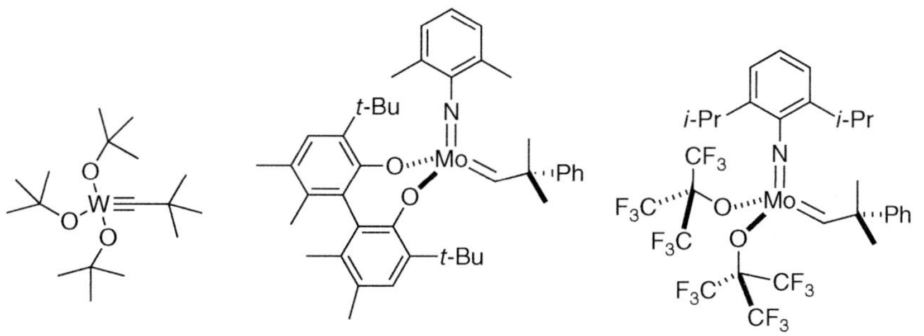

Figure 10.2 Popular molybdenum and tungsten catalysts.

10.2.2 Ruthenium Catalysts

Because of the robustness and functional group tolerance of ruthenium catalysts, the majority of industrial applications utilize ruthenium catalyst such as first Gen Grubbs [17], *N*-heterocyclic carbene (NHC) variants [18–20] (including second Gen Grubbs) [21], the Hoveyda–Grubbs [22], and the Grela catalyst [23] (Figure 10.3). The next few sections will present a few vignettes regarding the industrial uses of ruthenium catalyst for metathesis in pharmaceutical chemistry, fine chemicals, and materials and describe the various challenges one must consider when designing metathesis reactions for industrial use.

The types of industrial, metathesis transformations utilizing well-defined metathesis catalysts can be divided into three main reaction types: RCM, cross-metathesis (CM), and metathesis polymerization (ring opening and acyclic diene).

In order to understand the types of roadblocks that are unique to the three different reaction types, industrial case studies will be examined. For RCM,

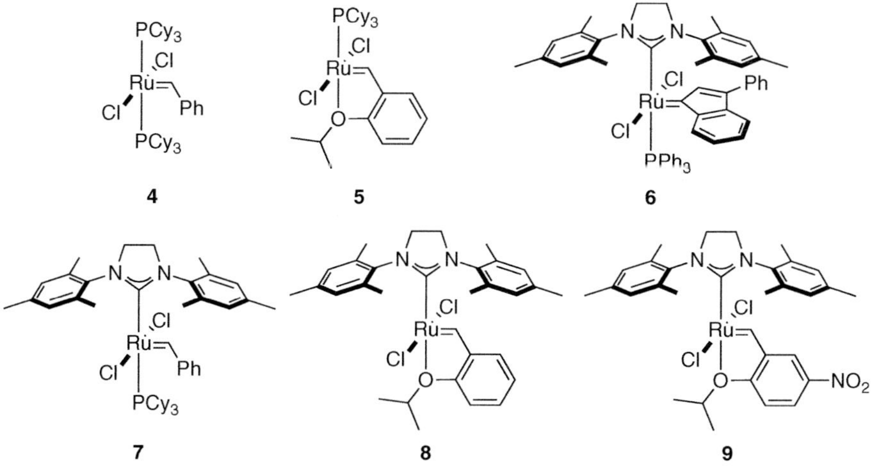

Figure 10.3 Popular ruthenium catalysts.

synthesis of the drug BILN-2061 will be reviewed and how the need for dilution and catalyst removal in the later stages of a synthesis can be problematic in the production of pharmaceuticals [24]. For fine chemicals, the joint work between Materia and Cargill which spun out the joint venture, Elevance, will be discussed and how substrate purity can impact catalyst loadings. A third example where metathesis has been utilized for the development of new materials and thermosets through the polymerization of DCPD will be examined and how catalyst selection and purity are crucial for commercial success.

10.3 Industrial Uses

10.3.1 Ring-closing Metathesis (RCM)

Industrial uses of RCM have primarily been relegated to the pharmaceutical sector for the synthesis of larger macrocycles (Figure 10.4). In the past five years, metathesis chemistry has seen significant use in the development of hepatitis C virus (HCV) drugs such as BILN-2061 [25], Simeprevir [26], and Vaniprevir [27]. Synthesis of drugs containing medium-sized rings such as SB-462795 [28] and rolapitant [29] has also benefitted.

The use of RCM for active pharmaceutical ingredient (API) synthesis presents its own synthetic challenges with the chemistry, but also other issues because of the tight regulation of the pharmaceutical industry. A key concern when designing a robust industrial synthesis utilizing a metal catalyst is the ability to remove the metal. This removal step has presented many challenges as the regulated

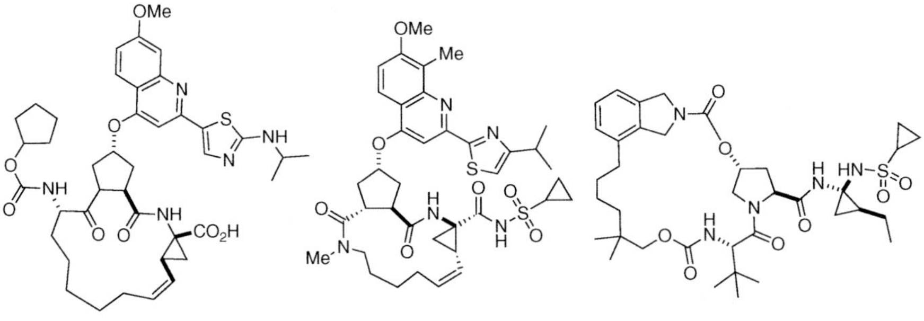

10
BILN-2061

11
Simeprevir

12
Vaniprevir

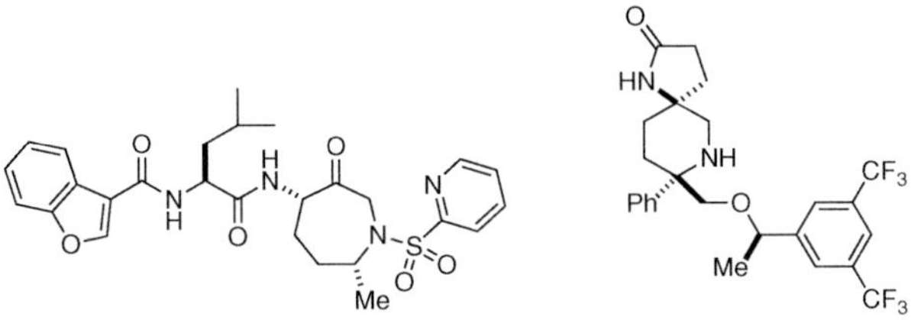

13
SB-462795

14
Rolapitant

Figure 10.4 Examples of pharmaceuticals that utilize metathesis for their synthesis.

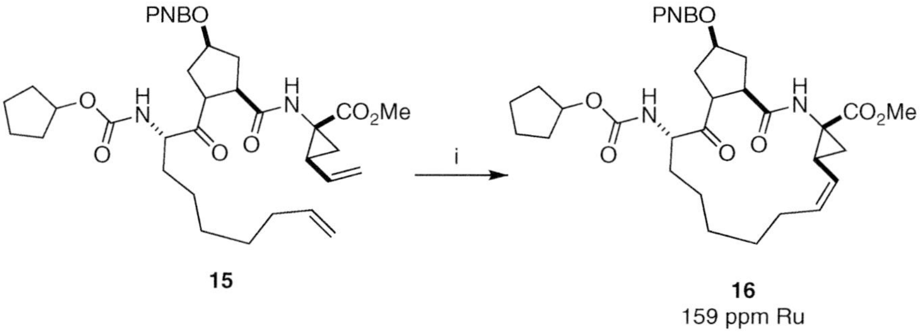

Figure 10.5 Initial synthetic route of BILN-2061. Conditions: (i) **5** (5 mol %) CH_2Cl_2 (10 mM), 40 °C, then sequential MNA/$NaHCO_3$ washes, silica pad filtration, and charcoal treatment (87%).

amount of ruthenium allowable in the final compound needs to be <10 ppm. Furthermore, other problems such as suppressing non-RCM reactions and enabling the RCM are also critical.

A good account of the hurdles overcome to exemplify RCM on a large scale is in the synthesis of BILN-2062, a 15-membered macrocycle developed by Boehringer Ingelheim for the treatment of HCV. To begin, early syntheses of BILN-2061 examined the use of first-generation Grubbs–Hoveyda (**5**), in which the catalyst was refluxed with diene **15** to yield the macrocycle **16** (Figure 10.5). After cyclization, the catalyst was quenched with the ruthenium catalyst chelator mercaptonicotinic acid (MNA) and isolated with MNA/$NaHCO_3$ washes followed by charcoal treatment and silica pad filtration. Unfortunately, after the workup, the residual Ru content was 159 ppm. This is can be attributed to the high catalyst loading of 5 mol%. Nevertheless, over four more steps, the authors were able to reduce the ruthenium content to <5 ppm; however, the initial RCM step still left some room for improvement with the high catalyst loading and long reaction time [30].

Because of the long reaction times and suboptimal ruthenium removal, the utilization of faster catalysts such as Hoveyda–Grubbs catalyst (**8**) and the Grela catalyst (**9**) was pursued for the synthesis of BILN-2061. The catalyst loading was also lowered to the range 0.05–0.1 mol% to facilitate ruthenium removal. The lower catalyst loading allowed for the use of less MNA and no silica or charcoal filtrations to give an RCM product with <50 ppm of residual ruthenium, which could be lowered further to <10 ppm after the final few steps. Previously, the removal stage focused on the use of a quenching reagent followed by a chelation moiety and subsequent silica filtration to remove the residual ruthenium. These steps lowered the levels of ruthenium contamination to 100–200 ppm, which could then be further reduced to the levels of <10 ppm, which are needed for the API by the final few steps to provide the pharmaceutical drug BILN-2061 [25].

Another option for the removal of ruthenium catalyst is the semicontinuous extraction with supercritical CO_2 (sCO_2). This process was compared to more classical methods such as the use of chelating agents and filtration. In summary, the use of the sCO_2 extraction method was found to be superior reducing residual ruthenium to 115 ppm with 90% recovery. However, this method was only demonstrated on a multigram scale and has yet to be used in a scale-up plant [31].

Not only were there issues with catalyst reactivity and ruthenium removal but problems with substrate dilution also plagued the scale-up of BILN-2061. The initial concentrations of the acyclic dialkene ranged from 0.1 to 0.01 M. Although reactions conducted at 0.01 M afforded >90% of the desired product, yields dropped to 45% if the acyclic dialkene concentration was increased to 0.1 M. However, these initial studies also used a high catalyst loading of 2–3 mol% with a first-generation catalyst. To be practical, the concentration of a substrate for RCM needs to be >0.2 M with catalyst loading <0.1 mol%. In later studies, the acyclic dialkene concentration was increased to >0.2 M with catalyst loadings as low as < 0.1 mol%. By making some adjustments to the nitrogen-protecting group, the alkene that the precatalyst initiates on can be controlled. This can be seen as an example of the "*N*-Boc-effect," where the *N*-Boc functional group not only acts as a protecting group but also as a directing group for the RCM as well (Figure 10.6) [32].

This example of the process to scale-up of BILN-2061 nicely highlights a few of the considerations one must think about when utilizing metathesis in the production of a pharmaceutical. These issues are catalyst selection, substrate concentration, and catalyst removal. As shown, all these obstacles can be overcome with the proper strategy.

10.3.2 Cross-metathesis (CM)

Another common industrial application of metathesis catalysts has been the preparation of fine chemicals by CM. This reaction class has been utilized

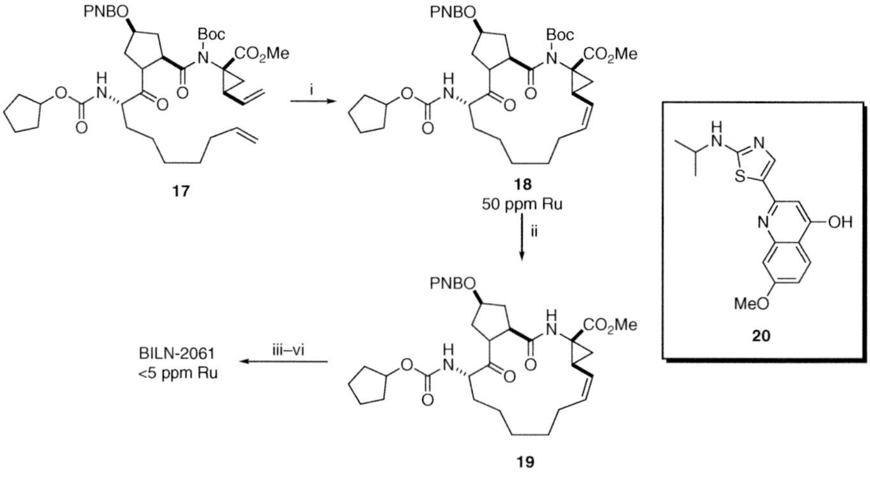

Figure 10.6 Improved route for BILN-2061. Conditions: (i) **9** (0.1 mol%), PhMe (0.2 M), 110 °C, then sequential MNA/$NaHCO_3$ washes (95%); (ii) $C_6H_5SO_3H$ (2 equiv), PhMe, 70 °C (95%); (iii) LiOH (1.3 equiv), THF, 0–5 °C (89%); (iv) *p*-$BrC_6H_4SO_2Cl$, Et_3N, cat. DMAP, CH_2Cl_2 23 °C (93%); (v) **20**, Cs_2CO_3, NMP, 50 °C, then crystallization (90%); (vi) LiOH (2 equiv), THF, H_2O, 40–45 °C (90%).

to synthesize small molecules with functions ranging from detergents to pheromones.

The synthesis of hydrogenated metathesized soybean oil (HMSBO) wax for candles was an early application of this technology. Initial work with partially hydrogenated soybean oil (soywax) was found to suffer from poor melting and solidification properties [33]. In an effort to improve the properties of soywax, Cargill and Materia codeveloped a procedure that utilized Grubbs' catalyst to metathesize the soybean oil. Interestingly, HMSBO wax was found to be firmer and less brittle than soywax and allowed for increased fragrance loading in candles compared to traditional paraffin-based systems. These property improvements in the HMSBO wax are presumably a result of the larger wax pieces that form as a result of the CM between triglyceride chains (Figure 10.7) [34].

CM has also been utilized in the synthesis of insect pheromones, which have become an attractive, environmentally friendly method for pest control. Insect pheromones are biologically active compounds that disrupt the communication between insect, thusly negatively impacting their mating cycles. Generally, pheromones consist of functionalized long alkyl chains, which contain 1–3 alkenes that can be constructed by CM with excellent stereoselectivity. Early efforts by Materia demonstrated the commercial feasibility of using metathesis for pheromones in the synthesis of pheromones for the Peach Twig Borer

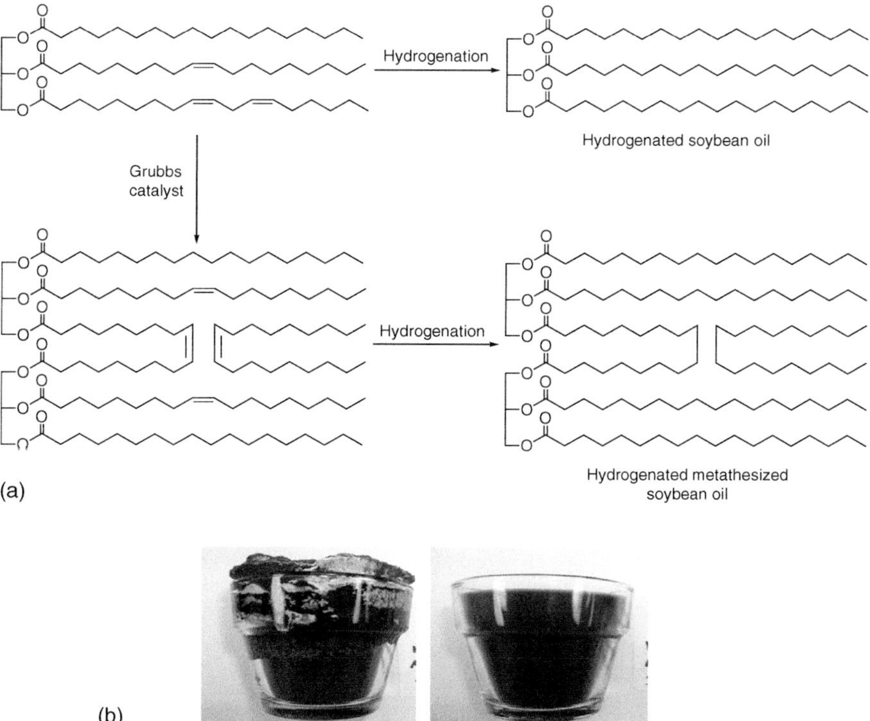

Figure 10.7 (a) Process for hydrogenated metathesized soybean oil. (b) Soywax and HMSBO (left to right).

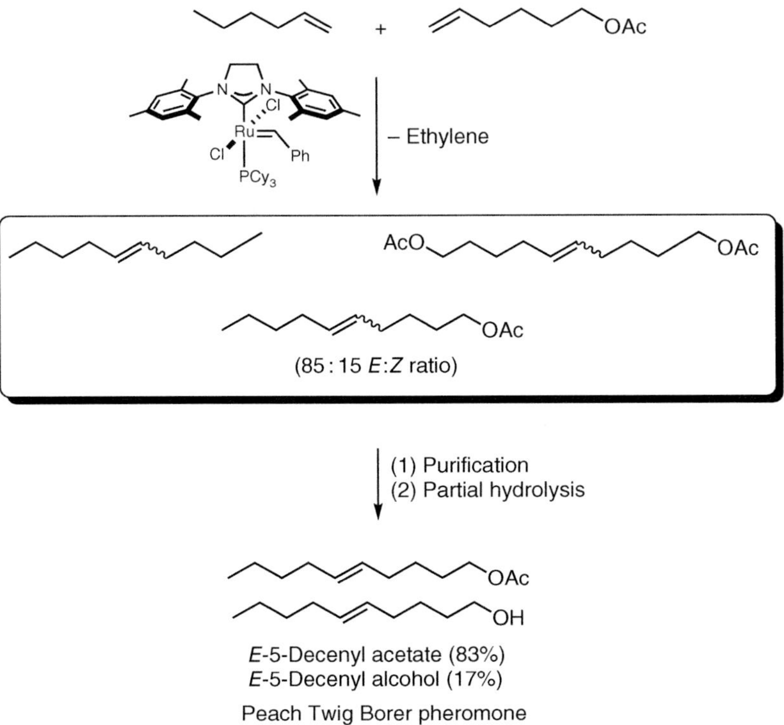

Figure 10.8 Synthesis of peach twig borer pheromone.

(*Anarsia lineatella*) [13]. The pheromones for Peach Twig Borer are composed of a 83 : 17 ratio of (*E*)-5-decenyl acetate and (*E*)-5-decenol and can readily be synthesized by the CM of 1-hexene and hex-5-en-1-yl acetate (Figure 10.8). Recently, Provivi has also shown interest in using metathesis for the production of insect pheromones; however, it is unknown if this process has gone into production [35].

The most commercial success with CM has come from the synthesis of new chemicals from seed oils. In this industrial sector, metathesis adds value to these renewable resources by modification of the triglyceride olefinic functionality (Figure 10.9). The use of CM with seed oils has contributed greatly to understanding the stability of the ruthenium catalyst under various conditions and the effect impurities have on the process.

Initially, the approach taken was to upgrade seed oils via ethenolysis, which would cleave the double bonds and provide a terminal alkene. This approach resulted in a mixture of the desired product as well as self-metathesis of the alkene portions. To begin, methyl oleate was chosen as a model substrate to examine the feasibility of ethenolysis. However, the ethenolysis yields were modest (Figure 10.10a). In order for this process to be viable on an industrial scale, a turnover number (TON) greater than 200 K needs to be achieved.[1]

---

1 TON definition: For well-defined substrates such as methyl oleate, TON = [(% conversion) * (% selectivity)] / (catalyst loading).

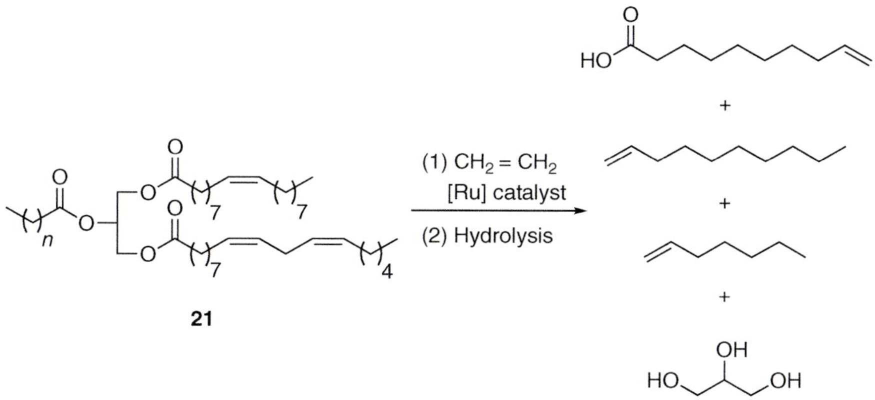

Figure 10.9 General scheme for ethenolysis of seed oils.

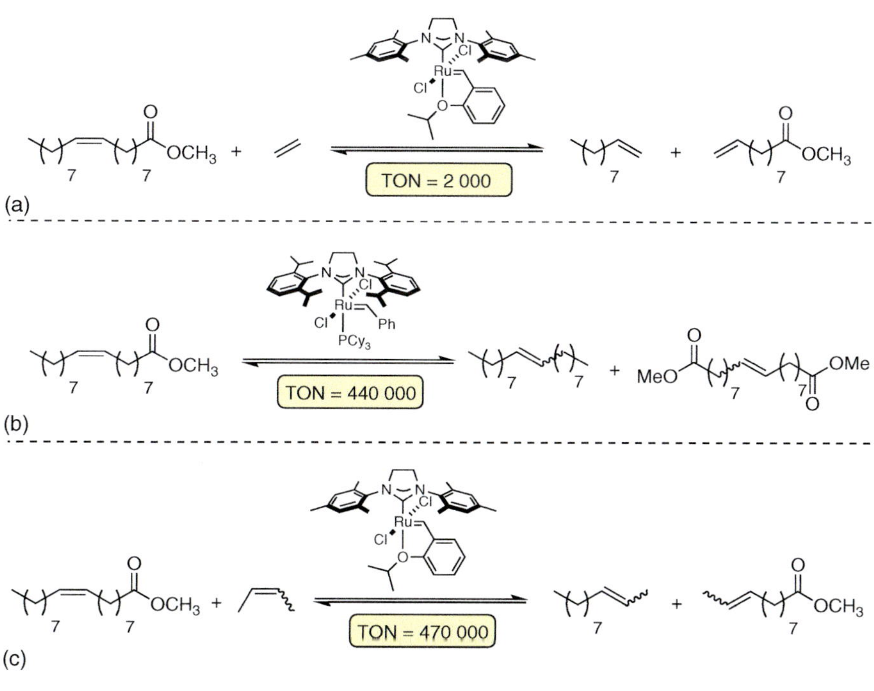

Figure 10.10 Differences in ethenolysis and alkenolysis

Needing to increase the TON for industrial use, a set of reactions were analyzed to tease out where in the metathesis process the catalyst was breaking down. From these experiments, it was discovered that using the thermodynamic CM of a substrate would deliver a TON > 400 000 (Figure 10.10b). Interestingly, these results were much higher than the previous ethenolysis reactions. Building on this success, the process was optimized to alkenolysis of the seed oils, more specifically butenolysis, which yield a high TON of 470 000 (Figure 10.10c) [36].

To better understand the issue with ethenolysis, a more involved look at the mechanism for CM is needed. As shown in Figure 10.11, when crossing with a terminal alkene, the ruthenium does not go through the methylidene because it is not part of the mechanistic pathway. By switching to an internal alkene, Elevance was able to increase the TON for the reaction and thereby get the desired product. In 2012, a plant was introduced in Indonesia, which has utilized this process [37].

Furthermore, the formation of a ruthenium methylidene can be problematic by undesired side reactions with the free phosphine ligand [38]. This undesired reaction can be avoided predominately in the presence of the alkylidene species. Examining the mechanistic pathway for the CM by alkenylosis, one can see how the methlyidene can be suppressed using alkenes vs ethylene. Formation of the methylidene is unfavored with alpha olefins because of steric interactions (Figure 10.11) [39].

10.3.3 Ring-Opening Metathesis Polymerization (ROMP)

Finally, the utilization of metathesis for polymers synthesis has long been a growing area in industrial chemistry. In general, this has been mostly on an industrial platform with cyclic olefins. A couple patents, however, use the metathesis to depolymerize larger polymers [40]. Lanxess has also subsequently hydrogenated these smaller polymers to provide saturated nitrile polymers [41].

Various cyclic monomers have also been used to synthesize polymers. Calderon and coworkers at Goodyear studied this approach heavily with the production of poly-cyclopentene. It was thought that poly-cyclopentene with a composition of trans <80 could be used as a substitute for polybutatdiene [4]. Another polymer made form cyclooctene, Vestenmar®, is produced by Evonik. The polymer is synthesized with a WCl_5 catalyst to initially yield large polymers (>500 000 g/mol) that are further treated with a secondary metathesis catalyst to a polymer of ~75 g/mol. These polymers have a high polydispersity index of about 5. By controlling the *cis*/*trans* content of the polymer yielded from the primary and secondary metathesis reactions, the polymer can be tuned to have different melting and glass transition temperatures. These polymers can then be used as a cross-linkable additive [42].

Besides cyclic monomers, highly strained bridged cyclic monomers, such as norbornene, make a good substrate for ROMP. One such polymer, Norsorex®, which is poly(norbornene) has been around since the late 1970s [43, 44]. This amorphous polymer has a high molecular weight of about 3 000 000 g/mol and has been applied toward the absorption of hydrocarbons [45].

Another strained cyclic bridged olefin, DCPD, has been used in a variety of polymer situations. One of the largest uses of metathesis in materials/polymers on an industrial scale has been the use to provide poly-DCPD. The polymerization of DCPD has been utilized for years by Telene and Hercules with molybdenum and tungsten metathesis catalysts [46]. This process is applied in reaction injection molding (RIM) to produce a variety of different parts ranging from cell covers to automobile side panels. In the past 10 years, this technology has been refined by Materia using ruthenium metathesis catalysts and commercialized as Proxima® (Figure 10.12). In general, poly-DCPD is a

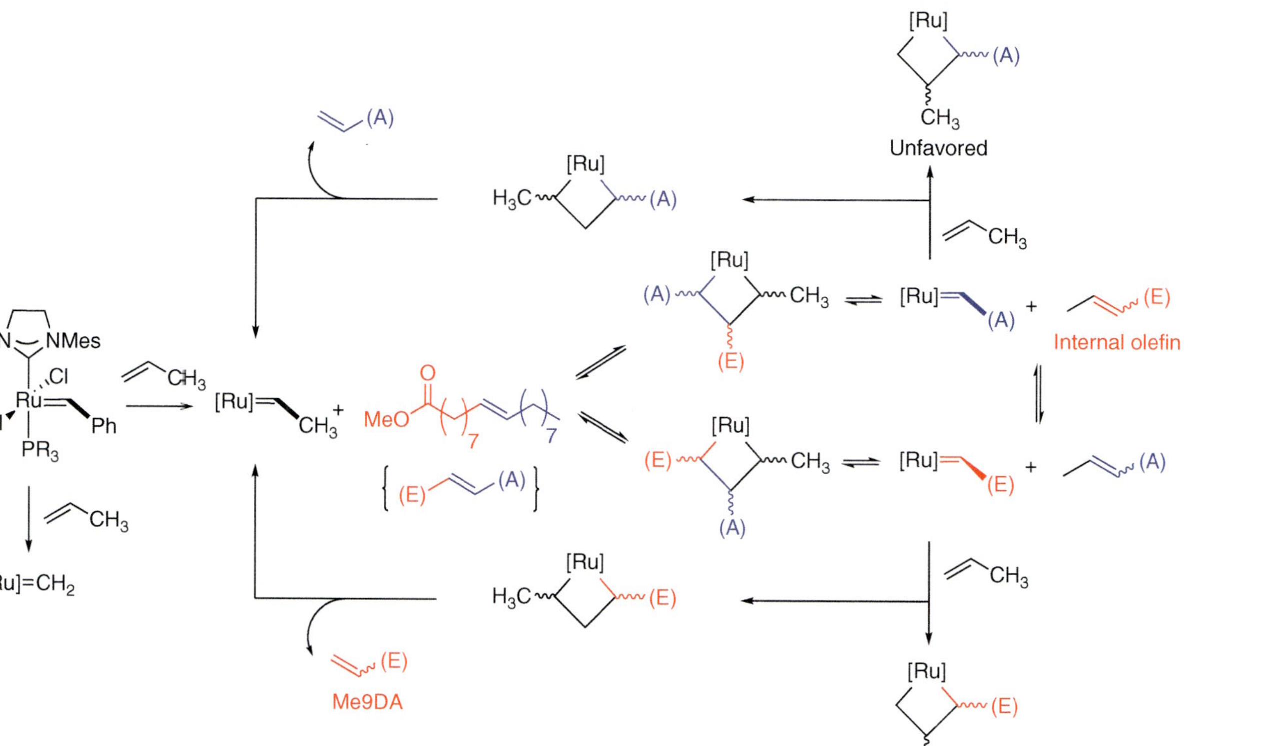

Figure 10.11 Mechanistic pathway fcr a-olefins.

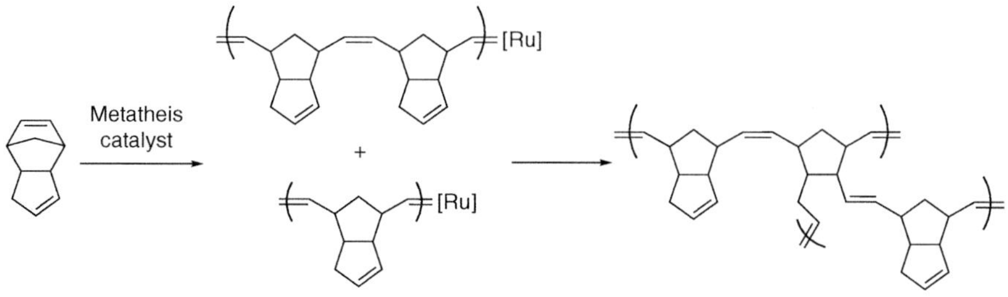

Metatheis catalyst

Figure 10.12 General polymerization scheme of DCPD.

thermoset that has interesting properties distinguishing it from other typical polymers such as epoxies or urethane. Poly-DCPD can reach Tgs that are above 180 °C and the polymer exhibits excellent toughness and impact resistance properties. Compared to epoxies, the poly-DCPD polymers have excellent moisture resistance and show slight loss in Tg with heated treatment in water. Secondly, the weight change after 4000 hours is less than 0.30% [47]. Furthermore, poly-DCPD is great for corrosive fluid handling and has been utilized for cell tops that are cast molded. Finally, in the past couple of years, poly-DCPD has been applied to the oil and gas industry to make centralizers and thermal insulation for the capturing of oil from underground [48].

In producing cast-molded parts with poly-DCPD, one needs to consider how to control the polymerization process so that the metathesis catalyst does not initiate rapidly yielding an out of control reaction. This has been solved by designing a catalyst with an initiation rate that matches the desired pot life. For instance, for producing materials with rapid injection molding, a catalyst needs to initiate within seconds. This problem can be solved by using a catalyst with fast thermal initiation. However, the same catalyst may not be applicable to produce a large pipe or casting a large part where the reactivity of the catalyst with the monomer needs to be suppressed so that the polymerization does not start before the mold or cast is filled with the monomer resin. Understanding the rate at which a catalyst will gel and then exotherm is extremely important as noted above with the polymerization with DCPD. Early efforts to control the polymerization of DCPD involved the addition of phosphine additives; however, this approach has its drawbacks, as the phosphine additives can adversely affect material properties [49]. More advanced methods for controlling ROMP with DCPD has been developed by Jeffrey Moore's group at the University of Illinois at Urbana–Champaign, which has implications for additive manufacturing processes [50].

10.4 Reaction Considerations

These three examples demonstrate how well-defined metathesis catalysts are currently utilized in industrial settings. As noted in these different vignettes, various roadblocks can be encountered with the different types of metathesis reactions: RCM, CM, and ROMP. Although in each example the approaches for solving setbacks differed, there are general approaches that can be applied to the utilization of ruthenium metathesis catalysts in general. Some of the topics that need to be considered when using metathesis are catalyst choice, catalyst loading, solvent, and reaction concentration.

10.4.1 Catalyst Choice

Depending on your end goal, whether it be a stereoselective bond or a fast/latent catalyst for polymer synthesis, there are some guiding principles to help in the decision-making process. In general, NHC-containing ruthenium metathesis catalysts can be divided into four sections: catalyst stereochemistry (*cis*/*trans*), the ruthenium carbene, the labile L ligand, and the NHC. To control the catalyst latency, the catalyst feature that provides the largest contribution is the overall stereochemistry of the catalyst itself. Typically, first and second Grubbs' catalysts are isolated as a *trans*-catalyst; however, Cazin and coworkers have shown that phosphite ligands can transform *trans*-ruthenium catalysts (**22**) into cis-isomers (**23**) [51]. These cis-isomers initiate significantly slower and need higher temperatures for initiation (Figure 10.13) compared to its *trans*-counterpart. This latency has been exploited for polymer synthesis by allowing the filling of molds or other items that would require a modest pot life. More recently, Lemcoff and coworkers utilized light activation to achieve this *cis*/*trans*-catalyst isomerization with the *cis*-thiol Grubbs–Hoveyda catalysts (**27**), which isomerizes a latent *cis*-thiol to a reactive *trans*-thiol via treatment at 350 nm [52, 53]. Although these thiol-Grubbs–Hoveyda catalysts have yet to be used on an industrial scale, they do show promise for applications where light activation is desired such as additive manufacturing or coatings.

Another method for controlling catalyst reactivity is by modifying the labile L-ligand. Typically, phosphorous ligands are utilized as L-ligands, and by varying the ligand, the dissociation rate can be manipulated, thus affecting the initiation rate of the catalyst. Switching from an aryl phosphine to an alkyl phosphine can substantially slow down the catalyst's initiation rate [54]. However, this change comes with a cost, as the alkyl phosphine-substituted catalysts have been observed to be more susceptible to decomposition. Conversely, catalyst initiation can be sped up to almost instantaneous with highly labile pyridyl ligands, which allows for ROMP synthesis of polymers with a low polydispersity index (PDI) [55]. Further control of catalyst initiation rates can be achieved by

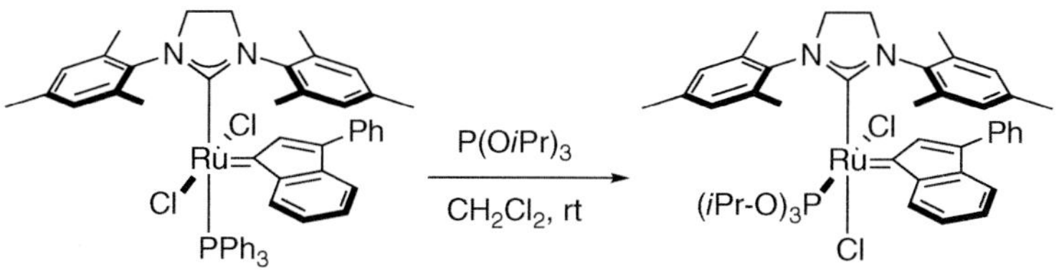

Ph
Cl
Ru
Cl
PPh_3
22
$P(O\textit{i}Pr)_3$
CH_2Cl_2, rt
Cl
Ph
Ru
$(\textit{i}Pr\text{-}O)_3P$
Cl
23

Figure 10.13 Formation of *cis*-catalyst.

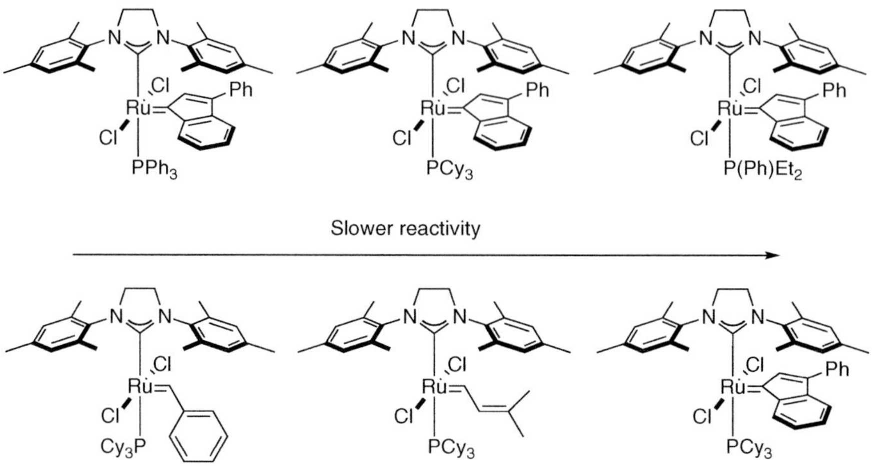

Slower reactivity

Figure 10.14 Effect of ligands on ruthenium metathesis catalyst reactivity.

varying the ruthenium carbene (Figure 10.14). In general, the reactivity goes from dimethylvinyl alkylidene as the fastest to phenylindenylidene being the slowest.

Finally the architecture of the NHC also has a role in the catalyst activity, although its effect is a little more difficult to understand [56]. Depending on the application, ruthenium metathesis catalysts have been designed to perform various transformations such as stereoselective reactions such as *cis*-alkene formation (**24**) [57]. Stereoretentive catalysts (**25**) [58, 59] have also been developed as well as sterically unhindered NHC groups (**26**) [60] and light activated (Figure 10.15) [53]. Finally, a few tests that are good for evaluating

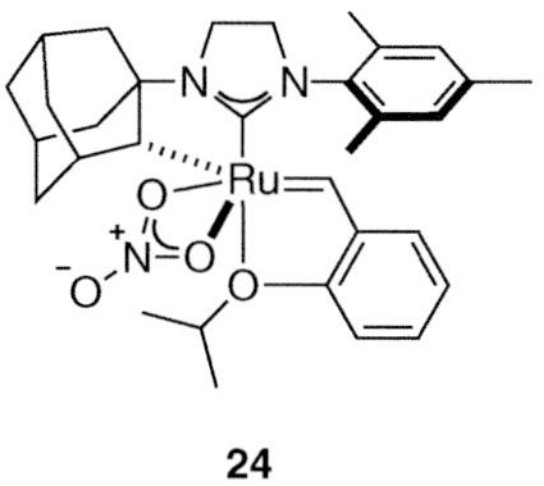

24 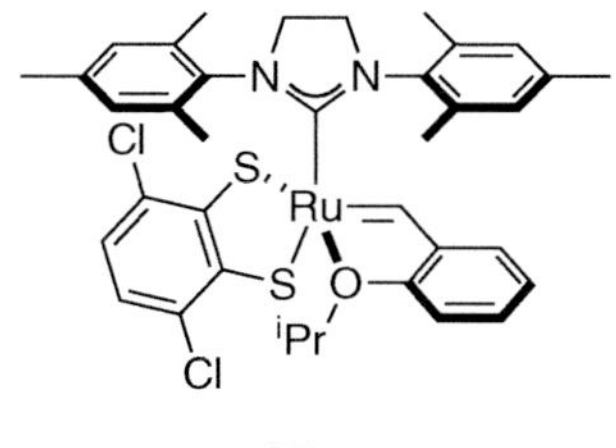 **25** 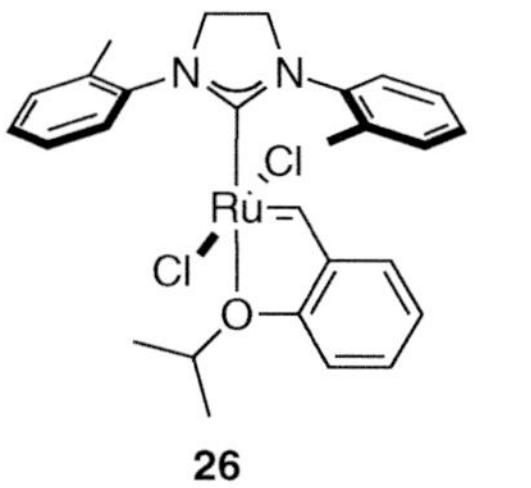 **26**

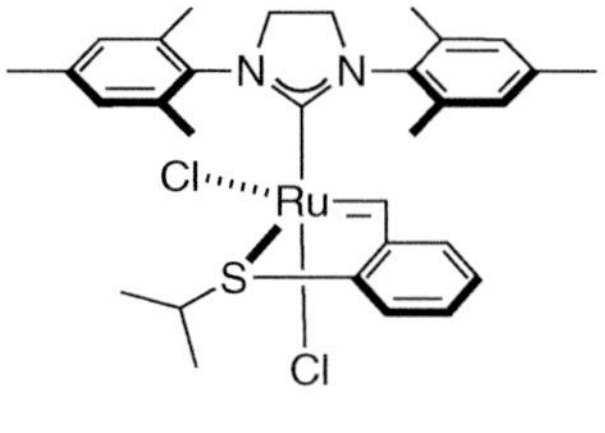

27

Figure 10.15 Ruthenium catalysts.

known catalysts were put together by the Grubbs group that can help in choosing the appropriate catalyst for an application [61].

In summary, the most common catalysts utilized for the synthesis of pharmaceutical APIs or fine chemicals are the Grubbs Second and the Hoveyda–Grubbs ruthenium catalysts. Catalysts containing alkyl phosphines or indenylidene complexes are currently experiencing traction in the material industry. Finally, some industrial interest in catalysts such as the *cis*-selective catalyst (**24**) or the sterically unencumbered catalyst (**26**) has shown recent interest in industry, but wide acceptance will depend on the economic viability of their syntheses.

10.4.2 Catalyst Loading

Generally, the loading amount of metathesis catalyst utilized in academic literature ranges between 1 and 5 mol%. Sometimes, these high catalyst loadings are needed to circumvent less than pure reactant streams, but this is financially problematic for most approaches and the catalyst loading needs to be lowered. Catalyst loading has been lowered as low as 1–5 ppm with TON > 150 000 [62]. However, these low loadings require highly pure reagent streams to be effective. Once an impurity enters the stream, it can greatly affect the loading and cause for higher temperatures and catalyst loadings that eventually lead to catalyst decompositions and other problematic side reactions.

10.4.3 Solvent

Ideally, solventless metathesis reactions such as Elevance's alkenyolsis of seed oils are optimal but cannot always be achieved. If a solvent is needed, noncoordinating solvents such as dichloromethane and toluene are excellent solvents for metathesis reactions and have seen success on the industrial scale. Other polar solvents that may weakly coordinate such as ethyl acetate or tetrahydrofuran can be used with limited success. However, polar solvents with strong coordinating potential with the catalyst such as dimethylformamide (DMF) and dimethyl sulfoxide (DMSO) should be avoided.

10.4.4 Reaction Concentration

For RCM reactions, unwanted side products arising from CM can be mitigated by controlling reaction concentration. Dilution is extremely critical with ring sizes greater than 8. In cases where the potential for unwanted CM is a competing reaction pathway, substrate concentrations are typically below 0.01 M. However, in an industrial plant setting, substrate concentrations need to be >0.2 M for a process to be feasible on scale. Nevertheless, as noted in the BILN-2061 example above, reaction conditions can be optimized for higher substrate concentrations [32]. Furthermore, recent efforts from Fogg and coworkers have demonstrated catalysts that favor RCM over CM in large macrocycles [63]. Although these advances are currently being pioneered in an academic environment, we may soon see efforts on the industrial scale.

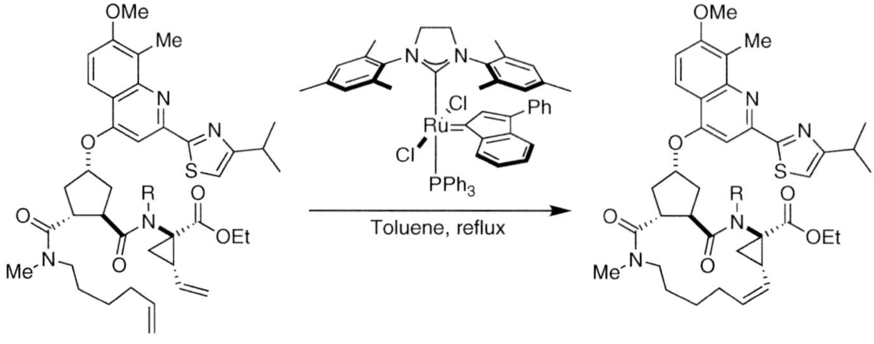

Toluene, reflux

R = H 28, $c = 0.01$ M
R = Boc 29, $c = 0.01$ M

R = H 30, 43%
R = Boc 31, 84%

Figure 10.16 Example of the "Boc effect" in macrocycle synthesis.

Finally, an approach similar to the one used in BILN-2061 was recently described by Janssen Pharmaceutica N.V. for the synthesis of Simeprevir. Utilizing the "Boc-effect," diene **29** was converted to macrocycle **31** (Figure 10.16). The RCM of the *N*-Boc-protected diene was twice as productive as the unprotected diene **28**. The authors noted that cyclization of the *N*-Boc diene was under kinetic control, whereas cyclization of the unprotected diene **28** was under thermodynamic control. Interestingly, the "Boc-effect" noted in the synthesis of BILN-2061 had the opposite effect where the *N*-Boc-substituted substrate underwent cyclization with thermodynamic control [64].

10.4.5 Overall Handling

In general, best practices for all catalysts are storage at 4 °C under an inert atmosphere; however, first- and second-generation Grubbs catalysts are relativity robust. For instance, the most first- and second-generation Grubbs can be handled on the benchtop in the solid form for short periods of time without adverse effects. A few exceptions are catalysts containing oxophilic ligands such as alkyl phosphine as an L-ligand or the C–H-activated catalyst (**24**), which are used for the synthesis of *cis*-alkenes. These *cis*-selective catalysts are sensitive to both light and air and have been slow to gain traction in the industrial setting.

10.4.6 Application Guide and Availability

In general, ruthenium catalysts are the most commonly used homogeneous metathesis catalysts for industrial applications because of their commercial availability on scale. Small research quantities of catalyst can be purchased through MilliporeSigma, whereas larger quantities can be procured from Umicore. The application chart below suggests ideal catalysts for common desired transformations. Recently, the commercial production of ruthenium catalysts was sold from Materia to Umicore; thus, catalysts are labeled with both the current Umicore name and the historical Materia name (Figure 10.17).

Reaction type	Catalyst			Product
	First choice	Second choice	Third choice	
Cross metathesis	M72	M2a	M1a	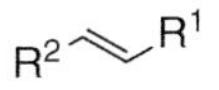
Ring closing metathesis	M72	M2a	M1a	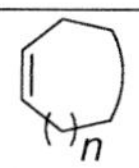
Sterically demanding cross metathesis	M72 SIPr	M72	—	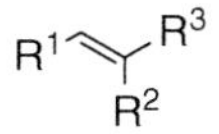
Sterically demanding ring closing metathesis	M72 SI(o-Tol)	M72	—	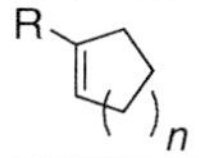
Z-Selective cross metathesis	M71c	—	—	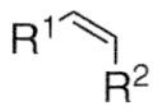
Ring opening polymerization	M2a	M22	—	Polymer

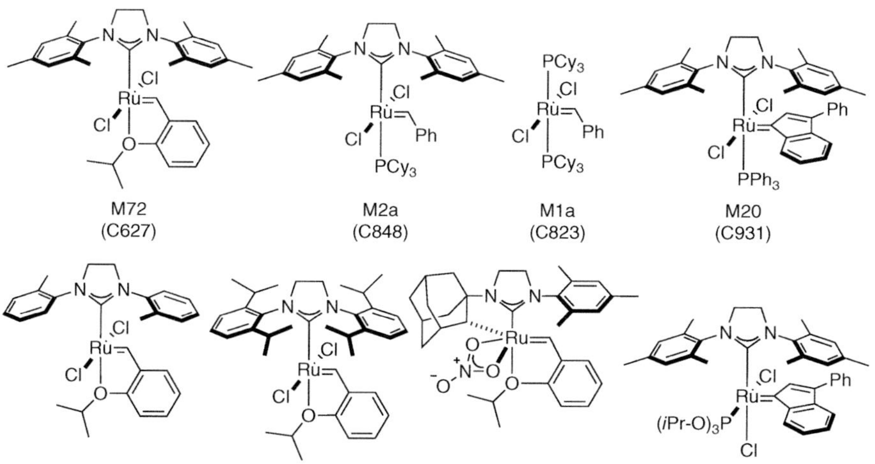

M72 (C627) M2a (C848) M1a (C823) M20 (C931)

M72 SI(o-Tol) (C571) M72 SIPr (C711) M71c (C633) M22

Figure 10.17 Application reference guide, Umicore naming (material naming).

10.5 Troubleshooting

10.5.1 Catalyst Removal

The removal of catalyst is crucial in pharmaceutical applications. For FDA approval, the final ruthenium content needs to be <10 ppm. As illustrated earlier, this can be accomplished by utilizing chelating reagents and silica gel. Furthermore, the combination of recrystallization and silica gel can lower the catalyst into the needed >10 ppm range [24]. Typically, the spent ruthenium catalyst can be removed by the addition of an ancillary ligand that can chelate to the metal catalyst and then removed (Figure 10.16). In the initial scale-up

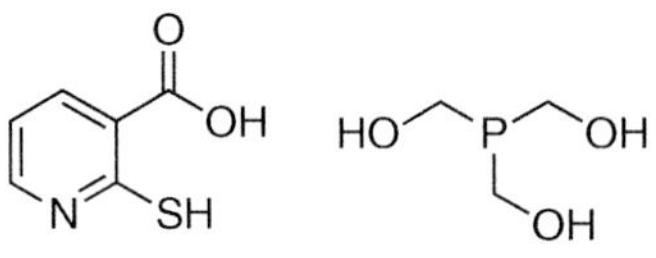

Figure 10.18 Examples of chelators used to remove ruthenium.

of BILN-2061, a variety of chelators were screened against the RCM of diallyl diethyl malonate; however, in the end, the use of MNA (**32**) was preferred. A couple of other methods that have been used on scale are the soluble phosphine chelator, tris(hydroxymethyl)phosphine (**33**), that is then separated via an aqueous extraction from the reaction mixture [65]. These methods work well with low catalyst loadings because >20 equiv of the THMP chelator is needed for efficient ruthenium removal. Other methods such as the process exemplified by Diver and coworkers utilize isocyanides to form a new ruthenium complex, which can be aqueously washed from an organic solvent [66]. However, the isocyanide used requires a multistep synthesis and has yet been proven on the industrial scale (Figure 10.18).

10.5.2 Functional Group Tolerance

In general, ruthenium metathesis catalysts are tolerant of a large range of functional groups. However, functional groups near the reactive site that are lone-pair containing heteroatoms such as amines, thiols, and alcohols are problematic. The presence of these groups can chelate with the catalyst's ruthenium center and retard the reaction rate. In order to overcome these challenges, protection of thiols, amines, and alcohols are often needed to improve the reactivity of the catalyst. The location of the alcohol in relation to the alkene is also critical because if the alcohol is nearby, it may chelate with the ruthenium center, forming a stable five- to six-member and thus suppressing reactivity. Furthermore, the use of acid to protonate and tie up the lone pair on the amine has been demonstrated [67].

10.5.3 Substrate Purity

For low catalyst loadings, substrate purity is critical. In general, oxygen has been observed to inhibit the reactivity of ruthenium catalysts. Small amounts of oxygen are believed to catalyze the formation of peroxides or other oxygen-containing functional groups. Peroxide value (PV) analysis can provide a good gauge on the purity of a reagent toward metathesis. In general, PV's lower than 10 ppm are desired for high conversion. Also, some trace contaminates are difficult to detect as evidenced by the ethenolysis by Grubbs and coworkers. Here, it was observed that the purity of ethylene was highly important and needed to be extremely pure for good reaction conversion [68].

In some cases, the reagent stream can be inherently problematic. For example, Elevance's alkynolysis of seed oils requires low PVs to obtain high TON but obtaining low PVs can be troublesome when utilizing a feedstock such as seed oils. To solve this problem, the pretreatment of the regent stream can be a feasible solution. The scrubbing of peroxides or other impurities from reagent streams can be achieved by treating with activated alumina or other oxygen scavengers. Other methods for improving the PV content of a reagent stream can be the addition of soluble deoxagenators such as magnesol or the use of small amounts of triphenyl phosphine [69]. One can also use a thermal treatment to clean up a resin stream [70]. There is also the use of soluble Lewis acids to remove the peroxides [71].

10.5.4 Catalyst Decomposition – Isomerization

The decomposition of ruthenium can be problematic. Again, in the case of CM used by Elevance, the catalyst loading was kept low not only for financial considerations but also that any decomposition of the catalyst cannot participate in isomerization side reactions. In some cases, ruthenium is believed to convert into a ruthenium hydride, which can facilitate isomerization of alkenes [72]. Other efforts by Grubbs and coworkers have shown that decomposition products can become dimers as well [73, 74]. To reduce isomerization, additives such as benzoquinone or phosphoesters can be added to the reaction [75–77]. In addition, Fogg and coworkers have looked at the effect of solvents and have shown that methanol can cause the ruthenium catalyst to decompose [78]. Furthermore, the presence of ethylene can cause problems with the catalyst stability. This can be troublesome for some RCM reactions and avoid by either pulling a vacuum to remove the ethylene or judiciously picking your metathesis substrates to avoid ethylene production.

10.6 Conclusion

Although the use of metathesis has been present in the industry for over 50 years, there is still much to be accomplished with metathesis chemistry. It is probably only time now until we will start to see metathesis become a daily part of the process chemistry toolbox in pharma and a way to design new materials with unique polymer applications.

Biography

John H. Phillips received his B.S. degree from the University of California, Berkeley, and his Ph.D. degree from the University of Wisconsin, Madison, with a focus on combinatorial chemistry. After graduation, he completed postdoctoral studies at the California Institute of Technology and the University of Michigan in the field of natural product synthesis and organometallic methodology. He joined Materia in 2010 as a research scientist in the Catalyst R&D group and has been engaged with the utilization of metathesis catalyst for new materials. He is the coinventor of patents in the areas of metathesis catalysts and functional polymers.

References

1 Truett, W.L., Johnson, D.R., Robinson, I.M., and Montague, B.A. (1960). Polynorbornene by coordination polymerization. *J. Am. Chem. Soc.* 82 (9): 2337–2340.

2 Peters, E.F. and Evering, B.L. (1960) Catalysts and their preparation. US Patent 2, 963,447, filed 31 October 1957 and issued 6 December 1960.

3 Banks, R.L. and Bailey, G.C. (1964). Olefin disproportionation. A new catalytic process. *Ind. Eng. Chem. Prod. Res. Dev.* 3 (3): 170–173.

4 Calderon, N. and Hinrichs, R.L. (1974). The way it was: polypentamer. *Chem. Technol.* 4: 627–630.

5 Bazan, G.C., Oskam, J.H., Cho, H.N. et al. (1991). Living ring-opening metathesis polymerization of 2,3-difunctionalized 7-oxanorbornenes and 7-oxanorbornadienes by Mo (CHCMe2R)(NC6H3-Iso-Pr2-2,6)(O-Tert-Bu)2 and Mo (CHCMe2R)(NC6H3-Iso-Pr2-2,6)(OCMe2CF3)2. *J. Am. Chem. Soc.* 113 (18): 6899–6907.

6 Nguyen, S.T., Johnson, L.K., Grubbs, R.H., and Ziller, J.W. (1992). Ring-opening metathesis polymerization (ROMP) of norbornene by a group VIII carbene complex in protic media. *J. Am. Chem. Soc.* 114 (10): 3974–3975.

7 Chauvin, Y. (2006). Olefin metathesis: the early days (Nobel Lecture). *Angew. Chem. Int. Ed.* 45 (23): 3740–3747.

8 Grubbs, R.H. (2006). Olefin-metathesis catalysts for the preparation of molecules and materials (Nobel Lecture). *Angew. Chem. Int. Ed.* 45 (23): 3760–3765.

9 Schrock, R.R. (2006). Multiple metal–carbon bonds for catalytic metathesis reactions (Nobel Lecture). *Angew. Chem. Int. Ed.* 45 (23): 3748–3759.

10 Trnka, T.M. and Grubbs, R.H. (2001). The development of L_2X_2RuCHR olefin metathesis catalysts: an organometallic success story. *Acc. Chem. Res.* 34 (1): 18–29.

11 Mol, J. (2004). Industrial applications of olefin metathesis. *J. Mol. Catal. A: Chem.* 213 (1): 39–45.

12 Higman, C.S., Lummiss, J.A.M., and Fogg, D.E. (2016). Olefin metathesis at the dawn of implementation in pharmaceutical and specialty-chemicals manufacturing. *Angew. Chem.* 128 (11): 3612–3626.

13 Pederson, R.L., Fellows, I.M., Ung, T.A. et al. (2002). Applications of olefin cross metathesis to commercial products. *Adv. Synth. Catal.* 344 (6-7): 728–728.

14 Nickel, A. and Edgecombe, B.D. (2012). *Industrial Applications of ROMP*, vol. 4, 749–759. Elsevier B.V.

15 Schrock, R.R., Murdzek, J.S., Bazan, G.C. et al. (1990). Synthesis of molybdenum imido alkylidene complexes and some reactions involving acyclic olefins. *J. Am. Chem. Soc.* 112 (10): 3875–3886.

16 Ondi, L., Nagy, G.M., Czirok, J.B. et al. (2016). From box to bench: air-stable molybdenum catalyst tablets for everyday use in olefin metathesis. *Org. Process Res. Dev.* 20 (10): 1709–1716.
17 Schwab, P., France, M.B., Ziller, J.W., and Grubbs, R.H. (1995). A series of well-defined metathesis catalysts–synthesis of [$RuCl_2(CHR')(PR_3)_2$] and its reactions. *Angew. Chem. Int. Ed. Engl.* 34 (18): 2039–2041.
18 Scholl, M., Trnka, T.M., Morgan, J.P., and Grubbs, R.H. (1999). Increased ring closing metathesis activity of ruthenium-based olefin metathesis catalysts coordinated with imidazolin-2-ylidene ligands. *Tetrahedron Lett.* 40 (12): 2247–2250.
19 Huang, J., Stevens, E.D., Nolan, S.P., and Petersen, J.L. (1999). Olefin metathesis-active ruthenium complexes bearing a nucleophilic carbene ligand. *J. Am. Chem. Soc.* 121 (12): 2674–2678.
20 Ackermann, L., Fürstner, A., Weskamp, T. et al. (1999). Ruthenium carbene complexes with imidazolin-2-ylidene ligands allow the formation of tetrasubstituted cycloalkenes by RCM. *Tetrahedron Lett.* 40 (26): 4787–4790.
21 Scholl, M., Ding, S., Lee, C.W., and Grubbs, R.H. (1999). Synthesis and activity of a new generation of ruthenium-based olefin metathesis catalysts coordinated with 1,3-dimesityl-4,5-dihydroimidazol-2-ylidene ligands. *Org. Lett.* 1 (6): 953–956.
22 Garber, S.B., Kingsbury, J.S., Gray, B.L., and Hoveyda, A.H. (2000). Efficient and recyclable monomeric and dendritic Ru-based metathesis catalysts. *J. Am. Chem. Soc.* 122 (34): 8168–8179.
23 Grela, K., Harutyunyan, S., and Michrowska, A. (2002). A highly efficient ruthenium catalyst for metathesis reactions. *Angew. Chem. Int. Ed.* 41 (21): 4038–4040.
24 Wheeler, P., Phillips, J.H., and Pederson, R.L. (2016). Scalable methods for the removal of ruthenium impurities from metathesis reaction mixtures. *Org. Process Res. Dev.* 20 (7): 1182–1190.
25 Farina, V., Shu, C., Zeng, X. et al. (2009). Second-generation process for the HCV protease inhibitor BILN 2061: a greener approach to Ru-catalyzed ring-closing metathesis. *Org. Process Res. Dev.* 13 (2): 250–254.
26 Horvath, A., Wuyts, S., Paul, D. et al. (2013). Improved Process for Preparing an Intermediate of the Macrocycle Protease Inhibitor TMC 435, WO2013061285, filed 26 October 2012 and issued 5 February 2013.
27 Kong, J., Chen, C.-Y., Balsells-Padros, J. et al. (2012). Synthesis of the HCV protease inhibitor vaniprevir (MK-7009) using ring-closing metathesis strategy. *J. Org. Chem.* 77 (8): 3820–3828.
28 Wang, H., Matsuhashi, H., Doan, B.D. et al. (2009). Large-scale synthesis of SB-462795, a cathepsin K inhibitor: the RCM-based approaches. *Tetrahedron* 65: 6291–6303.
29 Wu, G.G., Werne, G., Fu, X. et al. (2010). Process and Intermediates for the Synthesis of 8-rC1-(3.5-Bis-ftrifluoromethyl)phenvH-ethoxyV-methvn-8-phenyl-1.7-diazaspiror4.51decan-2-one Compounds, WO2010028232, filed 4 September 2009 and issued 11 March 2010.

30 Yee, N.K., Farina, V., Houpis, I.N. et al. (2006). Efficient large-scale synthesis of BILN 2061, a potent HCV protease inhibitor, by a convergent approach based on ring-closing metathesis. *J. Org. Chem.* 71 (19): 7133–7145.
31 Gallou, F., Saim, S., Koenig, K.J. et al. (2006). A practical method for the removal of ruthenium byproducts by supercritical fluid extraction. *Org. Process Res. Dev.* 10 (5): 937–940.
32 Shu, C., Zeng, X., Hao, M.-H. et al. (2008). RCM macrocyclization made practical: an efficient synthesis of HCV protease inhibitor BILN 2061. *Org. Lett.* 10 (6): 1303–1306.
33 Razaei, K., Wang, T., and Johnson, L.A. (2002). Hydrogenated vegetable oils as candle wax. *JAOCS.* 79 (12): 1241–1247.
34 Murphy, T.A., Tupy, M.J., Abraham, T.W., and Shafer, A. (2014). Candle and candle wax containing metathesis and metathesis-like products. US Patent 8, 911,515 B2, filed 29 January 2014 and issued 16 December 2014.
35 Chen, M.M., Coelho, P., Meinhold, P., and Lee, T.M. (2018). Agricultural phermone compositions comprising positional isomers. US Patent Application 2017/0135343 A1, filed 11 November 2014 and issued 18 June 2017.
36 Patel, J., Mujcinovic, S., Jackson, W.R. et al. (2006). High conversion and productive catalyst turnovers in cross-metathesis reactions of natural oils with 2-butene. *Green Chem.* 8 (5): 450–455.
37 https://www.elevance.com (accessed 14 July 2018).
38 McClennan, W.L., Rufh, S.A., Lummiss, J.A.M., and Fogg, D.E. (2016). A general decomposition pathway for phosphine-stabilized metathesis catalysts: Lewis donors accelerate methylidene abstraction. *J. Am. Chem. Soc.* 138 (44): 14668–14677.
39 Wenzel, A.G., Blake, G., VanderVelde, D.G., and Grubbs, R.H. (2011). Characterization and dynamics of substituted ruthenacyclobutanes relevant to the olefin cross-metathesis reaction. *J. Am. Chem. Soc.* 133 (16): 6429–6439.
40 Pawlow, J.H., Hergenrother, W.L., Dedecker, M.N., and Graves, D.E. (2013). Adducts of metathesis polymers and preparation thereof. US Patent 8, 378,034 B2, filed 04 August 2009 and issued 19 February 2013.
41 Obrecht, W., David, S., Liu, Q., and Wei, Z. (2015). Catalyst compositions and their use for hydrogenation of nitrile rubber. US Patent Application 2015/0025199 A1, filed 19 October 2012 and issued 22 January 2015.
42 Streck, R. (1990). Industrial aspects of olefin metathesis/polymerization catalysts. In: *Olefin Metathesis and Polymerization Catalysts* (ed. Y. Imamoglu), 439–516. Dordrecht, the Netherlands: Kluwer Academic Press.
43 Ohm, R.F. and Vial, T.M. (1978). A new synthetic rubber Norsorex® polynorbornene. *J. Elastomers Plast.* 10 (2): 150–162.
44 Marbach, A. and Hupp, R. (1989). The evolution of polynorbornene. *Rubber World*: 30.
45 https://youtu.Be/fzUBb7A0OFE (accessed 14 July 2018).
46 Slugovic, C. Industrial applications of olefin metathesis polymerization, *Olefin Metathesis: Theory and Practice*, Grela, K. Ed.; *Wiley*, 2014; pp 329 – 333.
47 Hu, Y., Li, X., Lang, A.W. et al. (2016). Water immersion aging of polydicyclopentadiene resin and glass fiber composites. *Polym. Degrad. Stab.* 124 (C): 35–42.

48 https://www.materia-inc.com (accessed 20 December 2018).
49 Woodson, C.S. Jr, and Grubbs, R.H. (2001). Polymeric composites including dicyclopentadiene and related monomers. US Patent 6, 310,121 B1, filed 08 July 1998 and issued 30 October 2001.
50 Robertson, I.D., Yourdkhani, M., Centellas, P.J. et al. (2018). Rapid energy-efficient manufacturing of polymers and composites via frontal polymerization. *Nature* 557 (7704): 223–227.
51 Schmid, T.E., Bantreil, X., Citadelle, C.A. et al. (2011). Phosphites as ligands in ruthenium-benzylidene catalysts for olefin metathesis. *Chem. Commun.* 47 (25): 7060–7063.
52 Aharoni, A., Vidavsky, Y., Diesendruck, C.E. et al. (2011). Ligand isomerization in sulfur-chelated ruthenium benzylidenes. *Organometallics* 30 (6): 1607–1615.
53 Sutar, R.L., Levin, E., Butilkov, D. et al. (2015). A light-activated olefin metathesis catalyst equipped with a chromatic orthogonal self-destruct function. *Angew. Chem.* 128 (2): 774–777.
54 Sanford, M.S., Love, J.A., and Grubbs, R.H. (2001). Mechanism and activity of ruthenium olefin metathesis catalysts. *J. Am. Chem. Soc.* 123 (27): 6543–6554.
55 Love, J.A., Morgan, J.P., Trnka, T.M., and Grubbs, R.H. (2002). A practical and highly active ruthenium-based catalyst that effects the cross metathesis of acrylonitrile. *Angew. Chem. Int. Ed.* 41 (21): 4035–4037.
56 Vougioukalakis, G.C. and Grubbs, R.H. (2010). Ruthenium-based heterocyclic carbene-coordinated olefin metathesis catalysts. *Chem. Rev.* 110 (3): 1746–1787.
57 Endo, K. and Grubbs, R.H. (2011). Chelated ruthenium catalysts for Z-selective olefin metathesis. *J. Am. Chem. Soc.* 133 (22): 8525–8527.
58 Khan, R.K.M., Torker, S., and Hoveyda, A.H. (2013). Readily accessible and easily modifiable Ru-based catalysts for efficient and Z-selective ring-opening metathesis polymerization and ring-opening/cross-metathesis. *J. Am. Chem. Soc.* 135 (28): 10258–10261.
59 Johns, A.M., Ahmed, T.S., Jackson, B.W. et al. (2016). High trans kinetic selectivity in ruthenium-based olefin cross-metathesis through stereoretention. *Org. Lett.* 18 (4): 772–775.
60 Stewart, I.C., Ung, T., Pletnev, A.A. et al. (2007). Highly efficient ruthenium catalysts for the formation of tetrasubstituted olefins via ring-closing metathesis. *Org. Lett.* 9 (8): 1589–1592.
61 Ritter, T., Hejl, A., Wenzel, A.G. et al. (2006). A standard system of characterization for olefin metathesis catalysts. *Organometallics* 25 (24): 5740–5745.
62 Schrodi, Y., Pederson, R.L., Kaido, H., and Tupy, M.J. (2013). Synthesis of terminal alkenes from internal alkenes via olefin metathesis. US Patent 8, 569,560 B2, filed 19 September 2012 and issued 29 October 2013.
63 Higman, C.S., Nascimento, D.L., Ireland, B.J. et al. (2018). Chelate-assisted ring-closing metathesis: a strategy for accelerating macrocyclization at ambient temperatures. *J. Am. Chem. Soc.* 140 (5): 1604–1607.
64 Horváth, A., Depré, D., Vermeulen, W.A.A. et al. (2019). Ring-closing metathesis on commercial scale: synthesis of HCV protease inhibitor Simeprevir. *J. Org. Chem.* https://doi.org/10.1021/acs.joc.8b03124.

65 Maynard, H.D. and Grubbs, R.H. (1999). Purification technique for the removal of ruthenium from olefin metathesis reaction products. *Tetrahedron Lett.* 40 (22): 4137–4140.
66 Galan, B.R., Kalbarczyk, K.P., Szczepankiewicz, S. et al. (2007). A rapid and simple cleanup procedure for metathesis reactions. *Org. Lett.* 9 (7): 1203–1206.
67 Stoianova, D.S., Yao, L., Rolfe, A. et al. (2008). High-load, oligomeric monoamine hydrochloride: facile generation via ROM polymerization and application as an electrophile scavenger. *Tetrahedron Lett.* 49 (29-30): 4553–4555.
68 Marx, V.M., Sullivan, A.H., Melaimi, M. et al. (2014). Cyclic alkyl amino carbene (CAAC) ruthenium complexes as remarkably active catalysts for ethenolysis. *Angew. Chem. Int. Ed.* 54 (6): 1919–1923.
69 Lemke, D.W., Uptain, K.D., Amore, F., and Abraham, T. (2015). Chemical methods for treating a metathesis feedstock. US Patent 8,642,824 B2, filed 11 August 2008 and issued 4 April 2014.
70 Uptain, K.D., Tanger, C., and Kaido, H. (2014). Thermal methods for treating a metathesis feedstock. US Patent 9,284,515 B2, filed 18 February 2014 and issued 15 March 2016.
71 Champagne, T.M. and Ung, T.A. (2016). Use of soluble metal salts in metathesis reactions. US Patent Application 2016/0039737 A1, filed 14 March 2014 and issued 02 November 2016.
72 Schmidt, B. (2004). Catalysis at the interface of ruthenium carbene and ruthenium hydride chemistry: organometallic aspects and applications to organic synthesis. *Eur. J. Org. Chem.* 2004 (9): 1865–1880.
73 Hong, S.H., Day, M.W., and Grubbs, R.H. (2004). Decomposition of a key intermediate in ruthenium-catalyzed olefin metathesis reactions. *J. Am. Chem. Soc.* 126 (24): 7414–7415.
74 Hong, S.H., Wenzel, A.G., Salguero, T.T. et al. (2007). Decomposition of ruthenium olefin metathesis catalysts. *J. Am. Chem. Soc.* 129 (25): 7961–7968.
75 Hong, S.H., Sanders, D.P., Lee, C.W., and Grubbs, R.H. (2005). Prevention of undesirable isomerization during olefin metathesis. *J. Am. Chem. Soc.* 127 (49): 17160–17161.
76 Formentín, P., Gimeno, N., Steinke, J.H.G., and Vilar, R. (2005). Reactivity of Grubbs' catalysts with urea- and amide-substituted olefins. Metathesis and isomerization. *J. Org. Chem.* 70 (20): 8235–8238.
77 Moïse, J., Arseniyadis, S., and Cossy, J. (2007). Cross-metathesis between α-methylene-γ-butyrolactone and olefins: a dramatic additive effect. *Org. Lett.* 9 (9): 1695–1698.
78 Beach, N.J., Lummiss, J.A.M., Bates, J.M., and Fogg, D.E. (2012). Reactions of Grubbs catalysts with excess methoxide: formation of novel methoxyhydride complexes. *Organometallics* 31 (6): 2349–2356.

11

Dehydrative Decarbonylation

Alex John

California State Polytechnic University Pomona, Chemistry and Biochemistry Department, 3801 West Temple Ave., Pomona, CA 91768, USA

11.1 Introduction

Conversion of biomass and its derivatives into useful chemicals has been deemed vital to the replacement of petroleum-based compounds with renewable and sustainable alternatives [1]. This could be relevant, particularly, for example, in the production of high-volume monomers (methacrylic acid [MAA], acrylic acid, and styrene). These are prepared from oil-based starting materials (acetone, propene, and benzene) by processes that require harsh conditions and hazardous chemicals as well as produce significant quantities of chemical waste. For example, the production of MAA from acetone by the acetone-cyanohydrin (ACH) process uses highly toxic raw materials (HCN) and forms significant amounts of by-products that need to be disposed [2]. The ACH process also results in an estimated emission of 5.5 kg of CO_2/kg MAA, which is much higher when compared to emissions from production of other platform chemicals. The use of biomass-derived starting materials to access these and other much-desired chemical feedstocks would circumvent some (or all) of the current undesired chemical transformations while supporting the chemical infrastructure in a sustainable manner. However, a key challenge in this direction is presented by the inherent structure of biomass and its derivatives, which, unlike petrochemicals, is highly functionalized (mostly oxygenated: hydroxyl groups and carboxylic acid derivatives) [3]. Hence, the realization of an efficient biorefinery is contingent upon the development of efficient, economical, green (nonpolluting and hazard-avoiding), and conceivably catalytic methods of converting bioderived molecules into chemical feedstocks by removing existing functionality.

With an annual global production that runs into billions of metric tons, biomass and its derivatives include lignocellulosic biomass (75%) as well as triglycerides such as vegetable oils and animal fats (Scheme 11.1) [1, 2]. This includes waste associated with current agricultural or industrial practices as exemplified by sugarcane bagasse, corn stover, and waste cooking oil/fat (from restaurants), etc. Cellulosic biomass is a source of glucose, which, through

Organometallic Chemistry in Industry: A Practical Approach, First Edition.
Edited by Thomas J. Colacot and Carin C.C. Johansson Seechurn.

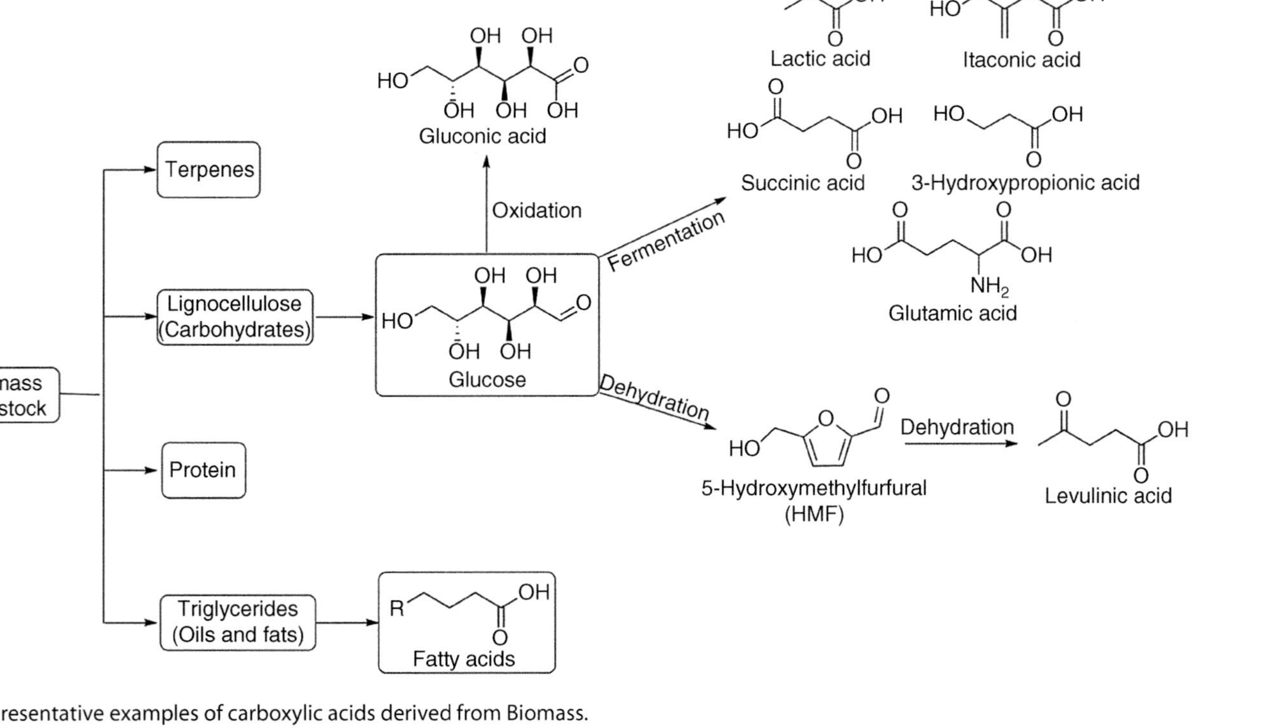

Scheme 11.1 Representative examples of carboxylic acids derived from Biomass.

various chemical and fermentative processes, is converted on industrial scales to a range of other chemical products, several bearing carboxylic acid functionality. Both oils and fats are similar sources of fatty acids (long-chain aliphatic carboxylic acids) upon acidic or basic hydrolysis. Given their easy availability, low toxicity, and structural diversity, these carboxylic acids represent a suitable class of biomass and its derivatives for conversion into chemical feedstocks. A unique reaction that is capable of reducing the functionality of such biomass-derived carboxylic acids is transition-metal-catalyzed dehydrative decarbonylation [4]. This reaction involves the formal removal of a molecule of carbon monoxide (CO) and water (H_2O) from an aliphatic carboxylic acid, thus transforming it to an olefin with one less carbon atom (Scheme 11.2). This is particularly interesting as a variety of biomass-derived chemical products (succinic acid, glutamic acid, levulinic acid, etc.) can be directly converted to value-added commodity chemicals (acrylic acid, acrylonitrile, methyl vinyl ketone [MVK], respectively) upon successful dehydrative decarbonylation.

Z–CH2CH2–C(=O)OH —Catalyst, −CO, −H_2O→ Z–CH=CH2

e.g. Z = Ph (styrene), CO_2R (acrylates), $COCH_3$ (MVK)

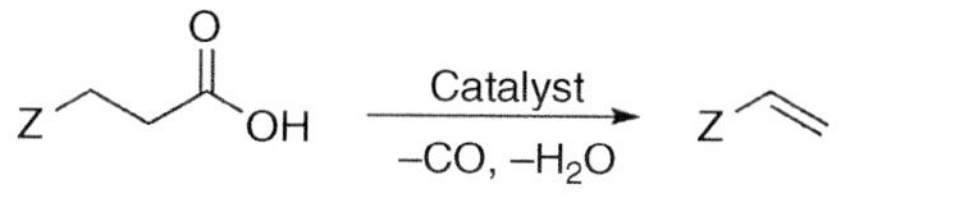

Scheme 11.2 Dehydrative decarbonylation of carboxylic acids.

Classical methods for conversion of fatty acids to olefins (including linear alpha-olefins [LAOs]) include $Pb(OAc)_4$-mediated oxidative decarboxylation [5], Kolbe electrolysis [6], and Ag-catalyzed decarboxylation [7]. These suffer from drawbacks such as either the use of stoichiometric amounts of a toxic reagent or generation of highly reactive intermediates (radical pathways) resulting in low yields. Transition-metal-catalyzed dehydrative decarbonylation offers a unique approach to LAO synthesis starting from fatty acids. LAOs are high-value chemicals used in plasticizers, lubricants, surfactants, and as comonomers in ethylene polymerization [8]. Currently, these are produced predominantly by ethylene (C_2) oligomerization [9] and hence contain an even number of carbon atoms (C_n; n is even). As naturally occurring fatty acids contain an even number of carbon atoms in the chain, LAOs derived from these via dehydrative decarbonylation would feature an odd-carbon number, resulting in a feedstock that is complementary to the one obtained from petrochemicals.

11.2 Use of Sacrificial Anhydride and Catalytic Mechanism

Owing to the low propensity of carboxylic acids to oxidatively insert a metal across the acyl–OH bond to initiate the reaction, activation of the carboxylic acid group is required. A prominent strategy to address this issue has been the use of carboxylic acid anhydrides. This was inspired by pioneering work by Trost and

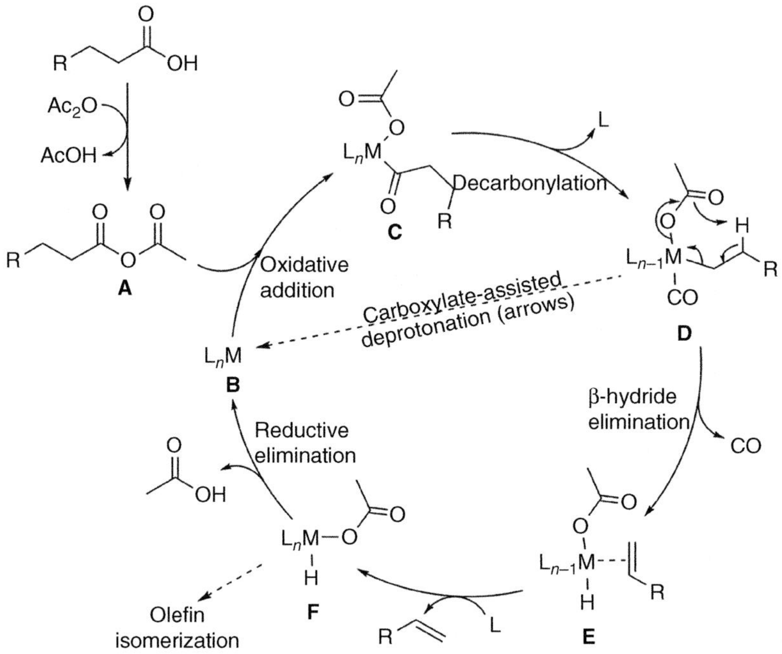

Figure 11.1 Proposed mechanism for transition-metal dehydrative decarbonylation using a sacrificial anhydride (Ac_2O).

coworkers, which documented the viability of anhydride decarbonylation using transition metals [10]. It has been demonstrated that the substrate carboxylic acid can react with the sacrificial anhydride to generate a "mixed anhydride" (**A**) that is subsequently decarbonylated by the transition metal [11]. After activation [12], the reaction is suggested to involve an initial oxidative addition of the mixed anhydride onto the low-valent metal center to generate an (acyl)M(carboxylate) (**C**) species (Figure 11.1). This is followed by a decarbonylation event accompanied by ligand (phosphine) dissociation if required to accommodate the CO ligand and yield an (alkyl)M(carboxylate) intermediate (**D**). This alkyl-metal species undergoes β-hydride elimination, resulting in a metal-bound alkene intermediate (**E**) that undergoes subsequent reductive elimination to release the alkene product as well as regenerate the catalytically active species. Alternatively, the alkyl-metal species could undergo a carboxylate-assisted deprotonation (**D** → **B**) with concomitant reduction to regenerate the active catalyst. Isomerization is believed to occur as a result of alkene re-coordination and insertion into an active M-hydride intermediate (**F**).

11.3 Rh-, Pd-, and Ir-Catalysis

11.3.1 Early Studies

The reaction was initially applied to the decarbonylation of fatty acids (stearic acid) using palladium and rhodium catalysts in combination with phosphine ligands by Foglia and Barr [13] This seminal work provided critical insights into the catalytic reaction. The use of preformed complexes such as $[(Et)_2PhP]_3RhCl_3$, Wilkinson's catalysts ($(Ph_3P)_3RhCl$) and $(Ph_3P)_2PdCl_2$, as well as simple metal salts such as $RhCl_3$ and $PdCl_2$ resulted in low turnovers in the absence of added phosphine ligand (PPh_3). The low reactivity was attributed to the reduction of the metal catalyst to the respective nascent metal under these conditions. The addition of exogenous PPh_3, which was deemed to be the best ligand additive, as well as the use of nitrogen (N_2) purge (120 ml/min) was found to greatly facilitate catalytic activity, presumably by facilitating CO removal from the system. The nuclearity of the metal precursor (mononuclear ($RhCl_3{\bullet}3H_2O$) vs polymeric (anhydrous $RhCl_3$)) as well as the reaction temperature (280–285 °C) were found to be critical for a successful reaction. Additionally, ligand-to-metal ratio was also found to be crucial with best activity observed at high ligand:metal (10 : 1) ratio (Scheme 11.3). The exogenous ligand (PPh_3) was proposed to aid in the generation of stearic anhydride from steric acid before onset of decarbonylation.

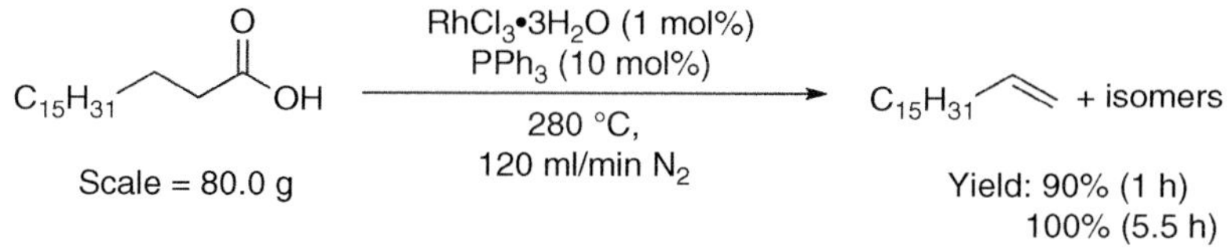

$C_{15}H_{31}$–CH₂CH₂–C(=O)OH → $C_{15}H_{31}$–CH=CH₂ + isomers

$RhCl_3{\bullet}3H_2O$ (1 mol%)
PPh_3 (10 mol%)
280 °C,
120 ml/min N_2

Scale = 80.0 g

Yield: 90% (1 h)
100% (5.5 h)

Scheme 11.3 Rhodium catalyzed decarbonylation of stearic acid.

Later studies invoked the use of a sacrificial anhydride to facilitate the reaction. The sacrificial anhydride of choice has been acetic anhydride (Ac_2O), given its low cost. Miller et al. [14] (Henkel Research Corporation) reported on the use of an equimolar mixture of the carboxylic acid and Ac_2O for Pd- and Rh-catalyzed decarbonylation to corresponding LAOs. The reactions were carried out at preparative scales (up to 500 g) and effected using $(Ph_3P)_2PdCl_2$ (0.01 mol%)/PPh_3 (0.50 mol%) or 1 mol% $(Me_2PhP)_2Rh(CO)Cl$ at 250 °C under a N_2 flow (Scheme 11.4). The procedure was performed using a short-path distillation setup (the distillation head was wrapped and heated with a heating

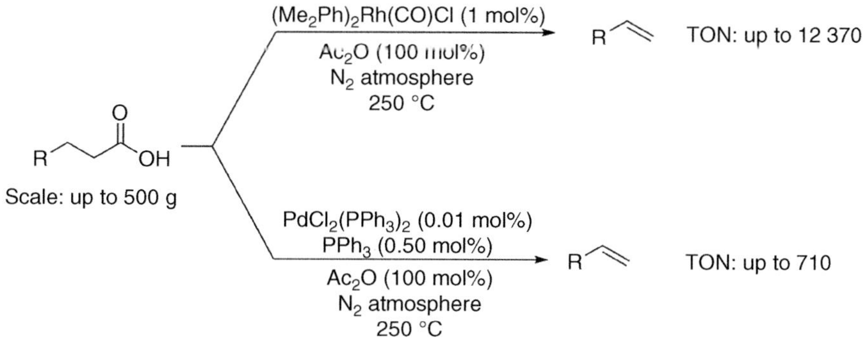

Scale: up to 500 g

$(Me_2Ph)_2Rh(CO)Cl$ (1 mol%)
Ac_2O (100 mol%)
N_2 atmosphere
250 °C
TON: up to 12 370

$PdCl_2(PPh_3)_2$ (0.01 mol%)
PPh_3 (0.50 mol%)
Ac_2O (100 mol%)
N_2 atmosphere
250 °C
TON: up to 710

Scheme 11.4 Pd- and Rh-catalyzed dehydrative decarbonylation of fatty acids.

tape), and the initial level of reactor contents were maintained throughout the course of the reaction by continuous addition of an equimolar mixture of the acid and anhydride. Decarbonylation initiated when the reaction temperature reached 180–190 and 230 °C for the palladium- and rhodium-catalyzed reactions, respectively. The product alkene was produced in high α-selectivity (up to 97%) and catalyst turnover numbers (TONs) as high as 12 370 could be achieved. Continuous removal of the olefin product from the reaction mixture was deemed essential for maintaining high-product selectivity. For carboxylic acids such as stearic acid (C_{18}) and myristic acid (C_{14}) that furnish higher boiling olefin products, high olefin selectivity could be maintained by carrying out the procedure under a reduced pressure (40–150 Torr). Overall, the palladium catalysts were found to be more efficient than the rhodium counterparts. Quite interestingly, Pd(II) catalysts were found to be more effective than the corresponding preformed Pd(0) compounds such as $Pd(PPh_3)_4$.

Miller and coworkers (Henkel Research Corporation) also disclosed practical aspects of the reaction with particular emphasis on high yield and α-selectivity in a related work [14]. The presence of an anhydride (preferably acetic anhydride) was deemed essential to maintain high catalyst TON, high yield, and high α-selectivity when the olefin was continuously removed from the reaction mixture by distillation (Table 11.1). Although the anhydride to carboxylic acid composition can vary between 0.5/1 to about 1.5/1, the optimal composition was suggested to be equimolar in both. The nature of the phosphine ligand was also found to be crucial with dimethylphenylphosphine (Me_2PhP) being especially effective. The catalyst lifetime and α-olefin selectivity were found to have a strong dependence on the nature of the phosphine ligand under reaction conditions (excess ligand). The presence of excess ligand was found to strongly inhibit conversion in the case of rhodium catalysts. Hence, optimal catalyst lifetime and selectivity are achieved by maintaining a steady concentration of a volatile phosphine such as Me_2PhP throughout the catalytic reaction by

Table 11.1 Effect of anhydride additive on TON and selectivity.

$C_7H_{15}CH_2CH_2COOH$ → $C_7H_{15}CH{=}CH_2$ [Catalyst (1 mol%), PPh_3 (10 mol%), Ac_2O (100 mol%), 255 °C]

Catalyst	Without Ac_2O TON	Without Ac_2O α-Selectivity	With Ac_2O TON	With Ac_2O α-Selectivity
$(PPh_3)_2NiCl_2$	24	25	50	33
$(PPh_3)_2PdCl_2$	16	17	520	92.5
$(PPh_3)_2PtCl_2$	27	58	130	79.4
$(PPh_3)_2Rh(CO)Cl$	105	90.2	310	94.0
$(PPh_3)_2Ir(CO)Cl$	270	54	1280	91.5
$(PPh_3)_3CuCl$	0	—	35	99.5

continuously introducing it either by itself (batch process) or as part of the feed (continuous process). In the case of palladium catalysts, the excess phosphine can be introduced at once. The process can be affected in any apparatus such as a batch reactor or a continuous reactor such as a plug flow reactor, a falling film evaporator, or an agitated thin film evaporator. As the reaction produces large volumes of the olefin in relatively short times, a continuous reactor is suggested as preferred. When the decarbonylation is effected under reduced pressure to facilitate removal of the olefin product, a higher boiling anhydride such as pivalic anhydride is used. In a representative example, decanoic acid (100 g, 0.58 mol), acetic anhydride (0.58 mol), $(PPh_3)_2PdCl_2$ (5.8×10^{-5} mol), and PPh_3 (5.9×10^{-3} mol) were heated in an oil bath maintained at 255 °C under a N_2 atmosphere. Acetic acid and/or acetic anhydride distilled out at reaching a reaction temperature of 130 °C under these conditions. The decarbonylation reaction initiated at 180–190 °C as evidenced by rapid bubbling, and the olefin (1-nonene) product distilled out. The initial level of materials in the reactor was maintained by continuous introduction of a feed mixture comprising decanoic acid and acetic anhydride throughout the reaction. Once the reaction ceased, the reactor was cooled, charged with more catalyst, and carried out again. A total of 11 consecutive runs were performed. The combined olefin product was washed sequentially with water, 3N NaOH solution, and saturated brine solution. It was then dried over anhydrous magnesium sulfate and purified by distillation (97.1% alpha; TON = 7885).

11.3.2 Recent Studies

Kraus and Riley [15] later developed modified conditions based on $PdCl_2$ (0.01 mol%)/PPh_3 (0.25 mol%) that lowered the phosphine:Pd ratio. Various carboxylic acids including fatty acids were converted to the corresponding alkenes in 62–91% yields using a combination of $PdCl_2(PPh_3)_2$ (0.01 mol%)/PPh_3 (0.5 mol%). Reactions were performed at 20–30 g scale using a short-path distillation apparatus, and Ac_2O as an additive. Alkene was removed readily by distillation to prevent olefin isomerization, and a reduced pressure (20–30 mmHg) was used for fatty acids containing more than 13 carbons. The olefin product was collected using a 100 ml collecting flask, which was kept cold using an ice bath. Utilizing an excess of the anhydride (4 equiv), diacids could be transformed to the respective α,ω-dienes in 55–65% yield. It is noted that excess anhydride is necessary in the case of diacids to avoid the formation of polymeric compounds. Kraus also disclosed the synthesis of a mixture of α-olefins using the reaction conditions under reduced pressure [16]. In a representative protocol, a 100 ml round-bottom flask was charged with a mixture of myristic acid (20.72 g), palmitic acid (27.23 g), and stearic acid (2.21 g), along with $PdCl_2$ (0.045 g), PPh_3 (1.338 g), and 50 ml acetic anhydride. The flask is attached to a 100 ml collecting flask (in an ice bath) using a short-path distillation apparatus and placed under weak vacuum (26 Torr). The distillate consisting of excess anhydride, acetic acid, and the olefin products is collected over a period of 30–40 minutes, while heating the reaction flask at 230 °C in an oil bath. The reaction is judged to be complete when distillation stops, and the reaction mixture darkens to black color, indicating inactivation of the catalyst.

Tolman and coworkers similarly reported palladium-catalyzed conditions for the synthesis of alkyl acrylates, styrene, and acrylonitrile starting from bioderived carboxylic acids [3]. The use of a high-boiling anhydride, pivalic anhydride (bp = 193 °C), was explored at a lower reaction temperature (190 °C) to avoid the need to continuously add anhydride (Ac_2O; bp = 140 °C) to the feed. DPEPhos and XantPhos were identified to be the most effective at promoting the reaction; however, the no stark differences in yield were noted compared to PPh_3. A preparative scale (25 g) reaction conducted using hydrocinnamic acid yielded ~70% styrene under these reaction conditions (Scheme 11.5a). The utility of Ir complexes in dehydrative decarbonylation was explored by Ryu and coworkers [17]. They demonstrated that Ir complexes such as Vaska's complex $[IrCl(CO)(PPh_3)_2]$ resulted in decarbonylation of carboxylic acids to alkenes (internal) in the presence of KI at 250 °C (Scheme 11.5b). Selectivity for α-olefins could be achieved by conducting the reaction at lower temperatures (160 °C) in the presence of both KI and Ac_2O. The Hapiot group recently demonstrated the utility of Ir catalysts ($[Ir(cod)Cl]_2$) in decarbonylating biomass-derived fatty acids to the corresponding olefins in good yields and selectivity [18].

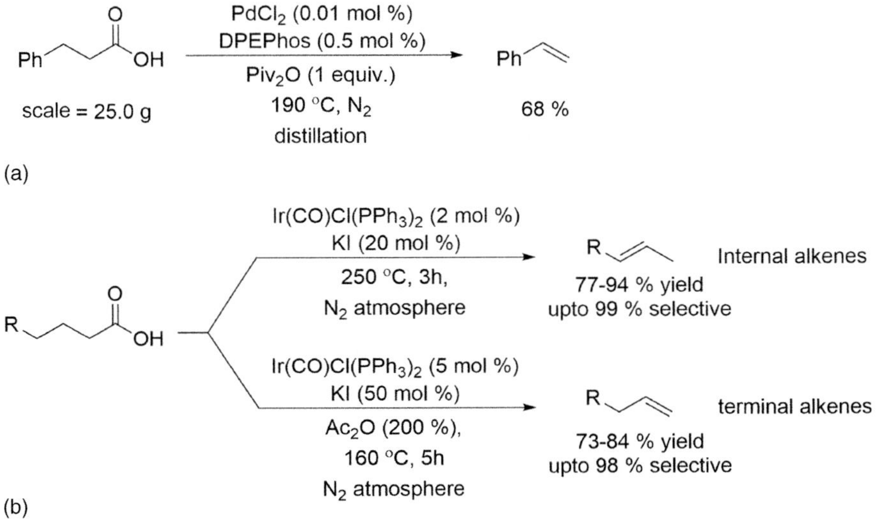

Ph–CH₂CH₂–C(O)OH → Ph–CH=CH₂

scale = 25.0 g

$PdCl_2$ (0.01 mol %)
DPEPhos (0.5 mol %)
Piv_2O (1 equiv.)
190 °C, N_2
distillation

68 %

(a)

R–CH₂CH₂–C(O)OH

$Ir(CO)Cl(PPh_3)_2$ (2 mol %)
KI (20 mol %)
250 °C, 3h,
N_2 atmosphere

Internal alkenes
77-94 % yield
upto 99 % selective

$Ir(CO)Cl(PPh_3)_2$ (5 mol %)
KI (50 mol %)
Ac_2O (200 %),
160 °C, 5h
N_2 atmosphere

terminal alkenes
73-84 % yield
upto 98 % selective

(b)

Scheme 11.5 Pd- and Ir-catalyzed dehydrative decarbonylations.

The generation of bio-based styrene by dehydrative decarbonylation of dihydrocinnamic acid was the basis of a recent disclosure by Hogan (Bridgestone Corporation) on its applications in the synthesis of copolymers, rubber compositions, etc. [19] In a representative procedure, bio-based dihydrocinnamic acid (1 equiv), $PdCl_2$ (0.25 mol%), pivalic anhydride (1 equiv), and phosphine ligand (2.2 or 4.4 mol%) are charged into an oven-dried round-bottom flask inside a N_2 filled glove box. The flask is fitted with a short-path distillation setup and removed from the glove box into an oil bath maintained at 160–170 °C under an inert atmosphere (N_2 or Ar). The oil bath temperature was then rised to 190 °C, when the initial heterogeneous reaction mixture changes to a yellow homogeneous solution. A strong effervescence is observed at ~185 °C, indicating the onset of decarbonylation. A clear distillate composed of styrene, pivalic acid, and unreacted pivalic anhydride is collected over two hours. The reaction is quenched by exposing to ambient atmosphere at the end of the reaction. The distillate collected was purified by basic workup followed by column chromatography to obtain pure styrene.

11.4 Milder Temperatures

Palladium-catalyzed reactions have received most attention for performing decarbonylation reactions and exhibit the best reactivity. A severe limitation to a majority of these procedures is the need for high temperatures (230–250 °C), and often, the olefin product has to be removed from the reaction flask by continuous distillation to avoid isomerization to the more stable internal olefin isomers. Processes that proceed at lower temperatures usually require a higher catalyst loading and high-boiling polar solvent such as DMPU (*N,N′*-dimethylpropyleneuarea). Gooben and Rodríguez showed that dehydrative decarbonylation of carboxylic acids can be carried out at lower temperatures (110 °C) in highly polar solvents (e.g. *N,N′*-dimethyl-*N,N′*-propyleneurea [DMPU]) and relatively higher catalyst loadings (3 mol%) [20]. Scott and coworkers demonstrated how cocatalytic amounts of simple trialkylamines, such as NEt_3, improve the selectivity for terminal alkenes under palladium catalysis ($PdCl_2$ [3 mol%]/DPEPhos [9 mol%]) in DMPU at 110 °C [21]. In the presence of 1 equiv of NEt_3, relative to the carboxylic acid, the α-olefin selectivity reached as high as 99 : 1, and allylbenzene was obtained in 97% yield (isolated) starting from 4-phenylbutyric acid. The alkene selectivity was retained even when a catalytic amount (9 mol%) of NEt_3 was employed. Changing the solvent to CH_3CN under these conditions, interestingly switched the selectivity in favor of the 2(*E*)-alkene.

The following section details two studies on palladium-catalyzed reactions under milder conditions at preparative scales (up to 100 mmol scale) in academic/industrial settings.

11.4.1 $PdCl_2$/XantPhos/(^{t}Bu)$_4$biphenol System

To tap into the enormous potential of the dehydrative decarbonylation reaction to streamline sustainable production of LAOs from fatty acids, Stoltz and coworkers developed a palladium-catalyzed protocol capable of effecting the transformation at relatively lower temperatures and without the need for polar high-boiling solvents [22, 23]. Preliminary screening that focused on identifying suitable ligands were based on the dehydrative decarbonylation of stearic acid (5 mmol) in neat acetic anhydride (2 equiv) as a solvent and dehydrating agent using $PdCl_2$(nbd) (0.1 mol%) at 132 °C over two hours under a N_2 atmosphere (Scheme 11.6). Yields were determined by ^{1}H NMR spectroscopy

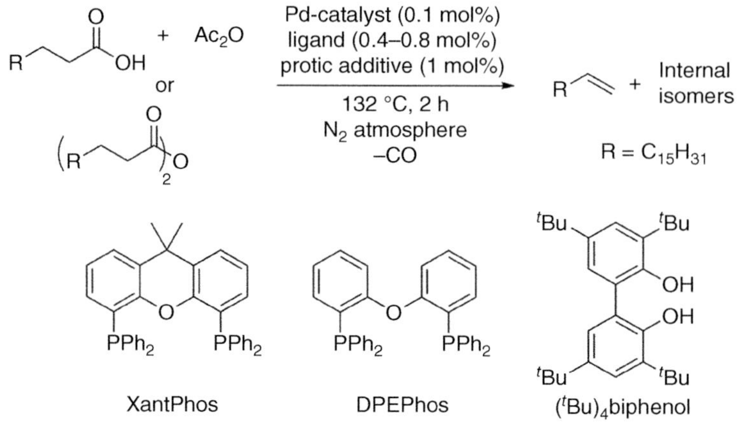

Scheme 11.6 $PdCl_2(PPh_3)_2$/XantPhos catalyzed dehydrative decarbonylation.

using methyl benzoate as an internal standard. Several monodentate (0.8 mol%; PPh_3, $P(p\text{-}MeOC_6H_4)_3$, $P(p\text{-}MeOC_6H_4)_3$, $P(2\text{-furyl})_3$, $P(o\text{-tolyl})_3$, PCy_3, and RuPhos) and bidentate (0.4 mol%; dppe, dppp, dppb, dppf, rac-BINAP, DPEPhos, and XantPhos) phosphine ligands were tested for supporting the palladium center during catalysis. However, only the bidentate ligands DPEPhos (43% yield; 59% alpha) and XantPhos (60% yield; 55% alpha) generated the olefin product. The bite-angle of the bidentate phosphine ligand is suggested to influence reactivity and selectivity, with optimal results obtained using ligands exhibiting bite angles in the range of 105–120°. When presynthesized neat stearic anhydride was used as the substrate, rather slow catalysis was observed (12% yield; 100% alpha) using the $PdCl_2(nbd)$ (0.1 mol%)/XantPhos (0.4 mol%) system. As this later reaction was missing an equivalent of acid as compared to the former ones, the role of acidic additives (1 mol%) was subsequently explored to evaluate their role in promoting reactivity. Marginal improvement in yield (22%; 96% alpha) was observed upon including isophthalic acid (1 mol%) in the decarbonylation of stearic anhydride. A significant improvement in yield (92%) was detected when the ligand/Pd ratio was dropped from 4 : 1 to 1.2 : 1, however, at the expense of α-selectivity (31%), suggesting that olefin isomerization was rampant under these conditions. After further optimization that involved screening other protic additives (p-TsOH•H_2O, salicylamide, 2,2′-biphenol and $(^tBu)_4$biphenol) and switching the palladium precursor to $PdCl_2(PPh_3)_2$, the best catalytic system was identified to include $PdCl_2(PPh_3)_2$ (0.1 mol%)/XantPhos (0.12 mol%)/$(^tBu)_4$biphenol (1 mol%) furnishing the olefin in 84% yield with 70% α-selectivity.

A major drawback of using presynthesized anhydrides in these dehydrative decarbonylation reactions is that the theoretical yield is capped at 50%; hence, the use of fatty acids is preferred instead. However, it was established that the build-up of acid in the reaction mixture leads to olefin isomerization, resulting in low selectivity. It was hypothesized that portionwise addition of the anhydride coupled with a continuous distillation of the acid by-product released could help circumvent this issue. It was demonstrated that the olefin can be obtained in 67%

Table 11.2 Protocol with portionwise addition of anhydride.

R–CH$_2$CH$_2$–C(=O)OH + Ac_2O (in portions) → [$PdCl_2(PPh_3)_2$ (0.05 mol%), XantPhos (0.06 mol%), (^{t}Bu)$_4$biphenol (0.5 mol%); 132 °C, 3 h; 1–5 mmHg distillation; –CO, –AcOH] → R–CH=CH$_2$ + Internal isomers

R = $C_{15}H_{31}$

Entry	Equivalent of anhydride	Yield (%)	α-Selectivity (%)
1	1 + 0.5 (once every 1.5 hours)	69	62
2	1 + 0.5 + 0.25 (once every hour)	67	86
3	1 + 0.14 + 0.12 + 0.1 + 0.09 + 0.08 (once every half hour)	68 (67)	89

Yield and α-selectivity determined by ^{1}H NMR spectroscopy.
Isolated yield in parentheses.

isolated yield (89% alpha) when acetic anhydride was added in six portions, once every half an hour, with concomitant distillation to remove the acetic acid generated (Table 11.2). With optimal conditions established, a variety of carboxylic acids, both naturally occurring and functionalized (ester, chloro, amide, keto, and silylether), were converted to the corresponding olefins in good to excellent yields (isolated, 41–80%) and selectivity (83–99% alpha) achieving TONs as high as 1600. A significant observation was that carboxylic acids featuring α- or β-substituents were relatively poor substrates (TON = 71–400).

The reaction was applied to the decarbonylation of two carboxylic acids, stearic acid and 10-acetoxydecanoic acid, at a 100 mmol scale using a 100 ml round-bottom flask. The scaled-up reactions afforded the corresponding olefins in comparable yields but slightly higher α-selectivity. It is noted that volatile olefins are distilled out along with the acetic acid, although distillation is not necessary to maintain high selectivity. The general procedure for a 20 mmol scale reaction involves loading a 15 ml round-bottom flask with catalyst (0.05 mol%), XantPhos (0.06 mol%), (^{t}Bu)$_4$biphenol (0.5 mol%), and the substrate (1 equiv). The flask is then attached to a distillation head and 25 ml receiving flask, which is held at −78 °C. (Other studies suggest using an ice bath to cool the collector flask. This should work adequately in this case as well unless the olefin being targeted is highly volatile.) The system is then connected to a vacuum manifold, evacuated, and refilled with N_2. Subsequently, the first portion of acetic anhydride (1 equiv) is added by syringe through the rubber septa that seals the distillation head. The flask is then lowered into an oil bath held at 20 °C and gradually heated to 132 °C (time required is 23 minutes). Vacuum (1–5 mmHg) is applied when the oil bath temperature reaches 122 °C, and the acetic acid produced is distilled out starting at 130 °C. After 30 minutes, the reaction setup is refilled with N_2 and the second portion of the anhydride syringed in. The process is repeated until all the anhydride has been added at the specified intervals. For reactions at a 100 mmol scale involving high melting substrates, the reaction was heated to 85 °C until all solid melted and then subsequently heated to 132 °C (time required is 40 minutes).

11.4.2 Well-Defined Pd-bis(phosphine) Precatalysts

Based on computational investigations that suggest that the phosphine ligand dissociates from the palladium center before the rate-determining step (RDS) and driven by the intent of evading the use of excess phosphine in palladium-catalyzed procedures, Jensen and coworkers reported the first example of dehydrative decarbonylation of fatty acids catalyzed by well-defined Pd(bis-phosphine) precatalysts. Initial screening to identify promising ligands for precatalysts design was performed using *in situ* generated complexes and stearic acid as a model substrate. This initial screen involved various palladium precursors ($PdCl_2$, $Pd(OAc)_2$, $Pd(dba)_2$, $Pd(PPh_3)_4$, $PdCl_2(PPh_3)_2$, and $[Pd(cinnamyl)Cl]_2$) and ligands (PPh_3, P^tBu_3, Xphos, ^{t}BuXPhos, DPPF, DPEPhos, XantPhos, and few imidazolium chlorides). Reactions were carried out at a 1 mmol scale of stearic acid using 0.5–3 mol% palladium catalyst, 0.5–9 mol% ligand, 200 mol% Ac_2O, and 9 mol% NEt_3 in DMPU (*N,N′*-dimethylpropyleneuarea) at 110 °C for 15 hours. Catalytic efficiency was evaluated as a function of yield as well as α-selectivity. Reducing the catalyst loading and phosphine/palladium ratio had a detrimental effect on catalysis. A combination of cinnamyl and DPEPhos ligands proved to be crucial for effective catalysis (65% yield; 96% alpha) even at low loading (0.25 mol%), yielding a TON of 130. Hence, a new air-stable precatalyst featuring the desired ligands was synthesized and crystallographically characterized.

Five different Pd–bis(phosphine) complexes bearing DPEPhos and XantPhos ligands were subsequently selected for further reactivity studies (Scheme 11.7). All of the discrete palladium precatalysts exhibited superior activity compared to the respective *in situ*-generated systems. Remarkably, complex **2** based on DPEPhos and cinnamyl ligands was eight times as active (88% yield; 99% alpha; TON = 176; TOF = 11.7 h^{-1}) as the DPEPhos/$PdCl_2$ (10% yield; 97% alpha) combination at 0.5 mol% loading. Precatalysts derived by combination of DPEPhos with the smaller allyl ligand (**4**) or replacing DPEPhos with XantPhos (**3**) were found to furnish inferior results. Further optimization of reaction conditions revealed that polar solvents (DMPU and NMP) are crucial for maintaining high activity and selectivity. These solvents are known to assist with charge separation during the alkene formation step during the catalytic cycle. The addition of organic bases (NEt_3) was found to improve the catalytic activity, presumably by facilitating catalyst activation. Inorganic bases (K_2CO_3, NaOH, KO^tBu, and Na_3PO_4) on the other hand inhibited catalysis.

The optimized reaction conditions were applied to several aliphatic carboxylic acids including naturally occurring fatty acids and other ω-functionalized long-chain carboxylic acids. Naturally occurring fatty acids including stearic, palmitic, myristic, undecylenic acids, etc. (C_{10}–C_{18}) were all converted to the corresponding LAOs in high yields as well as α-selectivity (yield = 25–88%; 95–99% alpha). High reactivity and selectivity were also observed for allylbenzene

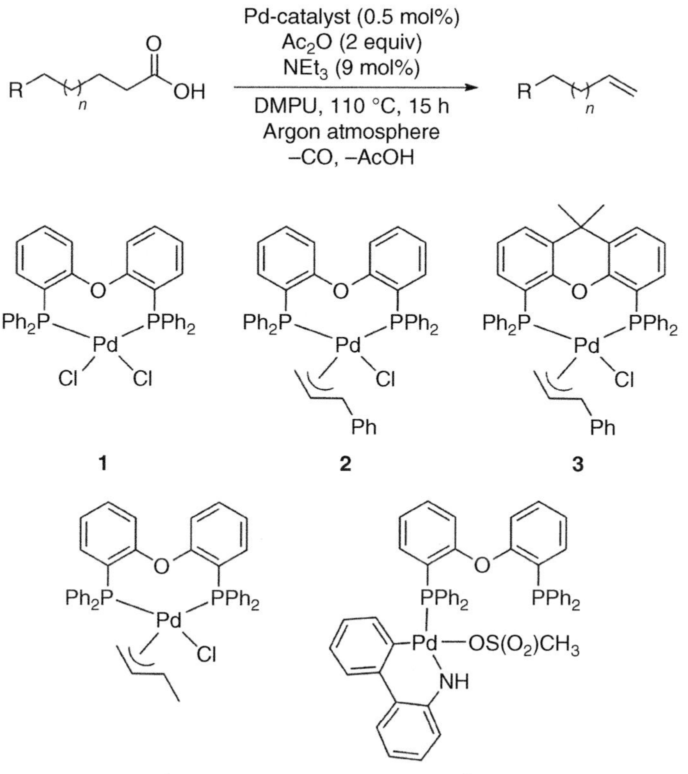

Scheme 11.7 Well-defined palladium precatalysts in dehydrative decarbonylation of fatty acids.

derivatives as well as ester (CO_2Et)-, hydroxyl (–OH)-, amine (–NH_2)-, and bromo (–Br)-functionalized carboxylic acids (10–64% yield; 84–97% alpha).

To demonstrate the practicality of the reaction conditions, a preparative reaction was carried out at 50 mmol scale of stearic acid. The reaction was carried out in a 250 ml Schlenk flask using 0.5 mol% catalyst. After charging the flask with the catalyst and fatty acid, it was sealed with a rubber septum and subsequently evacuated and refilled with argon (three times). Anhydrous DMPU (100 ml), acetic anhydride (2 equiv), and NEt_3 (9 mol%) were added in this order before heating the flak in an oil bath at 110 °C. The product was obtained as colorless oil with 80% yield and 99% α-selectivity following an ether workup and column purification using hexane as the eluent.

11.5 Nickel and Iron Catalysis

Over the recent years, efforts have been made at employing earth-abundant first-row transition metals such as Ni- and Fe- in lieu of the more expensive second-row metals such as Pd-, Rh-, and Ir-. Ryu and coworkers reported iron-catalyzed dehydrative decarbonylation for synthesis of α-olefins with high selectivity [24]. The presence of equimolar amounts of Ac_2O and KI was crucial for enabling catalysis. Heptadecenes were obtained in 79% yield staring from stearic acid using $FeCl_2$ (10 mol%)/PPh_3 (40 mol%) at 250 °C (Scheme 11.8). Internal alkenes were favored under these conditions (19 [terminal]/81 [internal]). Quite significantly, the use of CO atmosphere was found to influence both reactivity and selectivity, with the reactions finishing in 10 minutes. The use of a 5 atm. CO pressure increased the yield to 97% with improved selectivity (46 [terminal]/54 [internal]). The selectivity improved further (87 [terminal]/13 [internal]) when a higher pressure (20 atm.) of CO was used. Following extensive optimization involving several bidentate phosphine ligands (dppe, dppp, dppb, dpppent, dpph, etc.), dpppent (20 mol%) was identified as the best ligand yielding heptadecene in 91% yield with high α-selectivity (92%). The reaction could be performed to a 10 mmol scale without significantly compromising the overall yield and selectivity. This protocol could be successfully extended to other fatty acids including ω-functionalized ones. Consistently high α-selectivities (91–98 %) were recorded across all the substrates tested. Preliminary mechanistic studies ruled out the intermediacy of radicals in the reaction. The beneficial effect of CO atmosphere in promoting catalytic activity and efficiency has also been documented for Ni-catalyzed decarbonylation reactions in the presence of iodide (I^-) additives [25].

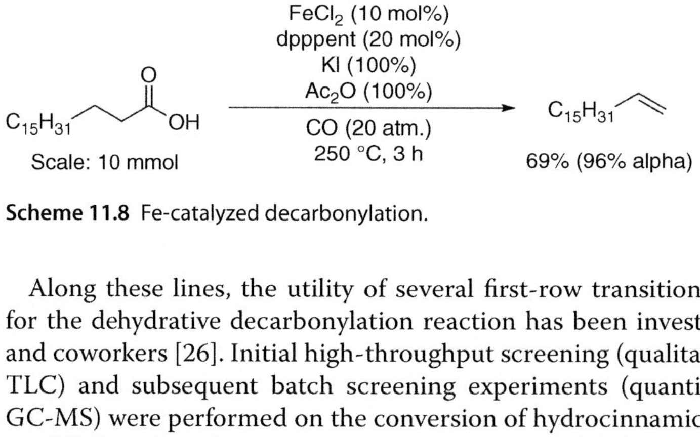

$C_{15}H_{31}CH_2CH_2COOH$ (Scale: 10 mmol) → [$FeCl_2$ (10 mol%), dpppent (20 mol%), KI (100%), Ac_2O (100%); CO (20 atm.), 250 °C, 3 h] → $C_{15}H_{31}CH{=}CH_2$, 69% (96% alpha)

Scheme 11.8 Fe-catalyzed decarbonylation.

Along these lines, the utility of several first-row transition metal complexes for the dehydrative decarbonylation reaction has been investigated by Tolman and coworkers [26]. Initial high-throughput screening (qualitative analysis using TLC) and subsequent batch screening experiments (quantitative analysis by GC-MS) were performed on the conversion of hydrocinnamic acid to styrene at 180 °C. Based on these screening experiments, Ni(II) (e.g. NiI_2) and Ni(0) (e.g. $Ni(COD)_2$) complexes of simple phosphine ligands such as PPh_3 were determined to be as effective (~30% styrene yield) as their more fancier counterparts (dppb, DavePhos, and DPEPhos). The low yields were attributed to styrene dimerization by a putative [Ni-H] intermediate. Mechanistic studies revealed that regeneration of the active catalyst was rate limiting at lower temperatures (up to 100 °C). Catalyst (Ni) poisoning by CO coordination is suggested to shut down catalysis; $Ni(CO)_2(PPh_3)_2$ was detected in the reaction mixture after two turnovers. The reaction conditions were also extended to include fatty acids as substrates. Using several bidentate phosphine ligands (Scheme 11.9a), nonanoic acid was conveniently converted to 1-octene in good yields (43–78%) and

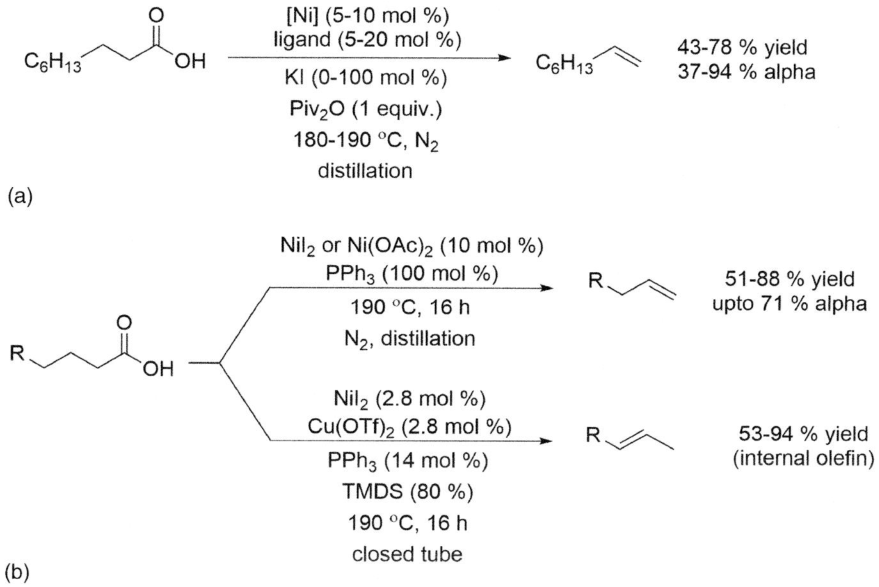

$C_6H_{13}CH_2CH_2COOH$ → [Ni] (5-10 mol %), ligand (5-20 mol %), KI (0-100 mol %), Piv_2O (1 equiv.), 180-190 °C, N_2, distillation → C_6H_{13} olefin; 43-78 % yield, 37-94 % alpha

(a)

Nil_2 or $Ni(OAc)_2$ (10 mol %), PPh_3 (100 mol %), 190 °C, 16 h, N_2, distillation → 51-88 % yield, upto 71 % alpha

Nil_2 (2.8 mol %), $Cu(OTf)_2$ (2.8 mol %), PPh_3 (14 mol %), TMDS (80 %), 190 °C, 16 h, closed tube → 53-94 % yield (internal olefin)

(b)

Scheme 11.9 Ni-catalyzed dehydrative decarbonylation by Tolman group.

α-selectivity (37–94%). Strongly coordinating phosphine ligands, such as dcype, were found to inhibit catalytic activity (10% yield).

Recently, the Tolman group also disclosed a Ni-catalyzed (NiI_2 or $Ni(OAc)_2$) dehydrative decarbonylation that does not require an anhydride additive [27]. It was found that in the absence of a sacrificial anhydride, PPh_3 served both as a ligand and a reductant (Scheme 11.9b). This chemistry is reminiscent of the Rh-catalyzed reaction developed by Foglia et al. Hence, a stoichiometric amount of PPh_3 is needed for a successful reaction. Fatty acids were converted to the corresponding olefins in up to 88% yields and 70% α-selectivity under the optimized reaction conditions ([Ni] (10 mol%)/PPh_3 (100%) at 190 °C for 16 hours). The reaction is rendered cocatalytic in PPh_3 by *in situ* reduction of the $OPPh_3$ by-product using TMDS catalyzed by $Cu(OTf)_2$.

11.6 Ester Decarbonylation

The Tolman group also investigated ester decarbonylation as a feasible strategy for converting carboxylic acid derivatives other than anhydrides to olefins. This was inspired by previous reports on Ni(0)-mediated [28] decarbonylation of phenyl propionates as well as studies on palladium-catalyzed [29] decarbonylation of *p*-nitrophenol esters of aromatic carboxylic acids. Tolman and coworkers studied decarbonylation of *p*-nitrophenol esters of aliphatic carboxylic acids using Ni and Pd catalysts at 160 °C [30]. Esters derived from simple alcohols were found to be inefficient at oxidatively inserting a metal into the acyl-O bond to initiate the reaction. Screening studies using the *p*-nitrophenol ester of hydrocinnamic acid and involving various phosphine ligands (PPh_3, PCy_3,

DPEPhos, dppe, dcype, and dppb) resulted in low yields of styrene (5–10%). When using $PdCl_2$ and in the absence of added ligand, styrene was produced in 46% yield after 16 hours. In the presence of alkali and alkaline-earth metal halides as additives, improved yields were observed. LiCl and $CaCl_2$ were the best additives, resulting in 78% and 71% yield of styrene, respectively (Scheme 11.10a). Optimized conditions were identified to involve $PdCl_2$ (2.5 mol%) and LiCl (100 mol%) in DMPU (c. 0.5 ml) at 160 °C for three hours. The optimized catalytic conditions were applied to *p*-nitrophenol esters of various aliphatic carboxylic acids, including fatty acids. Olefin isomerization was reported to be prevalent under these conditions, with the more stable internal alkenes being preferred.

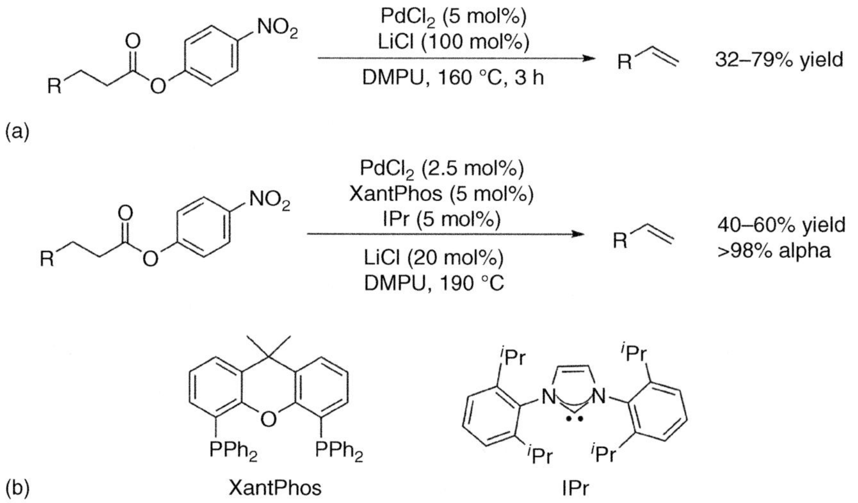

(a) $PdCl_2$ (5 mol%), LiCl (100 mol%), DMPU, 160 °C, 3 h → 32–79% yield

(b) $PdCl_2$ (2.5 mol%), XantPhos (5 mol%), IPr (5 mol%), LiCl (20 mol%), DMPU, 190 °C → 40–60% yield, >98% alpha

XantPhos   IPr

Scheme 11.10 Decarbonylation of esters.

With the intent of improving the α-selectivity of the olefins obtained by ester decarbonylation, the Tolman group explored the effect of ligand additives in the decarbonylation of *p*-nitrophenol ester of nonanoic acid. In the absence of any added ligand, octenes were produced with 14% α-selectivity. Although low conversions were obtained in the presence of ligand additives at 160 °C, higher conversions could be achieved at 190 °C. The presence of ligands was found to be beneficial with XantPhos yielding octenes with 82% α-selectivity [31, 32]. However, the $PdCl_2$ (5 mol%)/XantPhos (5 mol%) catalyst system was found to be ineffective in curbing isomerization in allylic systems (decarbonylation of 4-phenyl butyric ester). Increasing the XantPhos:$PdCl_2$ ratio improved the α-selectivity in this case. To avoid using large excess of an expensive XantPhos ligand, mixed ligand systems were explored. A high α-selectivity (92% allylbenzene) was realized using an equimolar combination of XantPhos and PPh_3 (10 mol% each). Upon further optimization, a combination of XantPhos and an *N*-heterocyclic carbene ligand (1,3-diisopropylphenylimidazo-2-ylidene; IPr) was found to be the most effective at curbing isomerization. The final optimized catalyst composition consisted of $PdCl_2$ (2.5 mol%)/XantPhos (5 mol%)/IPr (5 mol%). A variety of esters could be decarbonylated to the respective α-olefins in 40–60% yields and excellent selectivity (>98%) at 190 °C under the optimized conditions (Scheme 11.10b). The role of the two ligands in the catalyst system as probed by independent studies suggest that XantPhos maintains high α-selectivity while the IPr ligand helps maintain high conversions within short reaction times (up to 2.5 hours for 85–90% conversion).

11.7 Synthetic Utility: α-Vinyl Carbonyl Compounds

The Stoltz group demonstrated the utility of the palladium-catalyzed methodology they developed for the synthesis of α-vinyl carbonyl compounds featuring a quaternary stereocenter [33]. This is a common structural motif in a plethora of natural products including (+)-aspewentin-B, transtaganolide C, etc. A prominent approach to the synthesis of this motif is the α-vinylation of carbonyl compounds; however, known methods are limited in substrate scope, stereoselectivity, and require stoichiometric reagents [25]. Stoltz and coworkers showed that this privileged structural motif can be accessed by dehydrative decarbonylation of δ-oxocarboxylic acids, which can in turn be synthesized enantioselectively by either the (i) addition of an enolate nucleophile to an acrylate acceptor or (ii) by palladium-catalyzed allylic alkylation. Slightly modified catalytic conditions involving a higher catalyst, ligand, and additive loading were used (Scheme 11.11), and the reactions were performed at a 5 mmol scale. For smaller scale reaction, NMP was a used as a solvent along with the $PdCl_2$(nbd)

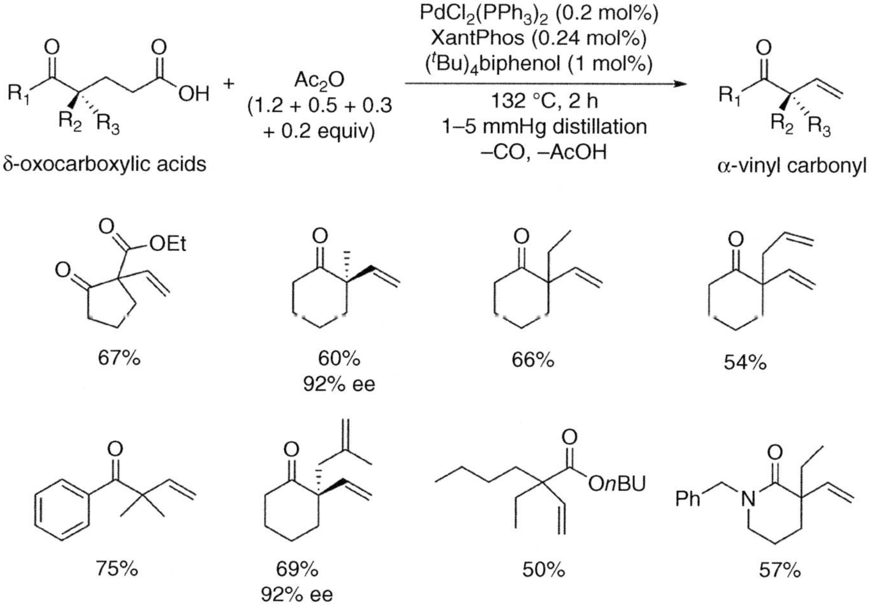

δ-oxocarboxylic acids + Ac_2O (1.2 + 0.5 + 0.3 + 0.2 equiv) → [$PdCl_2(PPh_3)_2$ (0.2 mol%), XantPhos (0.24 mol%), (^{t}Bu)$_4$biphenol (1 mol%); 132 °C, 2 h; 1–5 mmHg distillation; −CO, −AcOH] → α-vinyl carbonyl

67% | 60%, 92% ee | 66% | 54%

75% | 69%, 92% ee | 50% | 57%

Scheme 11.11 Synthesis of α-vinyl carbonyl compounds featuring a quaternary stereocenter.

(1 mol%)/XantPhos (1.2 mol%) catalyst system. The reaction was successfully applied to the synthesis of several α-Vinyl carbonyl compounds with quaternary stereocenters in moderate to good yields and excellent enantioselectivity.

11.8 Conclusions and Future Prospects

The dehydrative decarbonylation reaction catalyzed by transition metal catalysts provides a unique opportunity to convert carboxylic acids into olefins. In particular, the reaction provides a single-step route for conversion of fatty acids into LAOs. After generation of a mixed anhydride, mechanistically, the catalytic pathway is proposed to involve (i) initial oxidative addition, (ii) decarbonylation, (iii) β-hydride elimination, (iv) and reductive elimination sequence. The catalytic reaction has been the subject of avid research and has seen rapid development in the past decade. The reaction is catalyzed by a variety of transition metals including base metals such as Ni and Fe. Recent development allow for the reaction to be conducted under milder conditions, lower catalyst loading, and using lower ligand:metal ratios. The reaction exhibits broad substrate scope and functional group tolerance. Efforts to expand the scope of the reaction to include other carboxylic acid derivatives are ongoing. Future research in this area would unravel new modes for activation of carboxylic acids. Of special interest would be methodologies that involve catalytic amounts of an activator. Other significant improvements would revolve around the use of base metal catalysts. The developments of catalytic methodologies that operate under milder conditions as well as obviate the need to remove the olefin by distillation are highly desirable.

Biography

Dr. Alex John is an assistant professor of organic chemistry at California State Polytechnic University Pomona since Fall 2016. He received his Ph.D. degree in Chemistry from Indian Institute of Technology Bombay (IIT Bombay) working with Prof. Prasenjit Ghosh in the area of homogeneous transition metal catalysis. Subsequently, he did postdoctoral studies at the University of Oklahoma (Prof. Kenneth M. Nicholas) and as part of the Center for Sustainable Polymers at the University of Minnesota (Prof. William B. Tolman & Prof. Marc A. Hillmyer). His current research interest is in the area of sustainable chemistry with an emphasis on developing catalytic methods for converting biomass-derived molecules into platform chemicals.

References

1 (a) Dodds, D.R. and Gross, R.A. (2007). Chemicals from biomass. *Science* 318: 1250–1251. (b) Huber, G.W., Iborra, S., and Corma, A. (2006). Synthesis of transportation fuels from biomass: chemistry, catalysts, and engineering. *Chem. Rev.* 106: 4044–4098. (c) Kunkes, E.L., Simonetti, D.A., West, R.M. et al. (2008). Catalytic conversion of biomass to Monofunctional hydrocarbons and targeted liquid-fuel classes. *Science* 322: 417–421. (d) Zakzeski, J., Bruijnincx, P.C.A., Jongerius, A.L., and Weckhuysen, B.M. (2010). The catalytic valorization of lignin for the production of renewable chemicals. *Chem. Rev.* 110: 3552–3599. (e) Vennestrøm, P.N.R., Osmundsen, C.M., Christensen, C.H., and Taarning, E. (2011). Beyond petrochemicals: the renewable chemicals industry. *Angew. Chem. Int. Ed.* 50: 10502–10509.

2 Pyo, S.-H., Dishisha, T., Dayankac, S. et al. (2012). A new route for the synthesis of methacrylic acid from 2-methyl-1,3-propanediol by integrating biotransformation and catalytic dehydration. *Green Chem.* 14: 1942–1948.

3 (a) Chheda, J.N., Huber, G.W., and Dumesic, J.A. (2007). Liquid-phase catalytic processing of biomass-derived oxygenated hydrocarbons to fuels and chemicals. *Angew. Chem. Int. Ed.* 46: 7164–7183. (b) Corma, A., Iborra, S., and Velty, A. (2007). Chemical routes for the transformation of biomass into chemicals. *Chem. Rev.* 107: 2411–2502. (c) Petrus, L. and Noordermeer, M.A. (2006). Biomass to biofuels, a chemical perspective. *Green Chem.* 8: 861–867.

4 Miranda, M.O., Pietrangelo, A., Hillmyer, M.A., and Tolman, W.B. (2012). Catalytic decarbonylation of biomass-derived carboxylic acids as efficient route to commodity monomers. *Green Chem.* 14: 490–494.

5 Bacha, J.D. and Kochi, J.K. (1968). Alkenes from acids by oxidative decarboxylation. *Tetrahedron* 24: 2215–2226.

6 Kolbe, H. (1849). Untersuchungen über die Elektrolyse organischer Verbindungen. *Justus Liebigs Ann. Chem.* 69: 257–294.

7 (a) Anderson, J.M. and Kochi, J.K. (1970). Silver(II) complexes in oxidative decarboxylation of acids. *J. Org. Chem.* 35: 986–989. (b) van der Klis, F., van den Hoorn, M.H., Blaauw, R. et al. (2011). Oxidative decarboxylation of unsaturated fatty acids. *Eur. J. Lipid Sci. Technol.* 113: 562–571.

8 (a) Ittel, S.D., Johnson, L.K., and Brookhart, M. (2000). Late-metal catalysts for ethylene homo- and copolymerization. *Chem. Rev.* 100: 1169–1203. (b) Marquis, D.M., Sharman, S.H., House, R., and Sweeney, W.A. (1966). Alpha olefin sulfonates from a commercial SO3-Air reactor. *J. Am. Oil Chem. Soc.* 43: 607–614. (c) Franke, R., Selent, D., and Börner, A. (2012). Applied hydroformylation. *Chem. Rev.* 112: 5675–5732.

9 (a) Al-Jarallah, A.M., Anabtawi, J.A., Siddiqui, M.A.B. et al. (1992). Ethylene dimerization and oligomerization to butene-1 and linear α-olefins: a review of catalytic systems and processes. *Catal. Today* 14: 1–121. (b) Small, B.L. and

Brookhart, M. (1998). Iron-based catalysts with exceptionally high activities and selectivities for oligomerization of ethylene to linear α-olefins. *J. Am. Chem. Soc.* 120: 7143–7144.

10 Trost, B.M. and Chen, F. (1971). Transition metal mediated eliminations in anhydrides and thioanhydrides. *Tetrahedron Lett.* 12: 2603–2607.

11 Miller, J.A., Nelson, J.A., and Byrne, M.P. (1993). A highly catalytic and selective conversion of carboxylic acids to 1-alkenes of one less carbon atom. *J. Org. Chem.* 58: 18–20.

12 Ortuno, M.A., Dereli, B., and Cramer, C.J. (2016). Mechanism of Pd-catalyzed decarbonylation of biomass-derived hydrocinnamic acid to styrene following activation as an anhydride. *Inorg. Chem.* 55: 4124–4131.

13 Foglia, T.A. and Barr, P.A. (1976). Decarbonylation dehydration of fatty acids to alkenes in the presence of transition metal complexes. *J. Am. Oil Chem. Soc.* 53: 737–741.

14 Miller, J.A., Nelson, J.A., and Byrne, M.P. (1991). Process for making olefins. US Patent 5, 077, 447, filed 28 September 1990 and issued 31 December 1991.

15 Kraus, G.A. and Riley, S. (2012). A large-scale synthesis of α-olefins and α,ω-dienes. *Synthesis* 44: 3003–3005.

16 . Kraus, G. A., "Method for producing olefins" *US Patent App.* **2013**, US2013/0102823A1.

17 Maetani, S., Fukuyama, T., Suzuki, N. et al. (2011). Efficient iridium-catalyzed decarbonylation reaction of aliphatic carboxylic acids leading to internal or terminal alkenes. *Organometallics* 30: 1389–1394.

18 Ternel, J., Lebarbe, T., Monflier, E., and Hapiot, F. (2015). Catalytic decarbonylation of biosourced substrates. *ChemSusChem* 8: 1585–1592.

19 Hogan, T.E. (2015). Uses of biobased styrene. US Patent 2015/0274944A1, filed 8 November 2013 and issued 1 October 2015.

20 Gooben, L.J. and Rodríguez, N. (2004). A mild and efficient protocol for the conversion of carboxylic acids to olefins by a catalytic decarbonylative elimination reaction. *Chem. Commun.*: 724–725.

21 Nôtre, J.L., Scott, E.L., Franssen, M.C.R., and Sanders, J.P.M. (2010). Selective preparation of terminal alkenes from carboxylic acids by a palladium catalyzed decarbonylation-elimination reaction. *Tetrahedron* 51: 3712–3715.

22 Liu, Y., Kim, K.E., Herbert, M.B. et al. (2014). Palladium-catalyzed decarbonylative dehydration of fatty acids for the production of linear alpha olefins. *Adv. Synth. Catal.* 356: 130–136.

23 Liu, Y., Stoltz, B. M., Grubbs, R. H. et al. (2016) Palladium-catalyzed decarbonylation of fatty acid anhydrides for the production of linear alpha olefins. US Patent 9, 44,089,1B2, filed 27 November 2013 and issued 13 September 2016.

24 Maetani, S., Fukuyama, T., Suzuki, N. et al. (2012). Iron-catalyzed decarbonylation reaction of aliphatic carboxylic acids leading to α-olefins. *Chem. Commun.* 48: 2552–2554.

25 Suzuzki, N., Tahara, H., Ishihara, D., et al. (2016) US Patent 9, 26,679,0B2, filed 4 August 2011 and issued 23 February 2016.

26 John, A., Miranda, M.O., Ding, K. et al. (2016). Nickel catalysts for the dehydrative decarbonylation of carboxylic acids to alkenes. *Organometallics* 35: 2391–2400.

27 John, A., Hillmyer, M.A., and Tolman, W.B. (2017). Anhydride-additive-free nickel-catalyzed deoxygenation of carboxylic acids to olefins. *Organometallics* 36: 506–509.

28 Yamamoto, T., Ishizu, J., Kohara, T. et al. (1980). Oxidative addition of aryl carboxylates to nickel(0) complexes involving cleavage of the acyl-oxygen bond. *J. Am. Chem. Soc.* 102: 3758–3764.

29 Gooβen, L.J. and Paetzold, J. (2002). Pd-catalyzed decarbonylative olefination of aryl esters: towards a waste-free Heck reaction. *Angew. Chem. Int. Ed.* 41: 1237–1241.

30 John, A., Hogan, L.T., Hillmyer, M.A., and Tolman, W.B. (2015). Olefins from biomass feedstocks: catalytic ester decarbonylation and tandem Heck-type coupling. *Chem. Commun.* 51: 2731–2733.

31 John, A., Dereli, B., Ortuno, M.A. et al. (2017). Selective decarbonylation of fatty acid esters to linear α-olefins. *Organometallics* 36: 2956–2964.

32 Tolman, W.B., Hillmyer, M.A., John, A. et al. (2017). US Patent 9, 71,876,3B2, filed 7 January 2016 and issued 1 August 2017.

33 Liu, Y., Virgil, S.C., Grubbs, R.H., and Stoltz, B.M. (2015). Palladium-catalyzed decarbonylative dehydration for the synthesis of α-vinyl carbonyl compounds and total synthesis of (−)-aspewentins A, B, and C. *Angew. Chem. Int. Ed.* 54: 11800–11803.

Index

Organometallic Chemistry in Industry: A Practical Approach, First Edition.
Edited by Thomas J. Colacot and Carin C.C. Johansson Seechurn.

d

e

f

p

q

r